WHAT'S NEXT?

Eco Materialism & Contemporary Art

WHAT'S NEXT?

Eco Materialism & Contemporary Art

Linda Weintraub

intellect Bristol, UK / Chicago, USA

This edition first published in the UK in 2019 by
Intellect, The Mill, Parnall Road, Fishponds, Bristol, BS16 3JG, UK

This edition first published in the USA in 2019 by
Intellect, The University of Chicago Press, 1427 E. 60th Street,
Chicago, IL 60637, USA

A catalogue record for this book is available from the
British Library.

Copy-editor: Emma Rhys
Typesetting: Holly Rose and Sherry Williams
Production editor: Tim Mitchell

Color images can be viewed at www.lindaweintraub.com.

Support from Nomad/9 Interdisciplinary MFA,
Hartford Art School, University of Hartford.

Print ISBN: 978-1-78320-940-8
ePDF ISBN: 978-1-78320-942-2
ePUB ISBN: 978-1-78320-941-5

Printed by CPI / Anthony Rowe, UK.
This is a peer-reviewed publication.

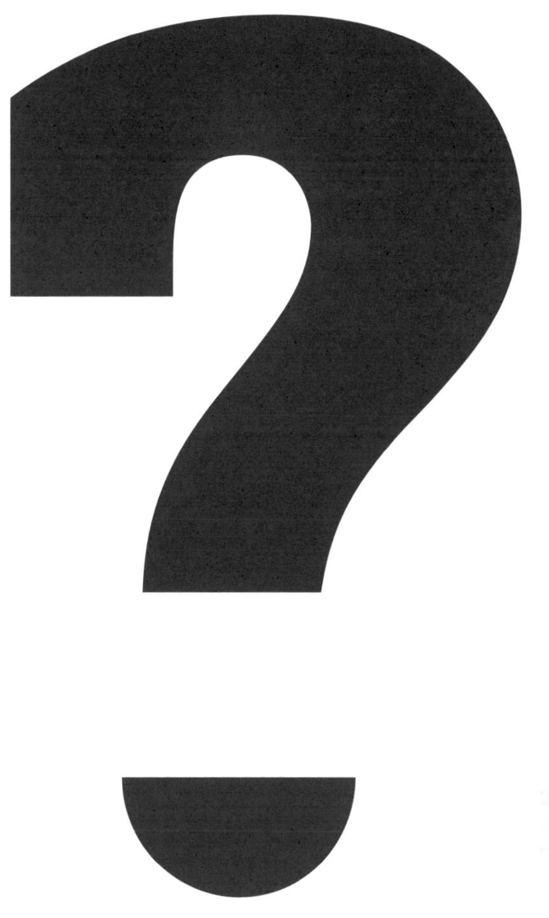

CONTENTS

Acknowledgments

Writing a book is typically a solitary procedure, requiring countless hours spent in lonely silence. I'm pleased to report that isolation never factored into the writing of this book. The process was socialized by the intention of forming a community of like-minded artists who were not yet acquainted, and had not yet discovered their commonalities. This book bestows a name upon their shared vision, and examines the cultural and technological conditions that permeate their shared concerns. The forty-three artists whose works are presented on these pages are among the founding members of this community. Their particular forms of creativity embodied the abiding principles of Eco Materialism even before this book was written. I am pleased to have this opportunity to formally acknowledge the timely forms of artistic ingenuity offered by Tim Hawkinson, Gu Wenda, Marc Quinn, Caitlin Pautler Albritton, Rose-Lynn Fisher, Amanda Cotton, Jess Dobkin, Sun Yuan, Peng Yu, Germaine Koh, Terence Koh, Maurizio Montalti, Amy Youngs, Walter De Maria, Hilda Hellström, Alexandra Regan Toland, Jon Cohrs, Laura Parker, Joe Scanlan, James Bridle, Thomas Thwaites, Mary Mattingly, Francis Alÿs, Rachael Mellors, Bruce Davies, Amy Franceschini, Markus Kayser, Qiu Song, Ren Nuo Ya, Kang Pengfei, Bai Ying, Guo Shen, Spurse, Simparch, Dove Bradshaw, Pierre Huyghe, Jason deCaires Taylor, Red Earth, Gebhard Sengmüller, Raphael Perret, Angelo Vermeulen, Alice Anderson, Tal Spelletich, Chim↑Pom, Zbigniew Oksiuta, and Natalie Jeremijenko.

While the artists are acknowledged for providing the book's rich and challenging content, converting content into a published volume was accomplished under the guidance of the conscientious and supportive staff at Intellect Books: the typesetter, Holly Rose; copy editor, Emma Rhys; the marketing team, Katy Dalli and James Campbell; and the production director, May Yao. Editor, Tim Mitchell, earns a special commendation for seeking sustainable publishing options that simultaneously reflect the content of this book and respect the limited budgets of a student readership. It is under Mitchell's guidance that the Intellect staff entered uncharted publishing territories regarding production and distribution. This book received a generous vote of confidence, even before the draft was completed from Nomad9, the Interdisciplinary Master of Fine Arts program at Hartford University, where I have been a guest artist since its inception. I am honored to acknowledge its generous grant that enabled Intellect Books to adopt environmentally responsible publishing strategies. In addition, I would like to acknowledge Carol Padberg, founder and director of this very special MFA program. While developing this text I was privileged to receive a measure of the guidance and encouragement she regularly bestows upon her students.

Julie Niemie accepted the responsibility of preparing the manuscript for publication by formatting endnotes and captions. Carlos Nazario generously provided technical assistance. Sherry Williams contributed her design acumen to developing the graphic style of the text. Their devoted attention to such details is greatly appreciated. In addition, The Crafts Council - Make:Shift; Ikon Gallery; David Zwirner Gallery; DIA Art Foundation; Pace Gallery; Futurefarmers; SEAD Network; Postmasters Gallery are acknowledged as arts organizations that either waived or reduced the fee for permission to reproduce images.

Informal conversations about this text as it was evolving were as vital to the outcome as the contributions of professionals . I am particularly grateful for the responses and queries offered by the following colleagues: Paula Turkon, Garry Hagberg, George Quasha, Aviva Rahmani, Lenore Malen, Ellen Levy, Nobohu Nagasawa, Michael Asbill, Matthew Friday, Iain Kerr, Dan Franck, Carlos A. Nazario, Carolyn Koppel, Matthew Burke and Victoria Vesna. Finally, a heartfelt expression of gratitude is extended to my husband, Andrew, whose unwavering support appeared whenever I wavered and needed support.

Preface
How Materiality Shaped This Book

Like other homesteaders, I create habitats on my property in upstate New York. These environments facilitate an organism's ability to reproduce and thrive. I provide sources of pollen for bees, stream-fed ponds for amphibians, nesting shelters for wild birds, meadows for wild plants, and compost for beneficial microbes. Likewise, this book has been created to provide such a congenial setting for a special group of contemporary artists who share a radical vision of the future, but who have never been granted a suitable context. In art, a habitat is known as a 'movement'. The art movement that is identified and named in this text is Eco Materialism. The analogy is apt. Ecologists define a habitat as a territory in which the organism can find food, shelter, protection, and mates. Artists define an art movement as a context in which shared values are cultivated, acknowledged, and explored. Without associating with a movement, pioneering artists tend to be viewed as oddities pursuing idiosyncratic impulses. But as contributors to a movement, their common beliefs and strategies gain significance as pertinent cultural assets. Thus, whereas an ecological habitat is made up of such physical factors as soil, moisture, temperature, light, and life forms, an art movement consists of a group of artworks that convey shared values, along with the analysis, discourse, and cultural traction that they inspired. The analogy also extends to the ever-changing nature of these phenomena. Habitats are subject to violent events such as earthquakes and gradual occurrences like rising ocean temperatures. Likewise, art movements emerge and fade in response to current technological, political, economic, and other cultural and environmental shifts. What is the impetus for the emergence of Eco Materialism as an art movement? Eco Material artists address the mounting threats to the planet's ability to support diverse forms of life on Earth. While they acknowledge humanity's role in causing this jeopardy, they are determined to avert the collapse of threatened eco systems. Their shared approach is to reassert the primacy of materiality, which has been displaced by contemporary technologies and manufacturing protocols. They honor the sensory, ethical, and pragmatic benefits of material interactions.

Some of the artists presented in this volume have international reputations. Others do not. Many have reported that they thought they were pursuing isolated courses until Eco Materialism was named as an emerging art movement. This book situates their individual endeavors at the fulcrum where three contemporary movements converge: the philosophy of Neo Materialism, the biological/ geological concerns of environmentalism, and the bold inventiveness of art.

The 'How' of Eco Materialism

Author's 'How'
Artists' 'How'
Publishers' 'How'
Readers' 'How'

Author's 'How': Materiality Shapes This Book's Content

As a contemporary art historian, I don't study past accomplishments. I may be referred to as a historian, but my task involves anticipating the way the current era may appear in future histories of art. I am, therefore, professionally obliged to venture into uncharted art territories, searching the perpetually shifting cultural scene for key indicators of emerging trends that offer this future-oriented promise. My colleagues and I often differ regarding the qualities that constitute significance. Some favor artists that excite the mind; others prefer art that lingers in their mind, or blows their mind, or opens their mind, or quiets their mind, or even boggles their mind. Nonetheless, we all delve into each season's abundant offerings with the intention of distinguishing fleeting trends from enduring contributions.

Such an exploration preceded the writing of this text. The process was fraught with anxiety. I fretted that the interest aroused by some artworks, and the indifference I felt toward others, was 'merely personal', and not 'culturally significant'. These doubts evaporated when I examined my bulging files and discovered that my favored artists shared a fixation on materiality that had congealed into a revolutionary twenty-first century philosophical movement, referred to as New Materialism. The 'Suggested Readings' that appear in this book's appendix testify to the cross-disciplinary nature of this pursuit. They include volumes with such titles as *Material Feminisms; Sociology and the New Materialism: Theory, Research,*

Action; Religion, Politics, and the Earth; The New Materialism (Radical Theologies and Philosophies); New Materialisms: Ontology, Agency, and Politics; Carnal Knowledge: Towards a 'New Materialism' through the Arts and *Religion, Politics and the Earth: The New Materialism (Radical Theologies).* This sample of texts written by feminists, sociologists, philosophers, political scientists, and theologians provided me with compelling evidence that Neo Materialism was a significant intellectual development, and that the artists in my files had independently invoked its principles. The tradition-defying artworks they were creating manifested the precise physical interactions that constitute the content of these texts. The timeliness of their efforts merited certification as a new art movement, and this movement would be best served by a book devoted to their materialist manner of anticipating 'What's next?' Furthermore, it seemed self-evident that the ideal readers of this future-oriented text would be young artists who are still formulating their career options.

There were far too many eligible candidates for inclusion in a single volume. Whittling down hundreds to the forty-three that you will encounter on the pages of this book required that I develop a selection process that gave all the artists in my files an equal opportunity to 'audition' for a role in this volume. The rigors of this curatorial selection reveals the author's 'how':

- First cut: Because Eco Materialism provides the thesis for this book, only those eco artists who introduce material interactions that convey an ecological theme were considered.

- Second cut: Of those Eco Material artists, only those who address one of the five categories of material interaction that comprise the book's organizational structure were considered. These categories reflect the breadth of Eco Material concerns, and their relevance to all human interactions with the material environment.

- Third cut: Of those who address these themes, only those whose work involves the particular material that is featured in that chapter were considered.

- Finalists: Of those who explore a chapter theme and utilize that chapter's featured material, only those whose work diversified this approach were considered.

- Accepted: Of the finalists, only those whose works promised to engage and instruct you, the reader, were selected.

These selections were collected, categorized, and assembled into the Table of Contents that summarizes the Eco Material reconstruction of cultural norms that are not confined to the practice of art. In an era of environmental peril, they are critical to all human activities.

Artists' 'How': Materiality Shapes Art's Form and Content

'What's next?' in art directs attention forward, beyond the present day. It may seem odd, therefore, that this book's futuristic orientation is launched by presenting three works that are ensconced in the annals of art history. However, these artworks were not selected for their endurance through the centuries. They appear here because, by depicting artists engaged in the process of creating art, they anticipate the focus of this contemporary art text. Like these early 'behind-the-scenes' studio depictions, as much significance is assigned to 'how' works of art were created, as to 'what' kind of artworks were produced. They introduce considerations of 'in what manner', 'to what extent', and 'for what reason' an artwork assumes its physical form.

Exploring the art of art-making typically yields insights into an era's defining relationship with the material environment. This is because studio practice evolves in tandem with the introduction of new forms of material interaction. Enea Vico (1523–1567), for example, adopted an engraving technique for creating prints on paper. While engraved designs have been found on prehistoric bones, stones, and cave walls, the technique that developed in the sixteenth century was dependent on the introduction of a copper plate and a steel tool called a burin. Because the plate could be reused, the technique facilitated duplication, enabling multiples of a single image to be produced. Engravers have been largely replaced by photographers today. Sargent's painting of Monet documents the introduction of another art-related technology: the paint tube. Invented in 1841 by the painter John Goffe Rand, the tube was transportable, making it possible for artists to leave their studios and paint outdoors.

The history of art can be written as a succession of 'hows'. This correlation gains special significance in this book because material interactions constitute its abiding theme. While the words digital, nano, robotic, laser, LCD, AI, and microprocessors are used to describe many contemporary artworks, this book diverges from the hi-tech arena to explore another influence upon contemporary material interactions: ecology. This is referred to as Eco Materialism.

Eco Materialism is both a philosophy that conveys a worldview, and a code of conduct that specifies how that worldview translates into material interactions. Eco Material artists express this philosophy through the issues they address, and they manifest this code of conduct through the protocols of art production that they adopt. As a result, special significance is attached to the creative insights conveyed by their work's medium, the tools that act upon these mediums, the settings where the creative acts occur, and the forms of energy that are expended in the art-making process. Eco Material mediums may include microbes and rot.

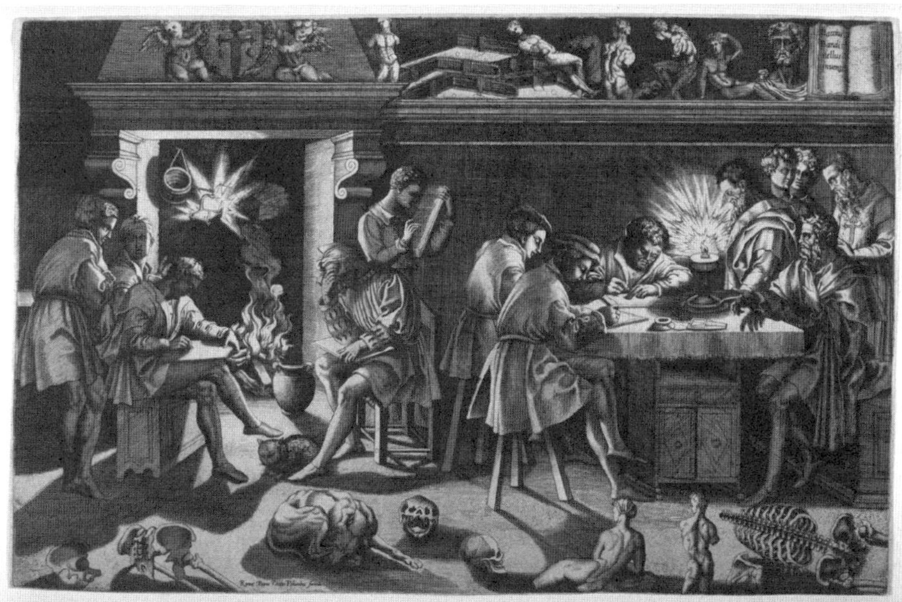

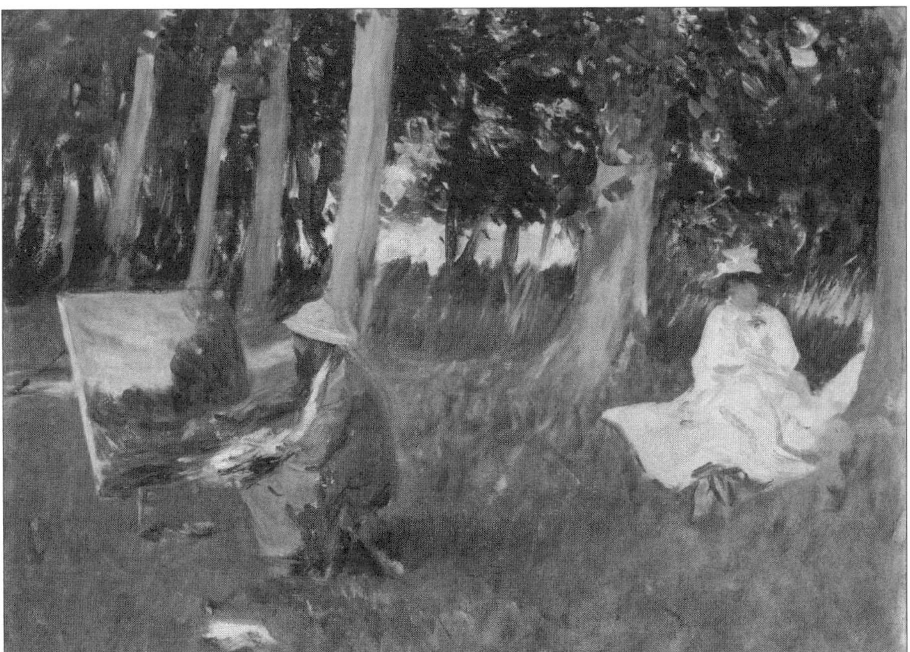

Figure 1: Enea Vico (Italian, 1523–1567), engraving of the studio of Baccio Bandinelli (Italian, 1493–1560). Image shows apprentice and master artists working. Published by Pietro Paolo Palumbo, c.1546. 309 x 479 mm (12 3/16 x 18 7/8 in.). Courtesy of Mr. and Mrs. Marcus Sopher Collection, de Young / Legion of Honor Fine Arts Museum of San Francisco.

Figure 2: John Singer Sargent (American, 1856–1925), *Claude Monet Painting by the Edge of the Wood* (1885), oil on canvas, 54 x 64.8 cm (21 1/4 x 25 1/5 in.). Courtesy of the Tate Gallery, London. Presented by Miss. Emily Sargent and Mrs. Ormond, Art Fund, 1925.

Tools may take the form of fire or flowing water. Settings where creativity occurs may be under water or in outer space. Forms of energy may include waste heat from machines or photosynthesizing algae. Ultimately, these issues of 'how' supply the 'whys' regarding the radicalism of Eco Material creativity. Indeed, Eco Material art practices deviate from conventional art practices as thoroughly as environmental reforms deviate from wasteful norms in consumption, or polluting norms in manufacturing. Thus, Eco Material artists propose three answers to the question 'What's next?'. Readers of this text will encounter examples of creative ingenuity embedded in the artists' material interactions; cultural significance conveyed through the artists' physical interventions; and survival strategies revealed through the artists' interactions with energy and matter.

Publishers' 'How': Materiality Shapes This Book's Form

Unlike political correctness in human interactions that is supported by consensus, ecological correctness in book publishing is a work-in-progress. Books impose environmental impacts whether they are published materially or electronically. While Intellect Books committed to minimizing the detrimental environmental impacts of the production of this text, determining 'how' it would be reproduced and distributed involved weighing ideals against practicalities. The 'specs' that resulted from these deliberations are as follows: paper sourced through environmentally responsible forestry; only black and white images appear in the print version to avoid chemical contamination from color inks; color images are available on the author's website and in the digital book formats; trim size is reduced to eliminate paper waste.

The editors acknowledged that focusing exclusively on renewable papers and non-polluting inks ignores the fact that resources are consumed and contaminants are generated during resource acquisition, printing and binding, transport, assembly, packaging, storage, distribution, warehousing, retailing, and disposal. Thus, the eco-friendly publishing strategy that was developed extends beyond the physical book, and includes:

1. Discrete orders for specialized readers: In addition to a bound book, eight sections of the text are available digitally as individual 'textlets'. This compartmentalization of the book's content allows instructors to make relevant portions of the book available to students at an affordable price. Furthermore, it provides flexibility to those educators who structure their own course syllabi from multiple texts instead of adopting an entire text book. The first textlet links philosophical materialism to ecology and contemporary art. Each of the six textlets that follow introduces a basic material (secretions, dirt, human effort, fire, water, and electronics). These substances convey six Eco Material themes (allures and repulsions; archetypes of material

interactions; operatives; tools; creativity; and technology). Instead of offering a specific forecast, the concluding textlet provides three possible 'What's next?' scenarios: desperation, relocation, and recuperation.

2. License to print files on campus: A license to print multiple copies of any chapter can be purchased through the Publishers' Licensing Services (plsclear.com). This strategy enables books to be printed from the PDF in the campus print shop, eliminating the financial and environmental costs of transportation, packaging, and warehousing.

Readers' 'How': Materiality Shapes This Book's Uses

This book not only offers instructive content for readers, it invites participation by readers. The Table of Contents serves as a menu of possibilities. It provides a range of topics from which to choose and sections to read in any order. Further inducements for reader engagement appear at the end of each section as Reader Interactions. By fostering personal engagement and self-reflection, these guided explorations enable readers to examine their current material interactions, imagine alternatives, and formulate plans of action. These material options apply, of course, to the creation of art, but they also apply to the imperatives imposed by being alive: eating, producing waste, relying on oxygen, and seeking security and comfort. Thus, these sections provide opportunities for readers to establish professional and extra-professional habits of acting on behalf of the environment.

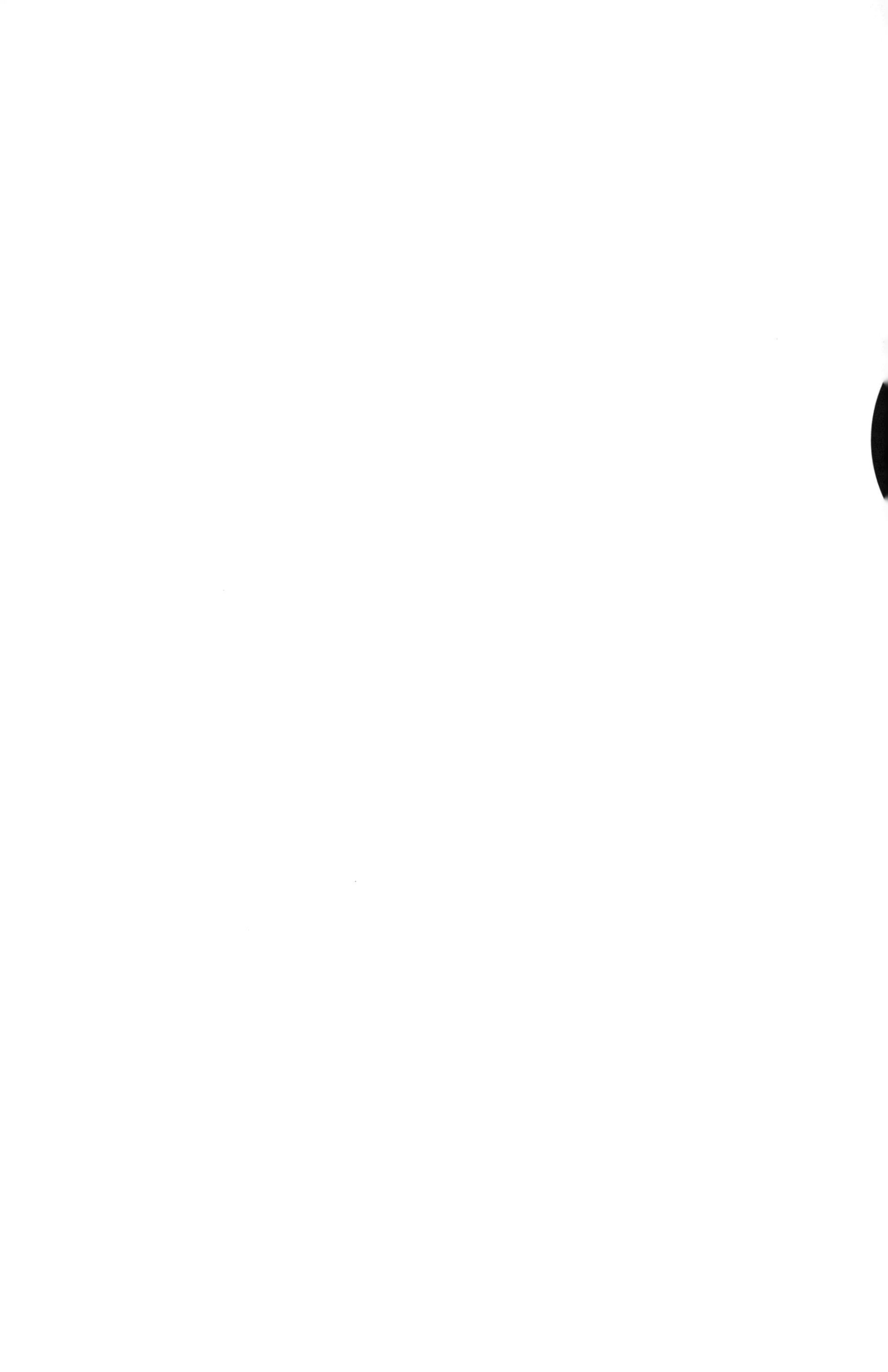

Introduction
Where to Begin?

Introduction to the Introduction

Enter an art studio and you have crossed the threshold from ordinary life, with its pesky distractions, into an enchanted zone where artists shed the encumbrances of necessity, liberate their creativity, and cultivate imaginative prospects... except that many contemporary art studios seem more like danger zones than havens where revelations are being envisioned. Although environmental consciousness is making significant inroads into construction, automobile design, food production, waste management, and many other sectors of society, it has been slow to infiltrate artists' classrooms and studios. Consider the following questions: Do you trust your art mediums? Would you eat them?

Artistic interactions carried minimal consequence in primeval times, when cave dwellers painted with earth (pigment), saliva (solvent), and animal fat (binder). Unlike these biodegradable substances, the ingredients listed on the labels of many manufactured paints, for example, portend ominous environmental outcomes. They are manufactured out of agglomerates, synthetic alkyd resins, biocides, coalescent solvents, defoamer additives, dispersants, polymerization reactions, thixotropic agents, and surfactants. While these concoctions are selected to give paint such desirable qualities as luminosity, stability, and affordability, they often impose undesirable health risks. 'Hazcom' warnings posted in studios are often accompanied by physical evidence of dangers: eye and hand protective gear, ventilation equipment, chemical storage, dangerous waste disposal units, emergency eyewash equipment, first aid kits, chemical spill kits, fire extinguishers, etc. The following signs posted in some art studios announce these hazards:

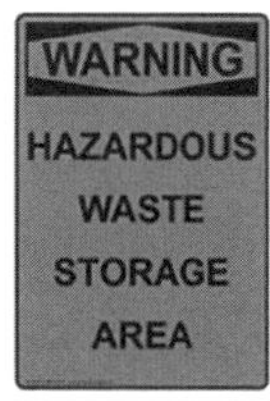

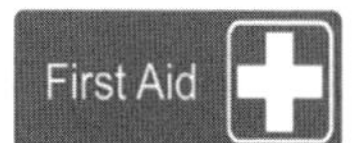

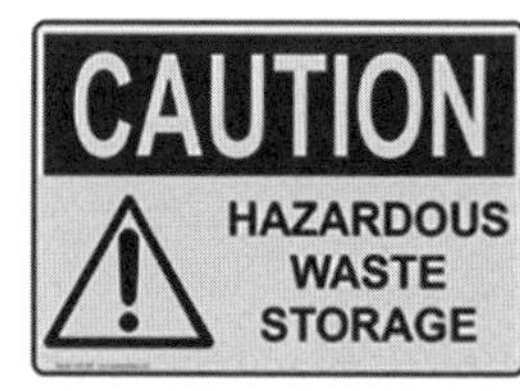

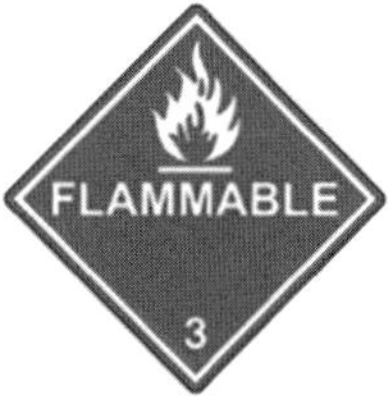

Figure 1: The MySafetySign Company, Brooklyn, NY. 'GHS Pictogram Signs: Global Harmonized System.

This extensive inventory of protective gear and treatments is still inadequate for dealing with the risks generated in many art studios. Artists may be protected by ventilation systems, but these systems merely transfer offending gases from interiors to exteriors. Likewise, toxic wastes, even those that are disposed of in compliance with safety protocols, are merely relocated from the studio to municipal waste facilities. Squirrels, blue jays, mushrooms, pine trees, and insects do not have gear to protect them from exposure to harmful substances. They are victims of artistic creativity whenever studio practice neglects environmental protections.

Introduction to the Introduction Explained

A pet peeve occupies the introduction to the introduction of this book. This peeve stems from the contradiction between art's esteem as a cultural beacon, and its disregard for a cultural challenge that defines the current era – the imperative to replace polluting, wasteful, and depleting material interactions with those that are environmentally beneficial, or at least harmless. This narrative earned the place of honor in the book because the entire project was sparked by indignation that so many contemporary art producers, consumers, instructors, and commentators ignore the material consequences of creating art. Visits to art classrooms and art galleries confirmed the misguided presumption that art is exempt from environmental responsibility.

Over time, my peeve escalated into a campaign because it was bolstered by professionals representing a great range of disciplines who were articulating comparable environmental considerations. The campaign developed into a crusade when I discovered that calls for sweeping material change constituted an emerging philosophy, referred to as New Materialism. This crusade became the book you are reading because I realized that artworks that were materially and conceptually allied with environmental mandates were being created by an international roster of artists that confirmed the mandates proposed by New

Materialism philosophies. Their timely innovations had never before been assembled for study and recognition.

This book was given the title *What's Next?* to highlight the fact that today's artists confront the age-old quandary of choosing between conformity and innovation. Conformists reinforce existing cultural norms; innovators introduce imaginative alternatives. In art, these norms refer to style, theme, format, medium, and tools. Because the tools and mediums that typically line the shelves of art studios are manufactured with little regard for their environmental impact, using them perpetuates the environmental disregard that accompanies Consumer Materialism. It requires a nonconformist to assert that every material choice by an artist is implicated in humanity's impact upon the planet's soils, waters, air, and life forms. These innovative artists locate life-affirming material choices at the top of their agendas.

Introduction: Where Are We Going?

If a map existed that could locate points of view presented in this book, it would appear as a large region without a clear boundary. Its edges would spill into three dynamic zones of contemporary exploration: emerging philosophical discourse, environmental strategies, and avant-garde art. Although this region's borders are fuzzy, the territory of philosophy–environmentalism–art that it charts marks the precise spot where a far-reaching shift in human consciousness is currently being formulated. Each component of this trio is exerting an influence upon the others. As a result, the creative contributions of individuals who identify as philosophers, ecologists, and artists comprise a rich blend of metaphysics, science, and culture. The new commons, referred to in this book as Eco Material art, is fortified by the intellectual rigor of philosophers, made credible by the exacting investigations of ecologists, and enriched by the creative imaginations of artists. This trio of contributors is introducing a sweeping alternative to the status quo by both embodying and envisioning human interactions that are rooted in the material substances and entities that comprise the ecosystems of the planet. It thereby diverges from reliance upon the digitized transmissions, communications, explorations, and entertainments, and the glut of merchandise that currently define contemporary lifestyles. This book honors the many artists whose fertile imaginations are seeding this Eco Material commons with heartening answers to the question, 'What's next?'.

By paying tribute to matter, materiality, and materialization, the artworks that are explored in the following chapters attend to the urgency of mounting environmental afflictions. These bold art initiatives reacquaint the public with the lapsed wonders of weight, texture, moisture, temperature, fragility, suppleness, elasticity, bulge, hollow, contour, and a host of other physical properties that are being neglected in favor of data, simulations, and digital transmissions, as well as subjected to the casual disregard that surrounds mass-produced commodities. While such fundamental qualities of materiality have been accounting for life on Earth for the past 3.5 billion years, currently they are accounting for extinctions, smog, pollution, industrial waste, water shortages, radioactive waste, oil spills, and an alarming litany of other planetary perils. The creative explorations among Eco Material artists present materiality as a strategy to convert society's environmental neglect into responsible stewardship. Their radical strategies and conceptualizations introduce three emerging frontiers of contemporary culture:

- Within the philosophical context, Eco Materialism diverges from the abstract conceptualizations that have dominated recent philosophy.

- Within the social context, Eco Materialism diverges from the dematerialized forms of communicating, working, shopping, learning, and playing.

- Within the art context, Eco Materialism diverges from the elimination of materiality from such recent art forms of creative expression as conceptual, relational, body and sound, as well as social practice, and performance art, and art that registers the dazzling antics of cyber technologies.

These deviations from current norms suggest that the answer to 'What's next?' may not be accounted for by more extensive automation, more advanced robotics, more powerful technologies, and more enticing commodities. Eco Materialism envisions 'What's next?' for humans in the twenty-first century in terms of respectful interactions with the tangible and measurable conditions of Earth systems. Within this book, this vision is provided by artworks that register the functional consequences of their materiality, and that cultivate responsible material interactions. The insights they convey may also be foresights that anticipate the material interactions of world politics, market economies and human fulfillment, as well as the production of art.

Before proceeding to the chapters that explore how philosophy–ecology–art coalesce, let us pause to dissect this trio of inputs so that the significance of each strand of innovation can be appreciated when it is encountered in the artworks presented in the following chapters.

New Materialism

New Materialists construct conscious relationships with all forms of matter, including such common objects as paperclips, coffee mugs, pennies, zippers, napkins, shampoo containers, paper cups, birthday candles, junk mail, pencils, and plastic spoons. These every day, manufactured objects acquire the capacity to enthrall when attention is paid to the elaborate network of professionals whose skills were invested in their production. Each exists as evidence of an extraordinary communal effort. These products represent the combined expertise of architects, builders, engineers, investors, shippers, accountants, bankers, regulators, marketers, distributors, designers, lawyers, mechanics, electricians, plumbers, miners, foresters, truckers, and more. Appreciation of these multidisciplinary achievements is enhanced by the fact that these facets of products and distribution can be widely dispersed across the globe, originating in countries that speak different languages, use different currencies, and represent different economic systems. Even more astounding is the fact that their diverse efforts are synchronized even though they are not mandated by a ruling authority. The stuff that lines market shelves and is typically taken for granted ceases to be ordinary when it is appreciated as the summation of such unfathomable complexity.

An object as ordinary as a wooden pencil provides a compelling illustration of the neglected material marvels that New Materialists savor.[1] Transforming a straight-grained tree into a log requires saws. Saws depend on the mining of metal ore. Transporting the log to the mill requires trucks and rope that are outputs of entirely separate industries. Shipping requires a communication system, roads, and railways. The graphite used in the pencil is mined with special tools and then packaged for transport to an assembly plant, which requires sacks and string and diesel-driven equipment. In the factory, the graphite is mixed with other ingredients to make it strong and smooth. Such complex scenarios also accompany the lacquer that coats the wood, the label that is made of film and applied with heat, the eraser that is inserted into the wood, the brass that holds the eraser in place, and the assembly of these diverse ingredients.[2] To New Materialists, a pencil is a wondrous human achievement.

New Materialists are equally inclined to revel in acorns, moss, nail clippings, banana peels, apple cores, chicken bones, peanuts, and egg shells, because each of these items encapsulates ongoing evolutionary struggles and genetic experimentations that originated in primeval times. By attending to these rich evolutionary histories, such banal objects epitomize successful responses throughout eons of shifting global conditions. These are the triumphant survivors!

Such awareness adds an 'extra' to the 'ordinary' stuff of our lives. Nail clippings appear extraordinary to New Materialists because they appreciate the advantage we primates acquired to perform delicate manual operations when nails evolved from claws about sixty-five million years ago! Egg shells are remarkable because the shells of the eggs of tens of thousands of species of insect, fish, reptile, amphibian, and bird are differentiated by size, color, thickness, and shape! Acorns are wondrous because they feed blue jays, pigeons, ducks, woodpeckers, mice, squirrels, rodents, ponies, pigs, bears, deer, moth larvae, and humans! Moss is marvelous because it was the first plant to evolve on Earth approximately 320 million years ago! Banana peels are amazing because they can soothe bug bites, stings, and other skin irritations; exfoliate and nourish your skin; and fade warts!

Along with the recovery of wonder, New Materialists seek ways to regain kinship between the physicality of human bodies and the physicality of the planet. Such accord was relinquished when industry and commerce began providing the means to be sheltered, clothed, entertained, and fed. These modern conveniences replaced personal connections with life-sustaining resources with mechanical and automated processes. People could survive without encountering plants with their roots still in the ground, animals grazing and birthing, insects pollinating fruit, and wool growing on sheep. As a result, the stuff of everyday life lost its connection with the substances and processes inherent to the planet. Instead today's stuff reflects human designs and technologies. By the time most items enter peoples' lives, the resources assembled to create them have been manipulated, dissected, and reconfigured out of recognition. New Materialism seeks to reverse this loss of sensory enrichment and rapport with our planetary home.

What Is New about New Materialism?

Since materialism factors so prominently in this book's exploration of 'What's next?', it seems imperative to contrast the word's newly acquired applications from its familiar associations.

New Materialism Is New Because It Reverses Anthropocentrism

Around the turn of the century,[3] three new terms emerged and were quickly adopted in scholarly and popular discourse. They all utilize the prefix 'anthro' meaning 'pertaining to humans'. 'Anthropocentrism' asserts that humans are more significant than other life forms. This mindset regards humans as not only separate from animals, plants, and mineral resources, but also superior to them. Non-human and non-living entities, therefore, are viewed as resources that exist exclusively for the benefit of humankind. The second and third terms address the consequences of Anthropocentrism. 'Anthropogenic' describes human-induced changes to the conditions and components of the planet. Agriculture, roads, and settlements are early examples of Anthropogenic influences. Hydroelectric installations, off-shore drilling, and mountain-top removal are current examples. This escalation accounts for the third term. 'Anthropocene' refers to the changes being imposed upon bio/geo/chemical planetary systems by the mega power of recent technologies, and the escalating demands for energy and consumption of resources that accompany them. While this assertion of human supremacy might be interpreted by some as a triumph, it also registers that humanity's takeover of the planet is the result of substances being claimed and exploited, forces being captured and squandered, and their potential for utility being depleted and exhausted. Concern that the resulting changes to Earth systems are so extensive and so enduring that they may constitute a new geological epoch, comparable to the introduction of the Paleozoic and the Mesozoic eras, is also revealed by this term. In August 2016, a team of scientists presented a formal recommendation to the International Geological Congress to officially name the new era the Anthropocene.

Rather than emphasizing human power and accomplishments, New Materialism positions humans within the teeming complexity of planetary events, not at the peak of the evolutionary ladder. The leveling of this entrenched hierarchy dissolves the assumptions of human privilege and status that have long been central to western thinking. New Materialists cultivate a hybrid view of reality that unifies nature and culture, matter and mind. According to Manuel DeLanda, a Mexican American writer, artist and philosopher who helped originate New Materialism, all beings exist within shared geological, social, biological, and psychological conditions. There is nothing exclusive about humanity. Even our bodies are composites of atoms, molecules, parasites, microbes, bacteria, and non-organic chemicals. Physically, we are participants in a confederacy of life forms. Behaviorally, we contribute to a reciprocal collective of planetary events. DeLanda extends this inclusiveness to non-living materials that, he believes, embody 'non-organic vitality.'[4] He expresses a view, widely shared among New Materialists, that inert forms of matter are as vital as living entities because they too are immersed in the Earth's flow of matter and energy. Whereas Anthropocentrism features the unique capacities and privileges of humans, New Materialism considers the 'inter' aspect of each human 'action'. Humans not only affect, but are affected by water, soil, stones, metals, minerals, bacteria, toxins, food, electricity, cells, atoms, plastic, and garbage. As a result, the presumption that humans can assert total control over the outcomes of their interventions is false. The term 'post-human' is often used to describe the dismantling of Anthropocentric assumptions.

New Materialism Is New Because It Is the Antithesis of Consumer Materialism

While the word 'materialism' originated in such geographically dispersed regions as India, Greece, and China between 800 BC and 200 BC, it was not until the

Industrial Revolution and the development of mass production that the term acquired its association with 'consumerism'. Consumerism is characterized by such behaviors as attending to shopping as a recreational activity; incurring debt to obtain luxury items; acquiring merchandise to earn affection and self-esteem; and believing that happiness and status are derived from material possessions. Consumerism is not dependent upon income. It is defined by attitudes and aspirations that may or may not be realizable.

The term 'conspicuous consumption' is often used to refer to Consumer Materialism that is wasteful and indulgent. Such purchases are induced by 'planned obsolescence', the policy instituted by manufacturers to design goods that intentionally break down or become obsolete, necessitating a new purchase. Both terms apply to businesses and customers that assume consumption is good, and more consumption is better. This lifestyle principle is evident in industrialized countries such as the United States, where the average home is filled with approximately three hundred thousand items.[5] Despite the tripling of home sizes in the past fifty years,[6] many residents find that there is still not sufficient space to store their accumulated stuff. Twenty-five percent of people with two-car garages don't have room to park cars inside them.[7] One out of every ten Americans rents offsite storage.[8]

Food, clothing, house wares, toys, hygiene products, and pharmaceuticals, indeed almost everything currently owned by the average family, is mass-produced. There is little incentive for frugality or repair when items are abundant, cheap, and replaceable. Consumer Materialism, because it invites careless disregard of material objects and neglect of the material environment, represents the opposite values promoted by New Materialism.

New Materialism Is New Because It Rejects Correlationism

In addition to the rejection of economic consumerism and Anthropocentrism, New Materialism rejects Correlationism, a philosophical stance that dominated European and Anglo-American philosophies throughout the twentieth century. This philosophical stance asserts that humans can only know about the material environment as it exists within the mind. As such, all knowledge depends upon the 'correlation' between thinking and being. The significance of material reality shrinks when existence is considered to be dependent upon human cognition. As Quentin Meillassoux, an influential French philosopher, explains, 'We can't know what the reality of the object in itself is because we can't distinguish between properties which are supposed to belong to the object and properties belonging to the subjective access to the object.'[9] Meillassoux highlights the role of mathematics in forming this correlation. This belief reduces materiality to a mere correlate of immaterial language and immaterial thought.

The denigration of materiality by Correlationists provides the springboard for the elevation of materiality by the new generation of materialists. New Materialists are rebooting contemporary culture's operating systems by rejecting the notion that thought, language, and representation are more significant than physicality and substance. This oppositional stance is expressed by Manuel DeLanda when he states, 'It is absurd to think that complex self-organizing structures need a "brain" to generate them.'[10] He cites atmosphere–hydrosphere conditions that do not require a brain to produce thunderstorms, hurricanes, and wind currents. Furthermore, he notes that no human brain was needed for bacteria to develop fermentation, photosynthesis, and respiration, which they accomplished two billion years before humans evolved. To New Materialists, the world in which humans exist is fundamentally independent of the human mind and human discourse.

New Material Principles Recapped

Timothy James LeCain, a historian of the environment and technology, summarizes the basic principles of New Materialism when he states that,

> *neo-materialist theory proposes that humans and their cultures are best understood as the products of their material environment, not its masters. Even more fundamentally, neo-materialism challenges the still dominant modernist belief that human culture is distinctly separate from the material world, suggesting that matter not only helps to create human intelligence, creativity, and culture, but may often be best understood as constituting these things. At its heart, this emerging neo-materialist theory challenges the modernist faith that the human intellect and culture have taken us out of nature, suggesting that humanists can build a powerful new methodological approach by adopting the contrary position: human culture must be understood and analyzed as a part and product of the material world, not its antithesis.[11]*

New Materialism: What's Next?

New Materialism originated in the second half of the 1990s. Credit for its emergence is shared between Manuel DeLanda, the theoretician and Rosi Braidotti, a philosopher and feminist. They developed New Material principles independently of one another. DeLanda proposed blurring entrenched dichotomies, such as the distinction between the natural and the artificial, and the living and the inert. Braidotti focused on the leveling of dichotomies by eliminating priorities. She rejected long-standing assumptions that the mind is more important than matter, and culture is superior to nature. The conceptual framework they established has been applied to a wide range of disciplines.

Scholars representing social studies, philosophy, political science, anthropology, and theology have discovered in New Materialism a conceptual framework for reversing the disregard for materiality that currently prevails in cultural, social, ethical, and pragmatic practices. Several of these scholars have developed distinct formulations of matter, materiality, and the processes of materialization that have become designated schools of New Material thought. Just as Baptists, Lutherans, Prespyterians, Episcopalians, and Catholics are all Christians, proponents of Thing Theory, Media Materialism, Object Oriented Ontology, Dark Vitalism, Feminist Materialism, and Transcendental Materialism are all New Materialists. This book introduces Eco Materialism: an application of materialism to ecological studies and environmental concerns. Eco Materialists are committed to replacing current industrial, technological, digital, and media norms that privilege short-term human convenience over long-term environmental vigor. Their agenda is grand, radical, and timely.

Eco Materialism

The 'Eco' in Eco Materialism can be traced to the origin of the word *ecology* – the Greek word *oikos*, meaning 'dwelling place'. Over time, *Ökologie* was coined in Germany, which became *ecology* in English. The 'dwelling place' referred to in the definition of ecology does not indicate huts, cottages, apartments, mansions, nests, dens, or any other lodgings. Ecology conceives of the entire planet as a single home shared by all living organisms on Earth. Ecologists respond to this domicile model of Earth by investigating its functional operations, while environmentalists respond by applying home-care regimes to maintain and improve it. Both regret that humans living in industrialized societies rarely include the planet within their concept of 'home'. Eco Materialists join ecologists and environmentalists by noting that the penalty of humanity's self-imposed exile from their planetary place of residence is evident in wasteful neglect that has made this 'home' sick. At the same time, we may be making ourselves 'homesick' by longing for a planet that invites our tender devotions.

The consequences of humanity's current material interactions are apparent in extinctions, erosion, deforestation, desertification, climate change, soil degradation, air pollution, acidified oceans, toxic spills, noxious emissions, vanishing coral reefs, melting glaciers, water contamination, and more. These accountings typically lead back to privileged, human-centric exclusiveness. Because disrupted ecosystems endanger all forms of life, Eco Materialists attend to the welfare of humans and non-humans. Multispecies inclusiveness is key to the merger of ecology and materialism.

Eco Materialism reunites people with their Earthly homeland by situating New Material principles within the context of ecology and environmentalism. Even though it acknowledges that industrial productivity and innovative technologies have delivered the material abundance they promised, it also recognizes that these operations are endangering the atmosphere as well as global forests, oceans, soils, and the life forms that inhabit these regions. Eco Materialism addresses the urgency of halting the binge of Consumer Materialism by promoting prudence and restraint. This field of inquiry, therefore, constrains the extended frenzy of material acquisition and misconduct among humans in industrialized societies. It not only challenges heedless human behaviors that disrupt or damage ecosystems, it strives to devise interactions that fortify the ability of ecosystems to support diverse forms of life. This task is daunting.

Eco Materialism Introduces New Social Norms

Rampant disregard that characterizes material interactions is entrenched in industrialized societies. Reversing this pattern begins by exposing contemporary consumers – who currently encounter goods only after they are fully formed – to evidence of how materials are extracted, hauled, loaded, poured, heated, shaped, assembled, manufactured, and packaged. Likewise, instilling Eco Material respect involves confronting consumers with the dismal fates of many unwanted items once they are classified as waste.

The takeover of human attention by digital networks is another source of material indifference that appears on Eco Material agendas of reform. Because interactions generated by keyboards and experienced on screens lack physicality, people who shop, work, and entertain themselves in cyberspace forgo opportunities for sensory contact with materiality. Terminology that defines cyber activities emphasizes their immateriality. We connect 'virtually'; befriend 'cyber' strangers; compute in a 'cloud'; connect with 'mobiles'; 'stream' programs; and 'tweet' messages. Eco materialists note that the devices that deliver these weightless, intangible, frictionless technologies nonetheless scar the physical body of Earth by emitting air contaminants (acid fumes, volatile organic compounds, doping gases); water emissions (solvents, cleaning solutions, acids, metals); and wastes (silicon, solvents).[12]

In addition to the challenges to Eco Material reform that are posed by negligent consumers and preoccupied cyber surfers, another hurdle lies with conscientious citizens who dwell on such dismal news reports as 'Tipping Point? Earth Headed for Catastrophic Collapse, Researchers Warn'.[13] It is difficult to motivate behavioral changes among people who have abandoned hope of redemption.

It entails countering a mounting litany of ecosystem woes: climate change, mass extinction, drought, flood, nitrogen deposition, rainforest collapse, overgrazing, acidified oceans, and so forth.

A new word has been added to the English language to convey ecological despair. In addition to 'suicide' (the murder of self), 'homicide' (the murder of a person), and 'genocide' (mass murder), environmentalists have added 'ecocide' (the murder of an ecosystem). 'Ecocide' fatalities are associated with dead lakes, species extinctions, and bee colony collapse because prospects for restoring these conditions are doubtful. In 2010, a proposal to classify 'ecocide' as an international crime was submitted to the United Nations International Law Commission. The proposal identifies human-induced environmental ills as criminal offenses. In addition, it establishes a legal duty of care for potential victims of ecocide.

A fourth hurdle confronted by Eco Materialists is posed by the close encounters with environmental collapse that pervade popular culture. Scenarios of eco disasters proliferate in books such as Paolo Bacigalupi's *The Water Knife* (2015) and Davis Schneiderman's *Drain* (2010); films such as *The Day After Tomorrow*

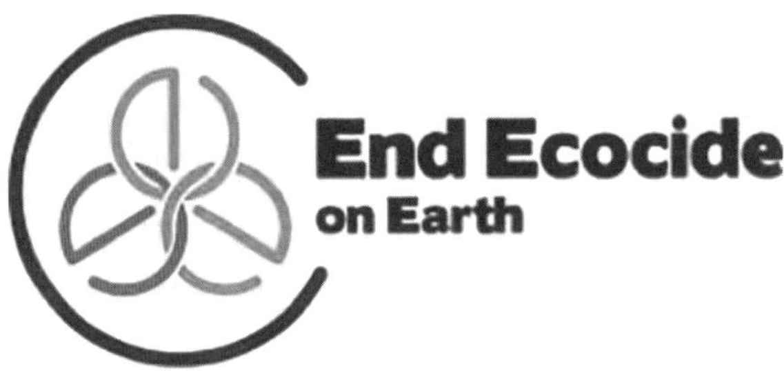

Figure 2: End Ecocide on Earth logo (2014).

(2004) and *The Impossible* (2012); games such as *Brink* (2011) and *Oil Rush* (2012); and television shows such as *Tokyo Magnitude 8.0* (2009) and *Earth 2100* (2009). Whether popular entertainments advance or retard the Eco Material agenda is difficult to determine. On the one hand, these powerful vehicles of mass communication may awaken the public's consciousness of needed reforms. However, consumers of these fictionalized eco disasters expect 'nick-of-time' reversals of doomsday scenarios. Virtual dangers with virtual outcomes are far more appealing than real-life alternatives that require sacrifice and patience.

Eco Material Resolve

Despite alarming evidence and psychological obstacles, futility and despair do not appear on the Eco Material radar. Its adherents track waves of resilience amid the ruins, and evidence of strategies that heal and enliven. A resolute commitment to hopeful outcomes concludes every Eco Material assessment of the current state and future prospects of the planet. Eco Materialists navigate this rocky zone of optimism, intent on halting society's short-sighted and destructive practices and establishing alternatives that are long-term and life-enhancing. For all these reasons, Eco Materialists are rewriting the operating manual for humanity's role in evolving events on the planet. The new manual is based upon the fundamental ecological insight that the planet is constituted of interrelated phenomena that extend to the functioning of the human organism, because approximately one hundred million non-human cells occupy the human body. This is ten times greater than the cells that share the person's DNA. Human existence, therefore, is not independent and self-determined; it as an ongoing, multispecies drama in which opportunity and emergency are continually being negotiated with myriad microscopic entities.

Furthermore, no precise border separates 'you' from your surroundings. While membranes enclose your organs, bones, and muscles, preventing you from blending into your environment, these surface layers are not sealed shut. Maintaining life requires many tiny pores and a few large portals that act as thoroughfares by providing entryways for resources and exits for wastes. In addition, sensory cells and nerves collect sights, smells, sounds, tastes, temperatures, and tactile sensations from your environs. They transmit this information to cells located within the recesses of your brain. In this manner, your body performs concurrent, contrasting functions: it separates by enclosing your physical organism, and it connects by providing channels for your physical organism to merge with its surroundings.

The fuzzy borders and fluid boundaries inherent to our physical existence challenge ordinary thought processes. These mental constructs rely upon accommodating the simultaneity of causes and effects within perpetually unfolding events. John Muir (1838–1914), an influential founder of the environmental movement, provided a vivid description of such fusion when he wrote, 'The sun shines not on us, but in us. The rivers flow not past, but through us, thrilling, tingling, vibrating every fiber and cell of the substance of our bodies, making them glide and sing.'[14] However, the popularity of protective sunglasses and sunscreens indicates that Muir's exuberance regarding sunshine is not the norm. Likewise, the proliferation of pools that treat water with chemically laden sterilizers is an example that Muir's call for immersion in wild waters is not being heeded. Avoiding sun and rejecting wild waters are two of many indicators of contemporary humans' willful alienation from the elemental substances and conditions of Earthly systems.

Such estrangement defies the sharing that is imbedded in our life blood, originating eons ago within the planet's primeval seas. The epic accounting of humanity's alliance with planet Earth's geological and biological record can be traced to these waters, which possessed a remarkable capacity to dissolve a wide range of mineral salts and gases. The first stirrings of life on Earth emerged out of this liquid fusion. In tiny increments, over vast spans of time, outer skins developed that enabled organisms to contain this life-sustaining 'sea' within their bodies. When internal pumping systems developed to circulate these internal seas, organisms took up residence on land. Saltwater became the protoplasm of cells; sap in plants; and blood in animals. Although humans are included in this narrative, awareness of this genetic mutualism rarely factors into legislative policies, manufacturing protocols, or consumer habits. Eco Materialists summon the conceptual framework for such pursuits.

Eco Systems as Models for Social Systems

Essentially, Eco Materialism eradicates the privileged status of a human-centered 'I', 'me', and 'we'; omits the possibility of 'mine'; and replaces these presumptions of special status with interspecies 'us' and 'ours'. This unifying worldview catalyzes Eco Materialism's rejection of humanity's self-serving exploits. The new notion of 'self' involves membership of a vast multispecies family that looks very different from ourselves. Membership also includes non-living environmental factors such as energy (light and heat), climate (wind, precipitation, temperature), and structure (topography and porosity). Such a composite entity, known as an ecosystem, consists of four units: habitats, watersheds, communities, and niches. These fundamental components of ecological study introduce the range and richness of Eco Material explorations.

Habitats are locations that provide the resources that a species requires to nourish and maintain its bodily functions, build shelter, and manage its wastes. These resources may consist of abiotic substances such as water, sunlight, minerals, oxygen, and carbon dioxide, or biotic resources derived from living and dead microorganisms, plants, animals, and their remains. Survival habitats of some species are hot and dry. Others species prefer cool and moist conditions. An organism's relationship with its habitat is always reciprocal. While habitats provide the substances and conditions that a species requires to survive, plants and animals also produce changes by depositing wastes, consuming food, casting shadows, compressing soil, and constructing shelters. Thus habitats and their populations engage in reciprocal processes of affecting and effecting. Eco Materialists apply this equation to human behaviors. They consider the resources gleaned to benefit humans against the resources diminished by excessive harvesting and toxic waste. They factor into their environmental accountings the long supply chains that liberate humans from dependence upon their local habitats by importing resources from afar. They might also interrogate humanity's ability to produce survival conditions in inhospitable deserts and polar tundra by constructing furnaces, air conditioners, insulation, waterproof boots, mittens, pipelines, reservoirs, and irrigation.

Watersheds are areas circumscribed by ridges from which water drains into streams, rivers, lakes, and/or oceans. Functionally, watersheds collect water from rainfall, channel rainfall runoff, recharge groundwater, transport sediment, influence local climate, and provide habitats for flora and fauna. A plethora of human inventions exist to bypass the constraints imposed by watersheds upon their occupation and development. Pipes, pumps, drains, gutters, reservoirs, culverts, irrigation, and sewage treatment plants harness water and direct its flow. When water appears from faucets far from where it originated, and when it disappears into drains without a known destination, it is taken for granted. Eco Materialists draw attention to the environmental costs of such disregard by seeking ways to re-engage people in industrialized nations with water as a precious, life-sustaining substance. They apply this mission to all the life-enhancing components of the planet, such as soil and microbes, from which so many people have become alienated.

Communities consist of all the living entities within a particular habitat. Their variety and abundance are determined by the conditions within their habitat, by the relationships among members of the same species, and by interactions between different species. These relationships can take many forms. They may be predatory, parasitic, competitive, cooperative, symbiotic, and so forth. Eco Materialists might consider the material consequences of human-to-human relationships that exceed such biological imperatives: professional colleagues; strangers in public spaces; virtual relationships in social media; celebrities, legislators, politicians; and so forth. But Eco Materialists are likely to pay particular attention to humanity's non-human relationships: communities of insects, weeds, molds, and germs, and microbes that are essential for sustaining planetary life.

Niches are not tangible components of the environment. Niche describes a population's response to, and effect upon, the distribution of resources in an ecosystem. The niche principle is observed when the population of a species increases because the resources it requires are abundant, and/or predators, parasites, and pathogens are scarce. Likewise, populations decline when their advantages diminish. These complementary interactions constitute the roles that species play as contributors to, or consumers of, a common pool of local resources. The niches for a fox, for example, are to eat small mammals, amphibians, and insects; supply blood that feeds flies and mosquitoes; and provide nourishment for scavengers and decomposers after it dies. Niche accounts for fluctuating populations that allow the system to replenish its resources, and thereby remain vital. Of the trillion species that are estimated to exist on Earth,[15] humans are uniquely capable of defying the constraints imposed by niche dynamics. Agriculture, irrigation, writing, architecture, mathematics, religion, urban planning, metallurgy, trade, money, printing, firearms, combustion engines, nuclear power, cybernetic technologies, and wireless communications fill this niche-defying roster. While these breakthroughs have accounted for humanity's proliferation and ascendancy, they tend to disregard humanity's dependence upon the finite pool of resources that we share with our multispecies neighbors. The goal of Eco Materialists' diverse creative enterprises is to construct niche roles for humans that sustain them and their fellow Earthlings.

Like ecologists, Eco Materialists engage flows of energy and matter between the organisms that occupy an area and the conditions and materials that surround them. Habitats and watersheds are accounted for whenever they address the non-living material aspects of an ecosystem; communities and niches underlie considerations of living entities. Most often, all four perspectives intermingle. Watersheds consist of diverse habitats; habitats support communities of plants, animals, insects, and microbes; communities occupy numerous niches; niches relate habitats to communities. By acknowledging such dynamic interconnections, Eco Materialists honor the intricate and evolving materiality of planet Earth, which may be unique in all the cosmos.

Eco Materialism and Contemporary Art

The preceding text surveys the radical reinterpretation of materiality that has propelled Eco Materialism into existence. It stakes its claim as an official twenty-first-century movement by overturning three social and cultural values that prevailed throughout the twentieth century: Consumer Materialism, Anthropocentrism, and Correlationism. While this investigation is reshaping numerous disciplines and pursuits, visual art has a special advantage that is not enjoyed by cultural theory, social studies, theology, and many other disciplines currently grappling with the implications of Eco Materialism. This is because manipulating matter is embedded in millennia of artists' operations. The innovative interactions with substances and forces that are currently being generated by an international group of artists perpetuate this distinguished heritage. Their efforts revive precedents whereby innovative modes of material creation are tested in art studios on behalf of the culture at large. For example, artists popularized ceramics in 24,000 BC in Czechoslovakia, copper in 9000 BC in the Middle East, and bronze between 3500 and 2500 BC in West Asia and Europe. These ancient artists explored the aesthetic potential of these new materials long before their practical applications were developed.

This tradition of material exploration was abandoned in recent decades by artists who pursued conceptual, relational, body and sound, as well as social practice and performance forms of creative engagement that excluded material interactions. Furthermore, the infatuation with the dazzling antics of cyber technologies reduce materiality to a vestige of a defunct era. Materials were retired from artists' toolboxes to accommodate virtual reality, video games, cyborg art, telecommunications, vector graphics, Internet art, video, computer animations, robotics, mathematically generated fractals, 3-D cameras, computational photography, sculpting in virtual space, hacking, digital painting, tactical media, Tumblr and Giphy transmissions, computer generated technology for film, nano technologies, and holography. Art departments in 182 college campuses in the United States alone established New Media programs to acquaint their students with the extravaganza of digital creativity that is being staged in the dematerialized zone of cyberspace.[16]

By reasserting the material basis of its creative enterprises, Eco Material artists are affirming art's time-honored role as an era's exemplar of material interactions. They are reacquainting the public with the lapsed wonders of weight, texture, moisture, temperature, fragility, suppleness, elasticity, bulge, hollow, contour, and a host of other physical properties that belong to Earthly matter. But Eco Material artworks are not merely explorations of physicality. Their materiality is elevated by the context of art whereby it acquires conceptual, pragmatic, and ethical significance. As such, Eco Material art exceeds reconnection with physical matter. It is also a catalyst for cultural transformation.

Here too, art history provides a compelling precedent. Approximately five hundred years ago, during the Reformation, some prescient European artists nudged public awareness back to the physicality of the everyday world from the otherworldly zones of heaven and hell. By shifting cultural consciousness away from prophecy and miracles relayed by the Catholic Church, these artists reinforced the upheaval induced by the Protestant Reformation in the sixteenth and seventeenth centuries. Artistically, this historic shift took the form of still life paintings and portraits that drew attention away from otherworldly visions to honor observable substances and entities. These art subjects were radical at the time because they announced that actual things and real people merited the attention formerly reserved for deities and the supernatural.

Whereas religious unrest inspired the Reformation artists, environmental jeopardy is motivating today's materially oriented artists. In both instances, artists reformulated culture so that attitudes and behaviors would be calibrated on a scale of material significance. Then, as now, artists are paying tribute to matter, materiality, and materialization. The contemporary artists who are manifesting these values actualize the Eco Material mandates by resisting the enticements offered by manufactured products and computer technologies. Nonetheless, their reassertion of materiality is not a nostalgic return to pre-industrial lifestyles. They pay as much attention to materials that take the form of iPhones as they do to sea shells.

Eco Material art also diverges from art's long-standing identification with original objects intended for display and commerce in the following ways: Eco Material art is not an enduring manifestation of an artist's skill; Eco Material art is not a representation or interpretation; Eco Material art is not an intellectual conceptualization or emotional expression; Eco Material art is not created with neutral mediums. In essence, Eco Material art relinquishes the centrality of 'form' to focus on 'matter'. Form is confined to static measures of shape and dimensions, whereas matter engages a multiplicity of physical qualities and the ever-changing continuum of conditions. Spatially, form and matter encompass the object and its environs; temporally, they expand 'now' to include each substance's lineage (how it came into being) and legacy (its lingering effects). In all these ways, materials are actively responding to and instigating change. This book acknowledges the dedicated efforts of Eco Material artists, the cultural transformations they are initiating to defend the planet by halting ecocides and promoting life-affirming renewable cycles of matter and energy.

By reinforcing the literary offerings at the forefront of contemporary intellectual discourse, Eco Material artists join a culture-wide initiative to replace the casual disregard that characterizes contemporary material interactions. Despite the growing fear of pandemics, famine, extinction, water shortages, social injustice, and war, Eco Materialists' faith in positive change propels their valiant attempts to ensure that humanity's fate is not fatal. Some accomplish this by drawing attention to the losses that have accompanied recent gains in speed, power, comfort, variety, and convenience. Others actively promote citizens' engagement with the water they drink, the air they breathe, the food they consume, the fibers in their clothing, the minerals in their tools, and the fuels they burn. Some reinstate opportunities for sensory appreciation that have been usurped by the current infatuation with disposable abundance and mediated experience.

Whether or not Eco Materialism assumes the dominant position in the twenty-first century that is comparable to the significance of Romanticism in the nineteenth century or Existentialism in the twentieth century, the spontaneous outpouring of strategies to connect humans with the physicality of the planet serves as a powerful transmitter of a developing intellectual, moral, and cultural climate. Significantly, for readers of this book, Eco Materialism opens a vast territory for vanguard artistic explorations. Eco Material attention to the world of solids, liquids, plasmas, and gases may originate in such neglected substances as sap, fur, bone, soil, stone, bark, and moss, as well as such familiar materials as Gortex, AstroTurf, lubricating oil, plastic, and Doritos. Every material invites consideration of the complexity of its formation, the efficiency of its design, the multiplicity of its functions, and the beauty of its form. At the same time, every material invites scrutiny regarding the resources and by-products of its production, the consequences of its use, and the outcomes of its disposal. Such timely initiatives replace the association of 'materialism' with consumerism and convenience, and secure it as an opportunity for participation and generosity. Ultimately, Eco Material artists respond to 'What's next?' by offering hope.

Your opinion: What's next?

Reader Interaction

GOOD BETTER BEST: An Eco Material Guide to Art Production

Cycle-logic is a new addition to the lexicon of English-speakers. It was invented to formalize the entry of a new ethical construct regarding human interactions with the materials of this planet. The term reverses the widespread assumption, made possible by industrialized and globalized manufacture, that material goods are cheap, disposable, and replaceable. Instead, it asserts that the molecules that comprise our bodies and all our material effects are inherited from our ancestors. This bequest does not belong to us. As living beings, we are bound to honor this loan by bequeathing it to future generations in order to enhance an ecosystem's ability to maintain vitality, cope with stress, withstand adversity, recover from a disturbance, and perform its own recycling strategies. Cycling Earth's finite supply of molecules is a process of material use (from acquisition to disposal), that applies to living species (from microbes to mammals), and incorporates all human behaviors (from mining metals to producing art).

Scrutinizing the environmental impacts of an artist's material interactions may offer the one indisputable standard of merit regarding art created in the 21st century. Material responsibility applies to performance art, conceptual art, and social practice art, as much as to painting, sculpture, and installation art. By challenging the presumption of art-for-art's sake, it introduces the principle of art-for-life's sake.

GOOD: Avoid manufactured art supplies that squander non-renewable resources and those that contaminate renewable resources by purchasing canvases, adhesives, charcoal, pigments, etc. from companies that produce these products according to "green" protocols.
BETTER: Use discarded and reprocessed mediums that are traded, found, refurbished, cultivated, bartered, scavenged, or mined.
BEST: Detoxify a polluted resource (e.g., decontaminate soil laced with heavy metals to make pigment) and/or utilize an excessive product (e.g., harvest an invasive weed to make paper).

GOOD: Reduce energy consumption and waste production during art production.
BETTER: Reduce energy consumption and waste production during transportation, packaging, and display of art, as well as art production.
BEST: Eliminate energy consumption and waste production during transportation, packaging, and display of art, as well as art production.

GOOD: The art work minimizes environmentally costly investments in climate control, archival papers, and storage.
BETTER: The art work eliminates environmentally costly investments in climate control, archival papers, and storage.
BEST: The artwork is either biodegradable or it is non-biodegradable but recyclable.

GOOD: Temporarily divert discarded materials from landfills (e.g., create sculpture with used plastic bags).
BETTER: Permanently divert discarded materials from landfills (e.g., disassemble objects made from multiple materials so that all the components can be recycled).
BEST: All wastes are retained and used as resources for future art or non-art activities.

GOOD: Ingredients are sourced from within a 100 mile radius.
BETTER: Ingredients are sourced from within a 50 mile radius.
BEST: Ingredients are sourced locally.

GOOD: Halt practices that decrease finite resources.
BETTER: Adopt practices that use renewable resources.
BEST: Invent practices that increase resources (e.g. cultivate a medium).

GOOD: Purchase a used tool or repair a broken tool.
BETTER: Purchase a used tool or repair a broken tool and share it.
BEST: Use scraps and discards to create your own tools. Share them.

GOOD: Consideration is given to the current impact of artwork on the immediate environment.
BETTER: Consideration is given to the long-term impact of artwork on the immediate environment.
BEST: Consideration is given to the long-term impact of artwork on the extended environment.

GOOD: Interference is temporary.
BETTER: Interference is avoided.
BEST: The artwork's presence improves the vitality of the environment (e.g., it removes toxins from the air).

Endnotes

1   Leonard E. Read, 'I, Pencil,' Foundation for Economic Education (FEE), March 3, 2017, accessed December 10, 2017, https://fee.org/resources/i-pencil/.

2   Leonard E. Read, 'I, Pencil: My Family Tree as Told to Leonard E. Read,' *Library of Economics and Liberty*, [1958] 1999, accessed December 13, 2017, http://www.econlib.org/library/Essays/rdPncl1.html.

3   The term 'Anthropocene' gained popularity after Nobel Prize-winning chemist and climate scientist Paul Crutzen began to use it in 2000.

4   Abstractgeologist, 'Geophilosophy and the Aesthetics of Emergence,' *Abstractgeology*, May 29, 2012, accessed August 11, 2017, https://abstractgeology.wordpress.com/2012/05/26/geophilosophy-and-the-aesthetics-of-emergence/.

5   Mary MacVean, 'For Many People, Gathering Possessions Is Just the Stuff of Life,' *LA Times*, March 21, 2014, accessed December 12, 2017, http://articles.latimes.com/2014/mar/21/health/la-he-keeping-stuff-20140322.

6   Margo Adler, 'Behind the Ever-Expanding American Dream House,' *LA Times*, July 4, 2006, accessed December 9 2017, http://www.npr.org/templates/story/story.php?storyId=5525283.

7   Gretchen Rubin, 'Good Stuff,' *New York Times*, August 18, 2012, accessed November 30, 2017, http://www.nytimes.com/2012/08/19/opinion/sunday/clutter-storage-and-happiness.html.

8   Jon Mooallem, 'The Self-Storage Self,' *New York Times*, September 2, 2009, accessed November 30, 2017, http://www.nytimes.com/2009/09/06/magazine/06self-storage-t.html?em&_r=0.

9   Quentin Meillassoux, 'Time without Becoming' (lecture given at Middlesex University, 2008), 6, accessed December 10, 2017, https://speculativeheresy.files.wordpress.com/2008/07/3729-time_without_becoming.pdf.

10   'Interview with Manuel DeLanda,' *New Materialism: Interviews & Cartographies* (University of Michigan Library: Michigan Publishing), accessed August 11, 2017, http://openhumanitiespress.org/books/download/Dolphijn-van-der-Tuin_2013_New-Materialism.pdf.

11   Timothy James LeCain, 'Against the Anthropocene: A Neo-Materialist Perspective,' April 23, 2015, accessed December 10, 2017, https://www.history-culture-modernity.org/articles/10.18352/hcm.474/.

12   Frans Berkhout and Julia Hertin, *Impacts of Information and Communication Technologies on Environmental Sustainability: Speculations and Evidence* (Brighton: University of Sussex, 1995), accessed August 13, 2017, http://www.oecd.org/sti/inno/1897156.pdf.

13   Stephanie Pappas, 'Tipping Point? Earth Headed for Catastrophic Collapse, Researchers Warn,' June 6, 2012, accessed December 9, 2017, https://www.yahoo.com/news/tipping-point-earth-headed-catastrophic-collapse-researchers-warn-171704844.html.

14   'Mountain Thoughts,' in *John of the Mountains*, ed. Linnie Marsh Wolfe (Madison, Wisconsin 1938) University of Wisconsin Press; 2 edition (May 15, 1979) accessed December 9, 2017, http://www.sierraclub.org/john_muir_exhibit/frameindex.html?http://www.sierraclub.org/john_muir_exhibit/writings/mountain_thoughts.html.

15   Peter Dockrill, 'The Largest Study of Life Forms Ever Has Estimated That Earth Is Home to 1 TRILLION Species,' *Science Alert*, May 3, 2016, accessed December 7, 2017, https://www.sciencealert.com/the-largest-study-of-life-forms-ever-has-estimated-that-earth-is-home-to-1-trillion-species.

16   Dr. Edgar Huang, *New Media Programs in the United States*, accessed August 13, 2017, http://www.iupui.edu/%7Ej21099/nmschools.html.

ECO MATERIAL ALLURES & REPULSIONS

Secretions

ECO MATERIAL ALLURES & REPULSIONS

Secretions

'Experience living hell!' could be the caption for the scene I observed while waiting for a bus in Manhattan on a brutally hot day in July. Workers wearing hard hats, work pants, leather gloves, and thick boots were jack hammering the sun-baked pavement, their bodies gyrating with each powerful pounding. The rat-tat-tat was deafening. The concrete dust was blinding. I thought, 'These unfortunate men must have the worst of all possible jobs!'

Then I noticed the phrase *Dirty Jobs* on a large poster that was plastered on the bus shelter where I stood, advertising the television program that had been airing on the Discovery Channel since 2004. Its host, Mike Rowe, travels the globe searching for the most disagreeable ways to earn an income. Rowe's smiling face dominated the poster. But it was the grime smeared all over his chin, cheeks, and forehead that was its most distinguishing characteristic. An arrow near each blob pointed to text identifying the job that produced this filth. Jackhammer operator was not among them. In fact, the poster included very few jobs associated with machines of any kind. Classifying a job as 'disgusting' was mostly reserved for interactions with organic matter. These jobs included bat-cave scavenger, catfish noodler, road-kill collector, pig farmer, chicken sexer, avian vomitologist, and worm-dung farmer.

The association of biological materials with 'living hell' seems to extend beyond the audience for this popular television series. Its producers did not invent the disgust aroused by the substances that account for life. They merely dramatized a cultural repulsion regarding interactions with digestion, excretion, sex, birth, death, and decay. The substances that perform these marvels of life are decidedly absent from experiences that are digitized, data-driven, abstracted, mediated, and simulated. Instead of paying tribute to organic substances, popular culture features them on 'worst jobs' lists.

The insights I gained on this summer's day in Manhattan instigated a multiyear investigation regarding the widespread preference for the rigid, hard, impermeable, stiff, dry, and odorless machines and electronic devices, as opposed to the mushy, spongy, slimy, oozy, gooey, and smelly organic substances associated with life. A list of common human inventions that are designed explicitly to separate the human body from organic matter includes tools like shovels, utensils like spoons, devices like thermometers, instruments like brushes, clothing like boots, apparatus like food blenders, infrastructures like sidewalks, implements like scissors, ointments like mosquito repellents, and appliances like garbage disposals. All these clever inventions block the body's sensory apparatus and obstruct sensory experiences.

Because interactions with industrial products and electronic technologies dominate current human practices, they are routine and commonplace. The domain that promises the excitement of discovery is less familiar. It consists of sensory interactions with sap, pollen, twigs, leaves, moss, lichens, and innumerable other materials that account for the wondrous uniqueness of planet Earth. These zones of excitement are scaled to the human body, which means forgoing remote perceptions accessed by microscopes and telescopes. Thus, one response to 'What's new?' activates the sensory receptors that are part of the standard equipment of every living human. Somatic interactions are 'new' because they are commonly disparaged as obsolete unreliable, and inadequate compared to machines that augment human physicality, and technologies that enhance mental powers. These human-scaled, personalized interactions might seem like nostalgic throwbacks to a romanticized version of past eras. In fact, they exist at the forefront of contemporary thought where material exploration is driving cutting-edge philosophical speculation, academic research, and social experiments. This book features contemporary artists' contributions to this bold exploration of physicality.

Our inquiry into Materialism commences by assembling artists who interact with materials that might appear on Tim Rowe's 'worst jobs' lists. The distaste these substances typically evoke is amplified because, in each case, they are not only generated by the human body; they are typically classified as waste discharged from the human body. In polite society, peoples' interactions with such substances occur in private. Like 'dirty jobs', touching them is typically followed by a hygienic scrubbing.

The artists who produced these works did not manipulate the secretions in the manner of paint, pastel, or any other conventional art medium. Because these substances retain their identity, physical matter serves double duty as the work's medium and its subject matter. The subject they introduce is the unfamiliar notion that secretions discharged from the body are not distasteful waste products; they are gratifying embodiments of life processes. In this manner, they urge viewers to loosen the grip of repugnance, reduce repulsion to squeamishness, and replace squeamishness with acceptance, then approval, then appreciation, and finally gratitude.

Tim Hawkinson (b. 1960, San Francisco, California, United States) created a meticulously realistic bird skeleton in miniature using his own fingernail clippings and superglue. *Bird* (1997) is tiny, but it is so meticulously formed that photographs of the work make it appear to be life-sized. Nail clippings, commonly considered bodily waste products, are salvaged to provide inspiration and facilitate creation.

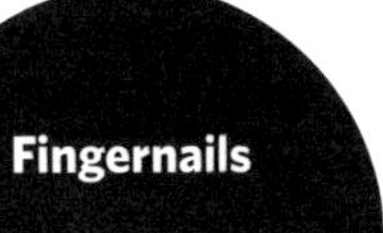

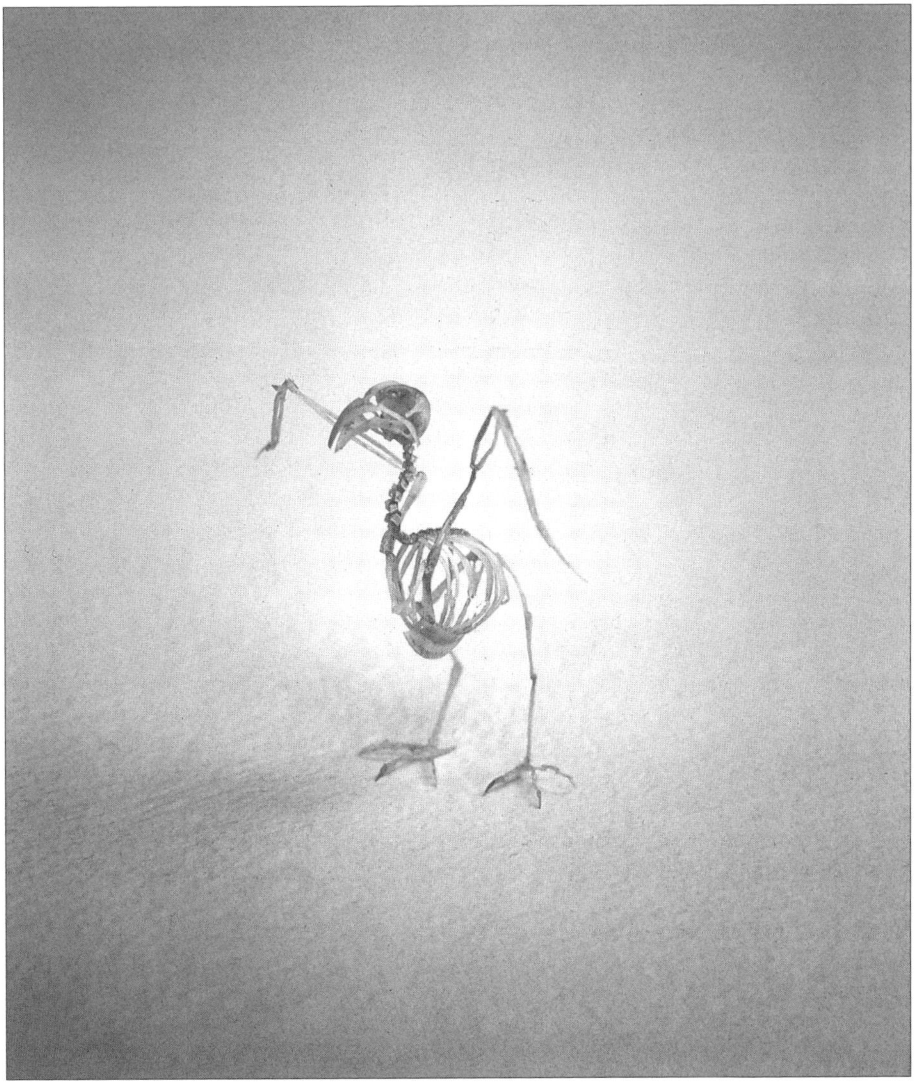

Figure 1: Tim Hawkinson, *Bird* (1997), artist's fingernail clippings and super glue, 5.1 x 5.7 x 5.1 cm (2 x 2 1/4 x 2 in.) © 2015 by Tim Hawkinson. Photo courtesy of the artist.

Marc Quinn (b. 1964, London, United Kingdom) collects ten pints of his own blood every five years to create a sculpted self-portrait. He begins by casting his head in frozen silicone, and then fills the mold with his blood. As such, the sculpture is both a material and a representational portrait of the artist. At the same time, it materially and metaphorically demonstrates that to survive, these sculptures must remain plugged into an external energy source. The series, entitled *Self*, commenced in 1991 and provides a vivid index of the artist's aging, and perhaps his dependence upon mechanized life-support.

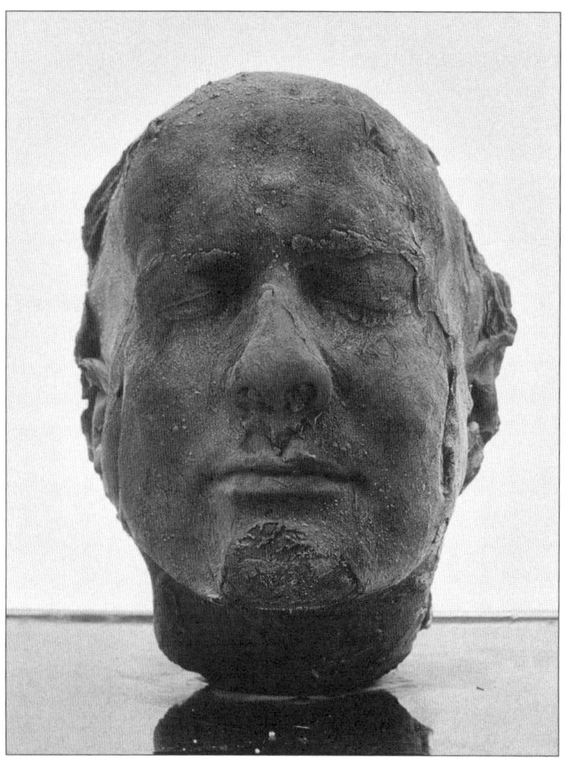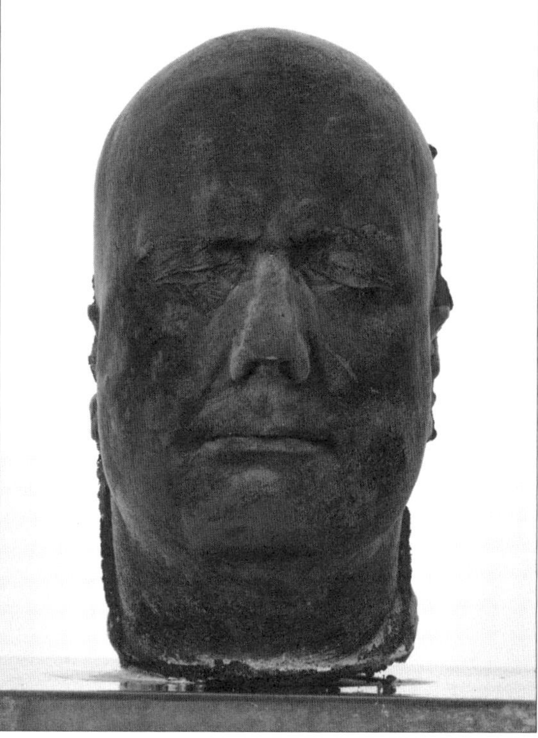

Figure 2a: Marc Quinn, *Self* (1991), artist's blood, stainless steel, Perspex, and refrigeration equipment, 208 x 63 x 63 cm © 1991 by Marc Quinn studio.

Figure 2b: Marc Quinn, *Self* (1996), artist's blood, stainless steel, Perspex, and refrigeration equipment, 208 x 63 x 63 cm © 1996 by Marc Quinn studio.

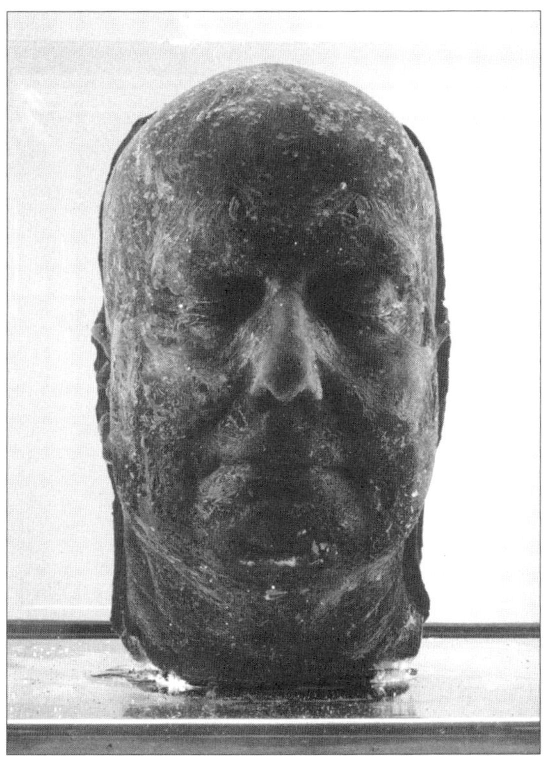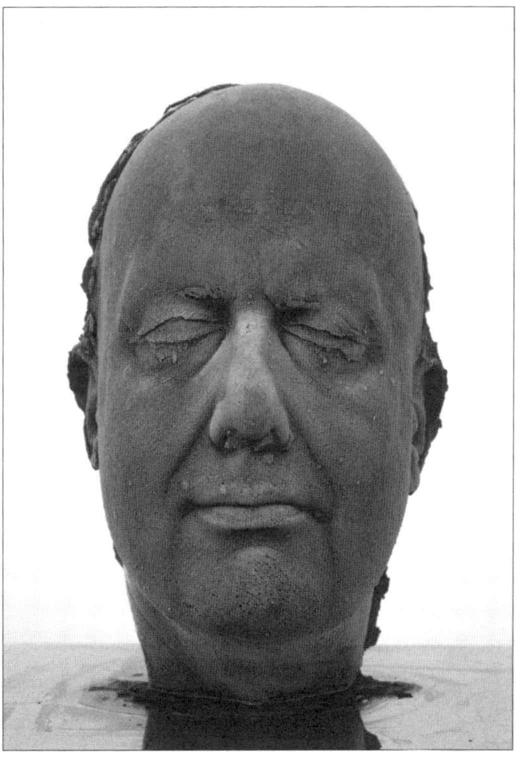

Figure 2c: Marc Quinn, *Self* (2001), artist's blood, stainless steel, Perspex, and refrigeration equipment, 208 x 63 x 63 cm © 2001 by Marc Quinn studio.

Figure 2d: Marc Quinn, *Self* (2006), artist's blood, stainless steel, Perspex, and refrigeration equipment, 208 x 63 x 63 cm© 2006 by Marc Quinn studio.

Gu Wenda (b. 1955, Shanghai, China) collected human placentas from four kinds of births: normal, abnormal, aborted, and stillborn. These by-products of human birth were given to this Chinese artist at a maternity hospital in China. The placentas were then ground into powders according to an ancient Chinese medical practice, because they were thought to enhance a person's vitality. Gu then placed the powdered placentas in cribs to form installations entitled *Oedipus Refound #2 and 3*. The artist explains that he chose this substance because it is highly prized in China, but discarded as waste in the West. 'These pieces narrate a polarized multicultural concern; the use of this material addresses highly charged issues in the West, but in China, its significance becomes elevated as the placenta can be made into a precious, medicinal tonic.'[1]

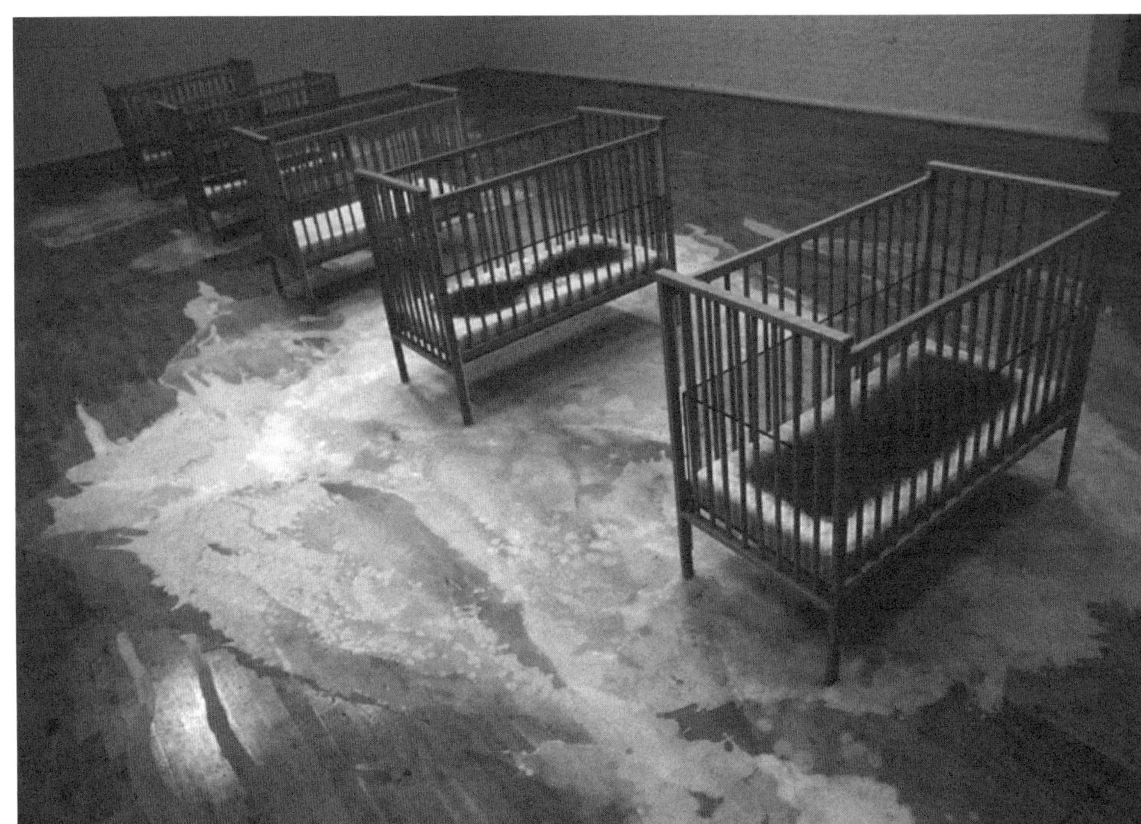

Figure 3: Gu Wenda, *Oedipus Refound #2, Enigma of Birth* (1993), site-specific installation. Four cribs containing human placenta powders (normal, abnormal, aborted, stillborn) produced according to Chinese ancient medical methods. Commissioned by the Modern Art Museum, Oxford, England. Courtesy of the artist.

Caitlin Pautler Albritton's (b. 1989, Tampa, Florida, United States) *Sweat Prints* (2016) are created out of cosmetics and exercise sweat on rice paper. The use of the female body as an athlete gains significance because of the prints' resemblance to Abstract Expressionism. This historic movement was dominated by powerful male artists. The artist explains, 'Though there is a bit of a poke at abstract expressionist works, the prints still relish in the beauty of simple mark making, implementing my own body as an artistic instrument and my sweat as the medium.'[2]

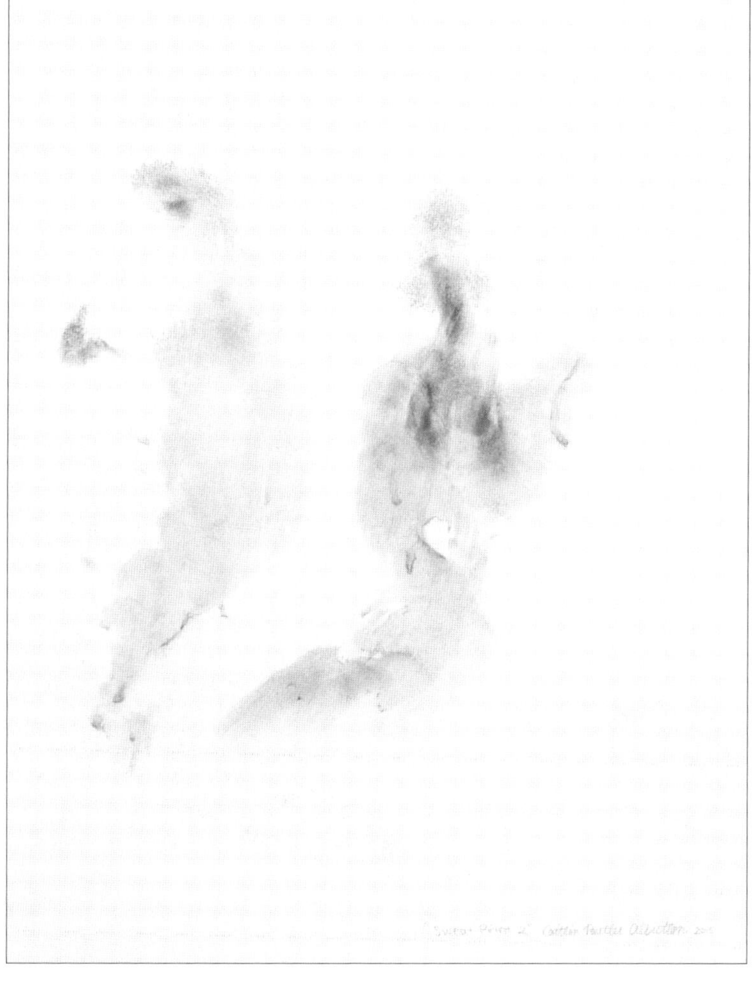

Figure 4: Caitlin Pautler Albritton, *Sweat Print 2* (2016), watercolor and workout sweat on rice paper, 15 x 12 x 12 in. Courtesy of Caitlin Pautler Albritton.

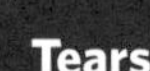

Rose-Lynn Fisher

Rose-Lynn Fisher (b. 1955, Los Angeles, California, United States) collected tears from one hundred different crying occasions that ranged from frustration, sorrow, rejection, laughing, yawning, and chopping onions. After drying the tears, she placed each under an optical microscope to discover if their structure changed according to their stimulus. She discovered remarkable differences. In a book about this project, she comments on the resulting artwork, *The Topography of Tears* (2008):

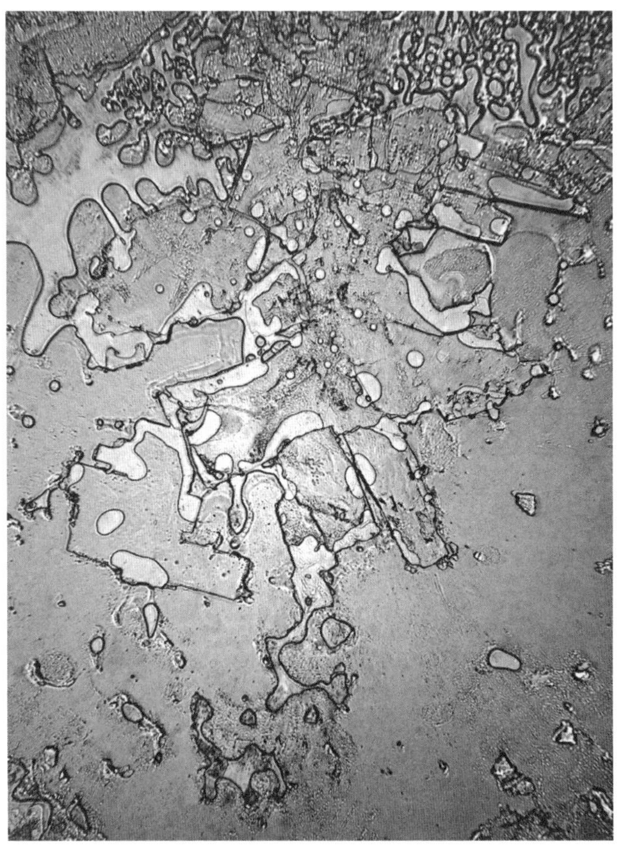

Figure 5a: Rose-Lynn Fisher, *Laughing Till I'm Crying* (2010), tears, photomicraphy, 6 x 4 1/2 in., Photo by Rose-Lynn Fisher. Courtesy of *The Topography of Tears* (New York: Bellevue Literary Press, 2017).

Tears are the medium of our most primal language in moments as unrelenting as death, as basic as hunger, and as complex as a rite of passage. They are the evidence of our inner life overflowing its boundaries, spilling over into consciousness. Wordless and spontaneous, they release us to the possibility of realignments, reunion, catharsis: shedding tears, shedding old skin. It's as though each one of our tears carries a microcosm of the collective human experience, like one drop of an ocean.[3]

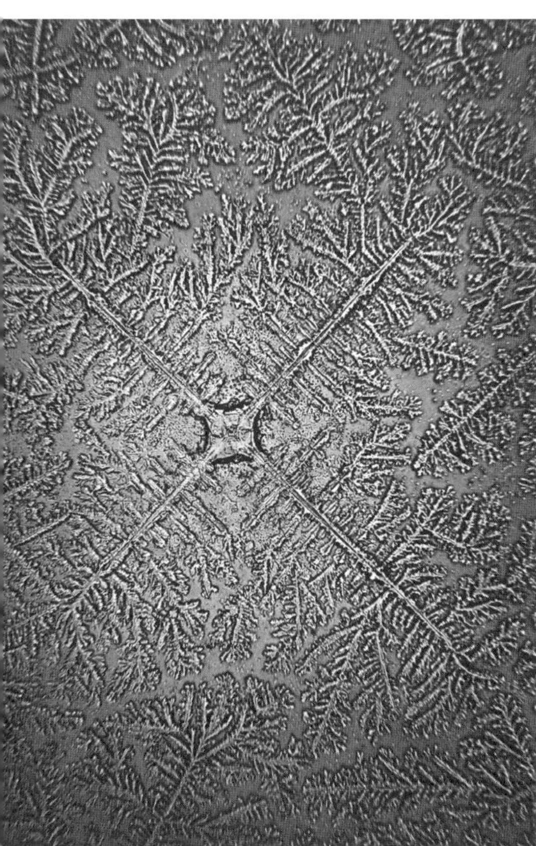

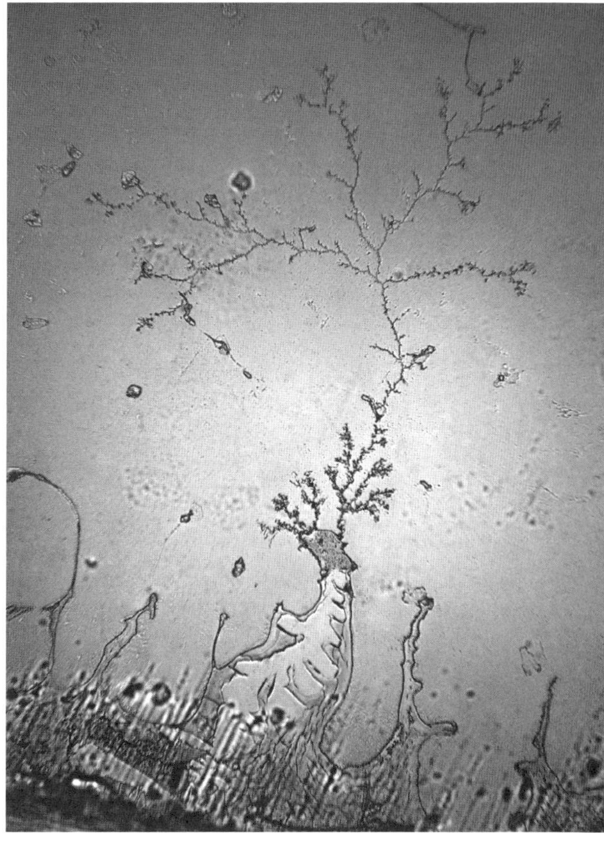

Figure 5b: Rose-Lynn Fisher, *Redemption* (2015), tears, photomicraphy, 6 x 4 1/2 in. Photo by Rose-Lynn Fisher. Courtesy of *The Topography of Tears* (New York: Bellevue Literary Press, 2017).

Figure 5c: Rose-Lynn Fisher, *Tears of Elation at a Liminal Moment* (2009), tears, photomicraphy. 6 x 4 ½ in. . Photo by Rose-Lynn Fisher. Courtesy of *The Topography of Tears* (New York: Bellevue Literary Press, 2017).

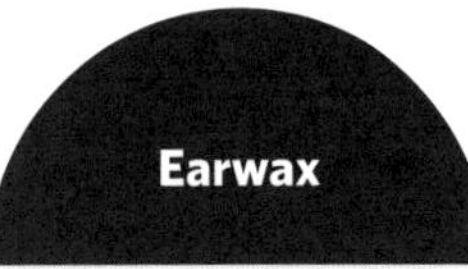

Amanda Cotton (b. 1988, Essex, United Kingdom) creates necklaces using earwax that she extracted over a period of many months. Because the wax resembles amber, she presents them as 'jewels' that she strings on chains created out of her own, plaited hair. Cotton describes earwax as 'a secretive bodily material produced to protect a precious sense.'[4] In this manner, the *Ear Wax Necklace* enables her to 'wear this material on display in an aesthetically pleasing manner.'[5]

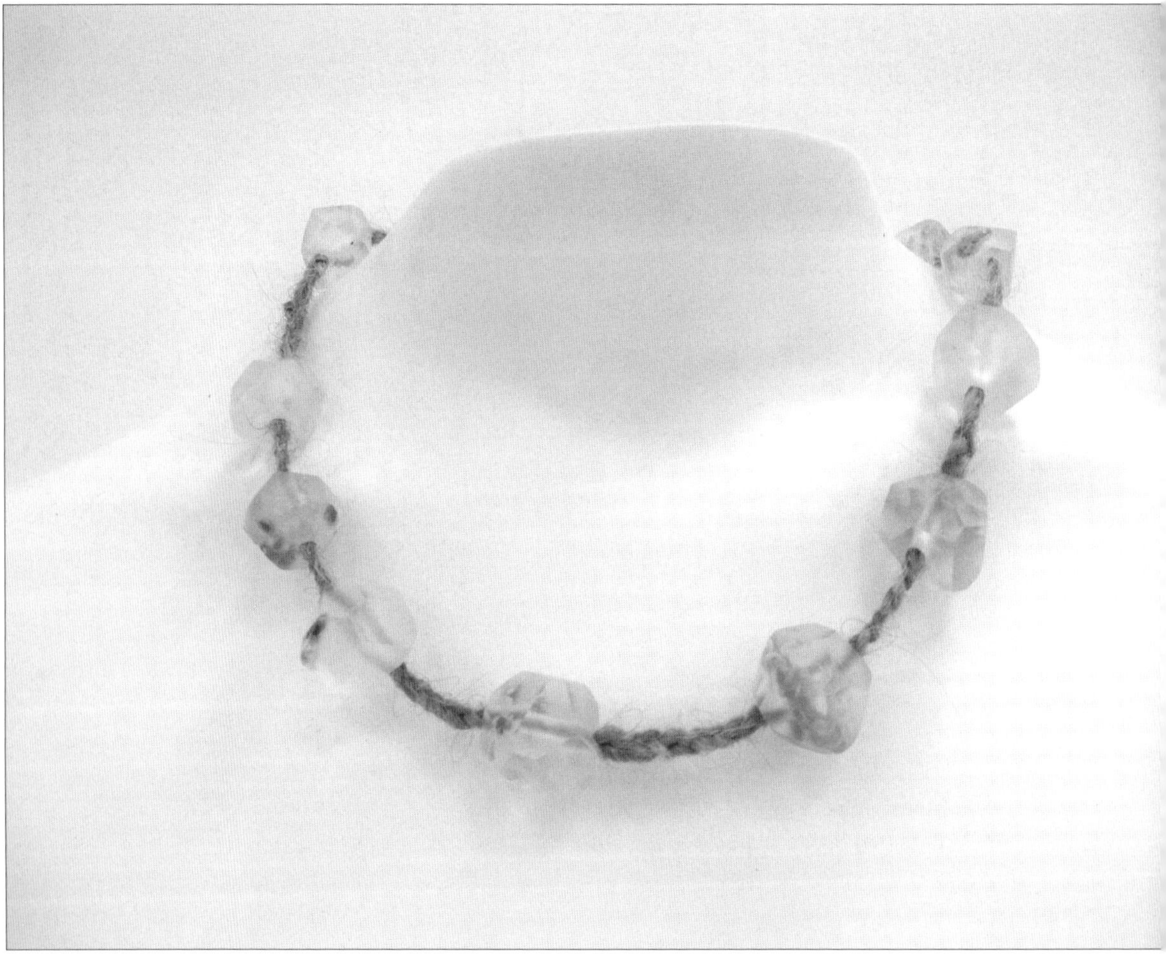

Figure 6: Amanda Cotton, *Ear Wax Necklace* (2012), earwax (jewel) and hair (chain), 40 x 10 x 17 cm. Courtesy of the artist.

The press release announcing Jess Dobkin's (b. 1970, Philadelphia, Pennsylvania, United States) exhibition at the Ontario College of Art and Design Professional Gallery in 2006 invited the public to 'quench their curiosity' by tasting samples of pasteurized human breast milk at *The Lactation Station Breast Milk Bar.* The pasteurized breast milk was donated by six lactating mothers.

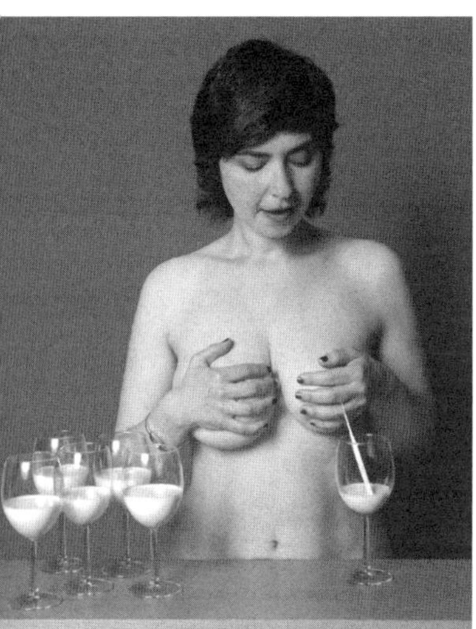

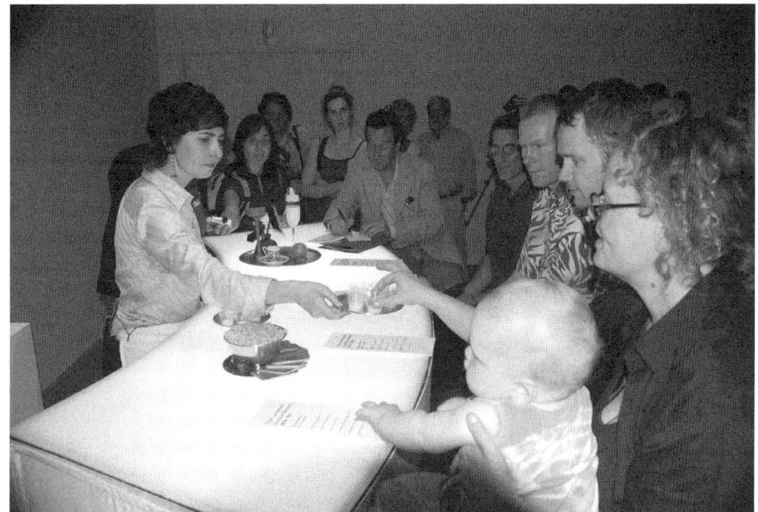

As part of the piece I interview the donors while I'm collecting their milk, asking them about their experience of breastfeeding, about their diet, about if they've tasted their milk, and if so, what they think it tastes like. And then as part of the performance I impart that information to the audience as they are sampling the milk.[6]

Figure 7a: Jess Dobkin, *The Lactation Station Breast Milk Bar* (2006). Breast milk. Promotional image advertising the public tasting. The Ontario College of Art and Design Professional Gallery. Photo by David Hawe. Courtesy of the artist.

Figure 7b: Jess Dobkin, *The Lactation Station Breast Milk Bar* (2006). Breast milk served at the public tasting. The Ontario College of Art and Design Professional Gallery. Photo by David Hawe. Courtesy of the artist.

Sun Yuan (b. 1972, Beijing, China) and Peng Yu (b. 1974, Jiamusi, China), who are married, created a towering thirteen-foot pillar in lurid yellow made from hardened human fat extracted by liposuction during cosmetic surgery. Titled *Civilization Pillar* (2001), the sculpture evokes disgust with gluttony, hedonism, excess, and decadence. The artists associate these indulgences with twenty-first century consumerism and capitalism.

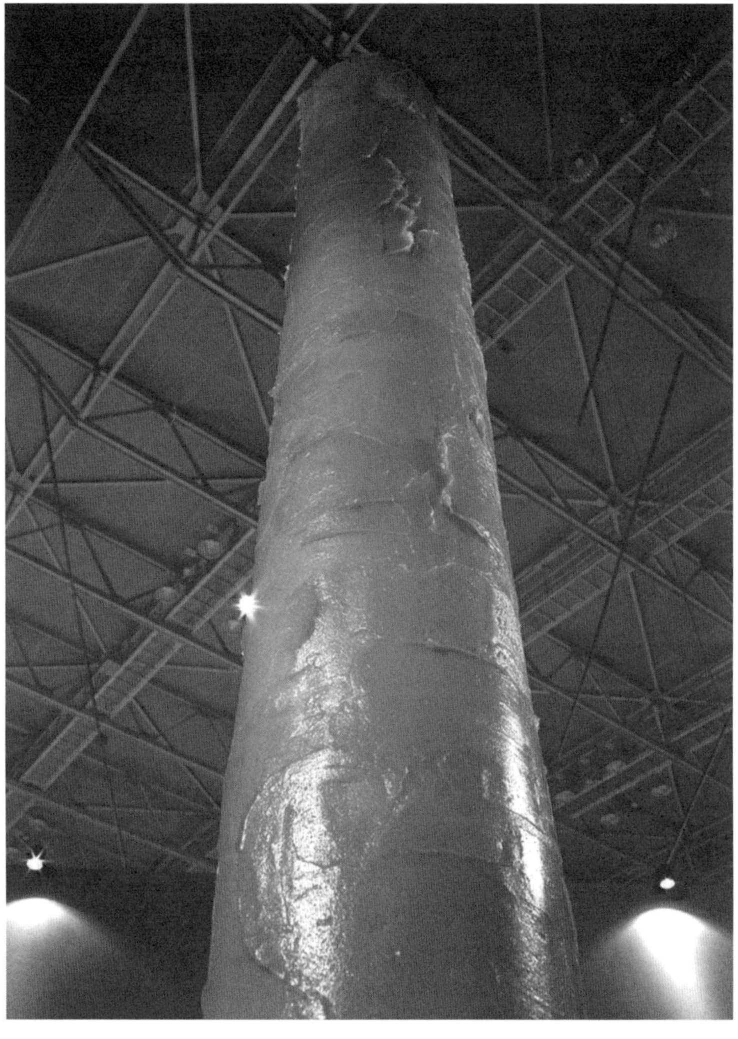

Figure 8: Sun Yuan and Peng Yu, *Civilization Pillar* (2001), liposuction fat, 3.8 m (13 ft.). Courtesy of Sun Yuan and Peng Yu.

Spurse

Spurse (formed in 2004) is a group of artists who identify themselves as a 'creative design consultancy'. Their whole-systems approach to eating dissolves cultural boundaries and allows all substances to be considered as ingredients for meals or activators for processing foods. This approach was applied using spit to raise dough for bread. One of the artists explains:

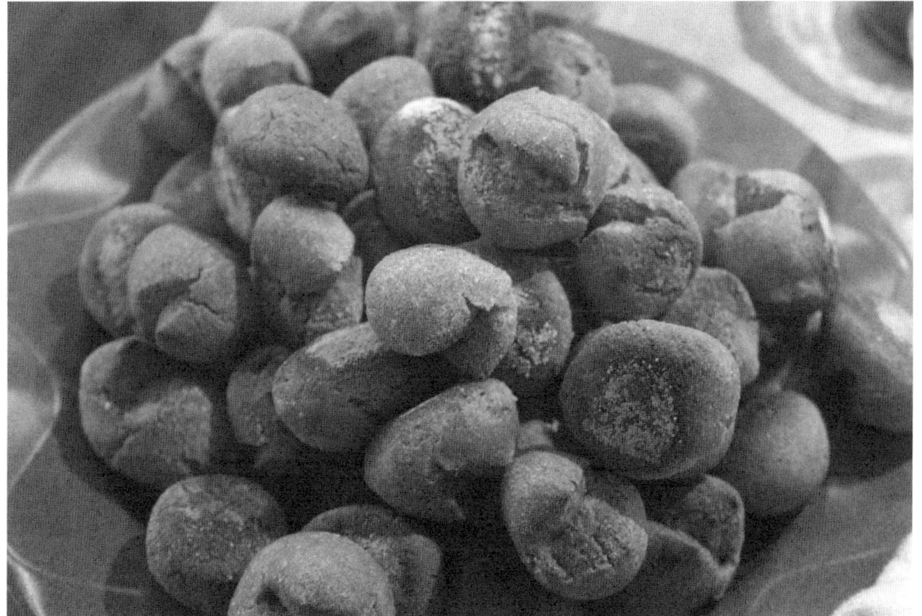

I am spitting into my baking. Every couple of days when I wake up, after putting an espresso pot on the stove, I take off the cloth covering the sourdough starter and I attempt to spit. I am adding to the community. After I spit, I add in a small handful of flour and water and give it a slow careful stir. I watch the bubbles rise, stretch and burst as I mix it.[7]

Figure 9: Spurse, *Spit Bread* (2017) Spit, wheat, water, fire. Each ball 1½". Photo by Lindsey Guile. Courtesy of the artists.

Fête is an ongoing project that the artist Germaine Koh (b. 1967, George Town, Malaysia) initiated in 1997. It consists of the artist's hair, sewn into swags, and hung in chronological order. The date of each haircut is embroidered on each section. Thus, the material changes in the hair, as it gradually turns from black toward grey, confirm the passage of time that is indicated by the dates.

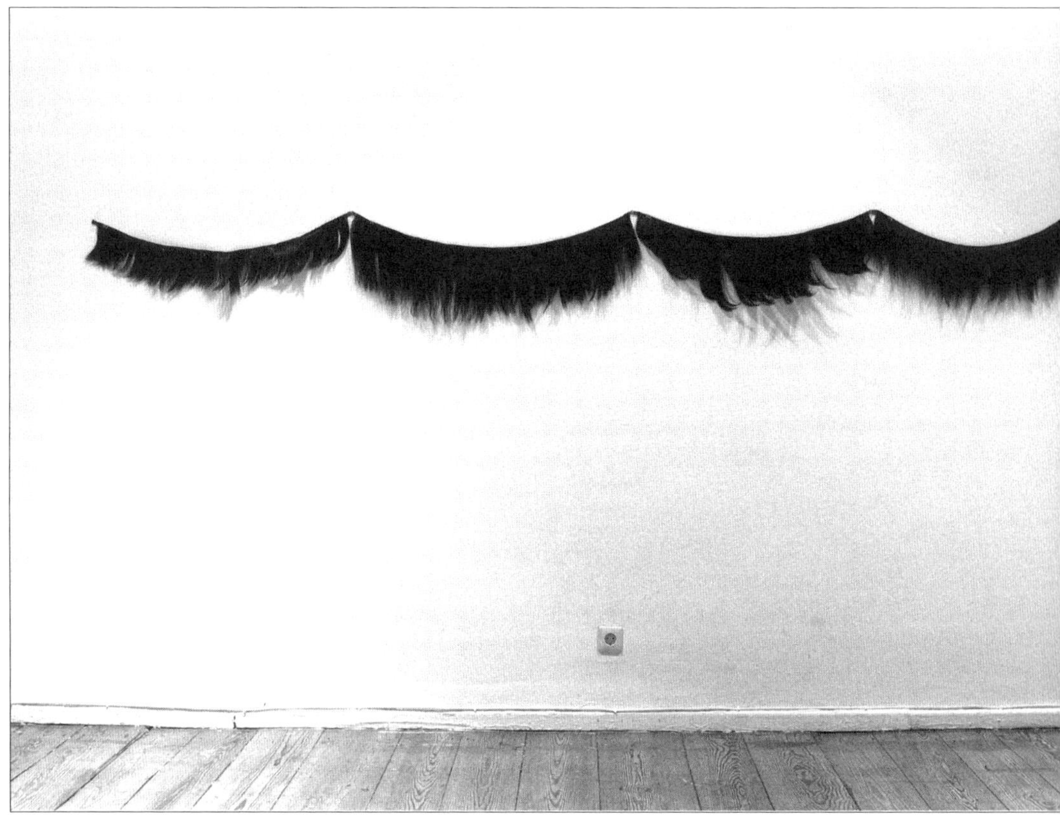

Figure 10: Germaine Koh, *Fête* (1997-ongoing), artist's hair, ribbon, binding. Courtesy of the artist.

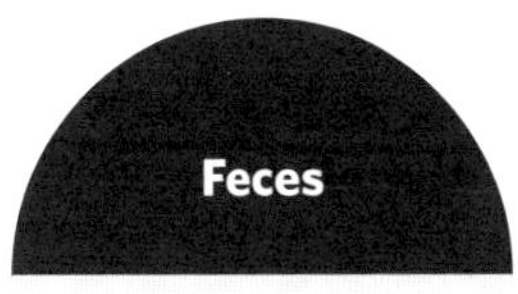

In 2007, the artist Terence Koh (b. 1977, Beijing, China) installed eighty-eight chunks of his own dried excrement that were plated with gold at Art Basel, a major international art fair. They were installed bathed in amber neon light and placed within glass cases to suggest that they were so valuable that they merited this protection – and a hefty price tag. It is reported that the artwork was sold for $500,000.[8]

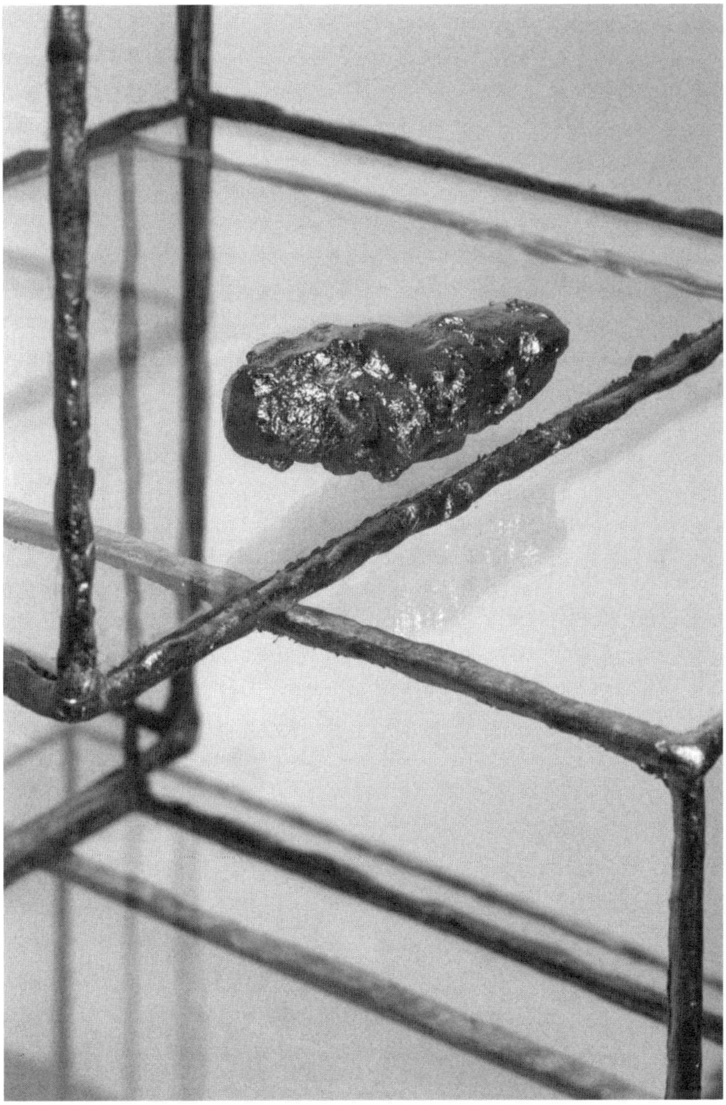

Figure 11: Terence Koh, *Gold Plated Turd* (2007), gold and artist's turd. Photo by Terence Koh. Courtesy of the artist.

Interrogating Repulsion

The impressive roster of contemporary art presented in the previous section represents New Material values. These disclosures can all be traced to the absence of 'the abject' from their interactions with human body secretions. The term 'abject' became affixed to art and literature in the 1990s in response to a surge in vivid depictions of filth, and a concurrent desire to disrupt the underpinnings of the social order. The Tate Modern, the eminent contemporary art venue in London, defines 'abject art' as that which 'explores themes that transgress and threaten our sense of cleanliness and propriety particularly referencing the body and bodily functions.'[9] The artists engaging secretions manifest New Materialism's seismic shift away from the abject. Their sanguine acceptance of the stuff of life purges them of repugnance. In each instance, a biological product is treated with respect, and even delight.

By selecting secretions as art mediums, these artists are flaunting an attitudinal shift away from industrial society's dominant material interactions: polyethylene terephthalate (beverage bottles, medicine jars, peanut butter jars, combs, bean bags, rope steel); high-density polyethylene (containers for milk, motor oil, shampoos and conditioners, soap bottles, detergents, bleaches); polyvinyl chloride (pipes, tiles, flooring); low-density polyethylene (plastic wrap, sandwich bags, plastic grocery bags); polypropylene (diapers, margarine containers, yogurt boxes, prescription bottles); Styrofoam (disposable cups, plastic food boxes, packing foam); polycarbonate (baby bottles, compact discs, medical storage containers). The absence of manufactured substances indicates a powerful reset of social values, as much as the presence of secretions in the artworks presented in this chapter,

It is just as significant that the artworks reject engineered items, which also dominate contemporary lifestyles. Consider, for example, that cell phones, computers, engines, thermostats, microwaves, LED bulbs, ID cards, cameras, game consoles, MP3 players, security systems, CAT scanners, MRI devices, alarms, printers, GPSes not only exclude living matter, most are designed and manufactured to suppress life functions.

Besides re-evaluating interactions with physical substances, Eco Material artists are scrutinizing the assumed advantages of contemporary interactions with material processes for signs of liabilities. They might ask, is it a privilege to proceed through life without regard for the substances that account for life, or for moon phases, seasonal shifts, animal migrations, and soil conditions? Do people in industrialized societies benefit by severing connection with the substances, rhythms, and patterns of planet Earth? Artists that welcome secretions as their mediums are implicitly challenging the functional and aesthetic foundations of contemporary experience.

Both New Materialists and Eco Materialists acknowledge that biological secretions tend to register in people's consciousnesses as revulsion instead of allure. However, this shared concern registers as the effect of alienation upon humans by New Materialists, and as an impact upon eco systems by Eco Materialists. In other words, New Materialists tend to address the emotional impoverishment that can afflict people who have lost rapport with the materials produced by biological processes; Eco Materialists focus on the fact that such alienation is contributing to the looming environmental crisis.

Converting Repulsion into an Eco Material Allure

The side-by-side collections of words in the following chart epitomize the turnaround of cultural norms that accompany Eco Materialism. That is because Eco Materialism attempts to banish 'repulsion' from the sphere of materiality, and embrace every instance as an 'allure'. Thus, both sets of words inspire Eco Material respect.

Rotting. Decaying. Dying. Disintegrating.	*Developing. Growing. Beginning. Connecting.*
Dissolving. Spoiling. Breaking. Crumbling.	*Integrating. Repairing. Binding. Uniting.*
Festering. Molding. Diminishing. Wasting Away.	*Flourishing. Erecting. Enlarging. Blooming.*
Breaking. Deteriorating. Ruining. Purging.	*Fixing. Strengthening. Constructing. Maintaining.*
Destructing. Putrefying. Disembodying.	*Constructing. Purifying. Embodying.*
Eroding. Dilapidating. Atomizing. Oxidizing.	*Accreting. Embellishing. Uniting. Stabilizing.*
Dissecting. Disassembling. Depreciating.	*Combining. Assembling. Appreciating.*
Disfiguring. Dismantling. Infecting.	*Configuring. Mantling. Curing.*
Evaporating. Reducing. Perishing. Polluting.	*Precipitating. Expanding. Flourishing. Cleansing.*
Fouling. Fading. Pulverizing. Mortifying.	*Sweetening. Appearing. Assembling. Vitalizing.*
Shriveling. Waning. Withering. Dwindling.	*Unfurling. Dawning. Growing. Augmenting.*
Decreasing.	*Expanding.*
Declining.	*Ascending.*
Dissolving.	*Integrating.*
Decomposing.	*Composing.*
Extinct.	*Extant.*

The elaborate collection of words that appear on the preceding page was assembled by the Department of Decomposition at the University of Illinois, Chicago.[10] The department invites its painting students 'to explore anything and everything that has to do with decay.'[11] Ironically, by stating that explorations of decay involve the 'abject', the academic announcement identifies decay as something that is foul, worthless, contemptible, and therefore unrelated to the Eco Material respect for decaying organic matter as a procreative resource. The following text tracks an example of what is arguably the most extreme example of the abject: rotting corpses.

Data is lacking that supports the assertion that the majority of people in industrialized societies are disgusted by decomposing corpses, and that they strenuously avoid contact with the fungi, viruses, worms, maggots, turkey vultures, carrion beetles, and bacteria that devour dead bodies. Thus, let us turn to Heavy Metal bands to support the contention that the common perception of the decomposition of organic matter is gross! These bands flaunt their counterculture perversity by selecting this grizzly theme to inform their names: Lust of Decay, Man Must Die, Mortal Decay, Decomposed, Still Life Decay, Putrid Pile, Garden of Decay, The Years of Decay, Garden of Bones, Our Kingdom of Decay, Internal Decay, Stench of Decay, Display of Decay, House of Decay, Years of Decay, Sounds of Decay, and Divine Decay. It seems significant that Heavy Metal bands rarely refer to death, presumably because it merely evokes fear and sorrow. It is the post-mortem state of the body that they relish because, by evoking disgust, it scorns social decorum.

Eco Material artists have little in common with Heavy Metal bands besides their mutual embrace of decay. While the bands situate decay within the context of social taboos, Eco Materialists view decay through the functional lens of ecology. The narrative that emerges shows that, despite the ongoing search for extraterrestrial life, Earth remains the only known celestial body where the cycling of organic and inorganic matter occurs. Decay is essential to the ongoing phase changes of matter. To the microorganisms that process organic molecules, a rotting carcass is as enticing as a well-stocked pantry. They relish the residues and excretions of corpses. In the process of feasting, they dismantle the complex organic molecules of the deceased body, reducing them to their primary components so they can reconstitute new forms of life. Decomposition is creative destruction. It is a reassuring sign of the continued health, resilience, and vitality of an ecosystem. When Eco Materialists proclaim the uniqueness of this planet, they not only cite the presence of life; they bestow equal admiration upon decomposition, because Earth would not be a living planet if it were not for the complementary processes of life/growth and death/decomposition. Maurizio Montalti extends the application of this principle to the life-enhancement

Maurizio Montalti (b. 1981, Cesena, Italy) founded the Amsterdam-based multidisciplinary design studio, Officina Corpuscoli, in 2010. It provides consultant services for creative practices that work with living systems and organisms. Additionally, Maurizio is co-founder and forming partner of the Amsterdam-based WNDRLUST collective. In 2010, Montalti received his MA in Conceptual Design in Context) at the Design Academy Eindhoven in the Netherlands. He currently co-heads the Materialization in Art and Design program at Sandberg Instituut.

Maurizio Montalti demonstrates that the last breath of an organism does not conclude its biography. This is because the molecules that once constituted that person's body when he or she was alive persist after death. This post-mortem state transforms the physical organism's niche role. When it was alive, the body was a consumer of resources (food, oxygen, sneakers, MP3 players, and so forth). After it died, it became a donor of resources (molecules from the corpse are released as it is dismantled by microorganisms). Montalti created *Bodies of Change* (2010) to materialize the donor-half of this equation. This sculptural project, which was undertaken when he was a student at the Design Academy Eindhoven in the Netherlands, integrates the actual organisms that accomplish decomposition. They are intended for interactions with actual corpses upon which they would dine.

Montalti is fully aware that his project will distress many members of the public. He addresses this discomfort by asking:

> *The human being seems to feel repulsion for some kinds of organisms that are supposed to be dangerous and harmful for the human body [...] How could I develop something beneficial and attractive out of what we generally tend to associate with disgust?*[12]

His answer is suggested in the following quote:

> *I'm extremely fascinated by the human body, considered in its whole incredible complexity, a universe in itself, and at the same time so incredibly ephemeral and temporal. In my research I investigate the role of the body during life, at the end of life, and beyond.*[13]

Montalti offers an Eco Material interpretation of humanity's perennial quandary regarding how to make sense of life within the context of the inevitability of death. Conventional funeral rites and mortuary laws rarely offer comfort to the mourners or guidance for the terminally ill because conventional mortuary practices assume humans cease existing after death. Within these contexts, the afterlife is reserved for otherworldly locations. By manifesting the ecological role of cadavers,

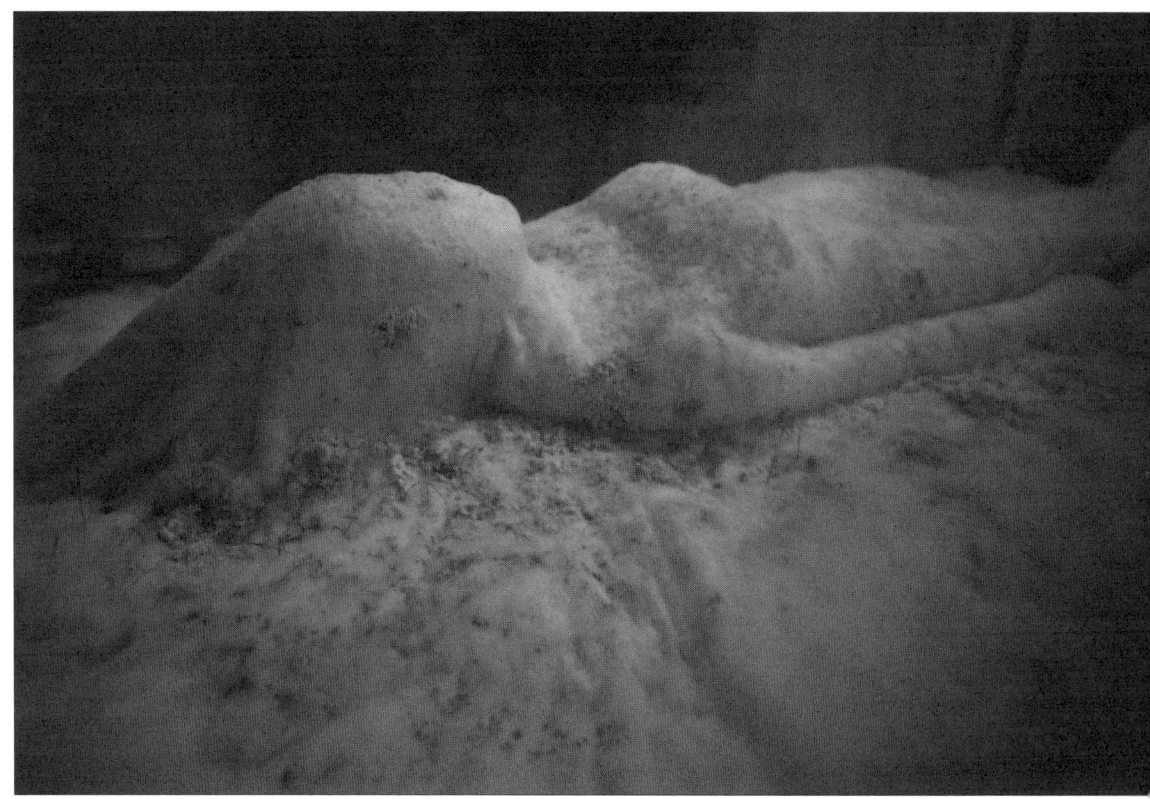

Figure 12a: Maurizio Montalti, *Bodies of Change* (2010) from the *Continuous Bodies* series, felted wool and selected fungal cultures, 180 x 80 x 25 cm. Photo by Officina Corpuscoli/Maurizio Montalti. Courtesy of Officina Corpuscoli/Maurizio Montalti.

Bodies of Change upends this belief. It demonstrates that the afterlife of every living creature exists right here on Earth. The ecological funeral methodology it introduces presents the passage from life to death – and the ensuing physical decay – as natural processes without which there could be no new life. Montalti explains:

> *I aim to explore and demystify denial and anxiety, related to the acceptance of the loss of a loved somebody, by transporting the process of decomposition of human remains to a more natural level, through an ecological connection with our changing environment.*[14]

The research Montalti conducted while he was planning this project identified numerous members of the multispecies that comprise the clean-up crew for corpses. These sanitation workers include turkey vultures, carrion beetles, some maggots, and bacteria. Despite the beneficial services they all provide for replenishing an ecosystem, fungi became the focus of his project because they are the champion

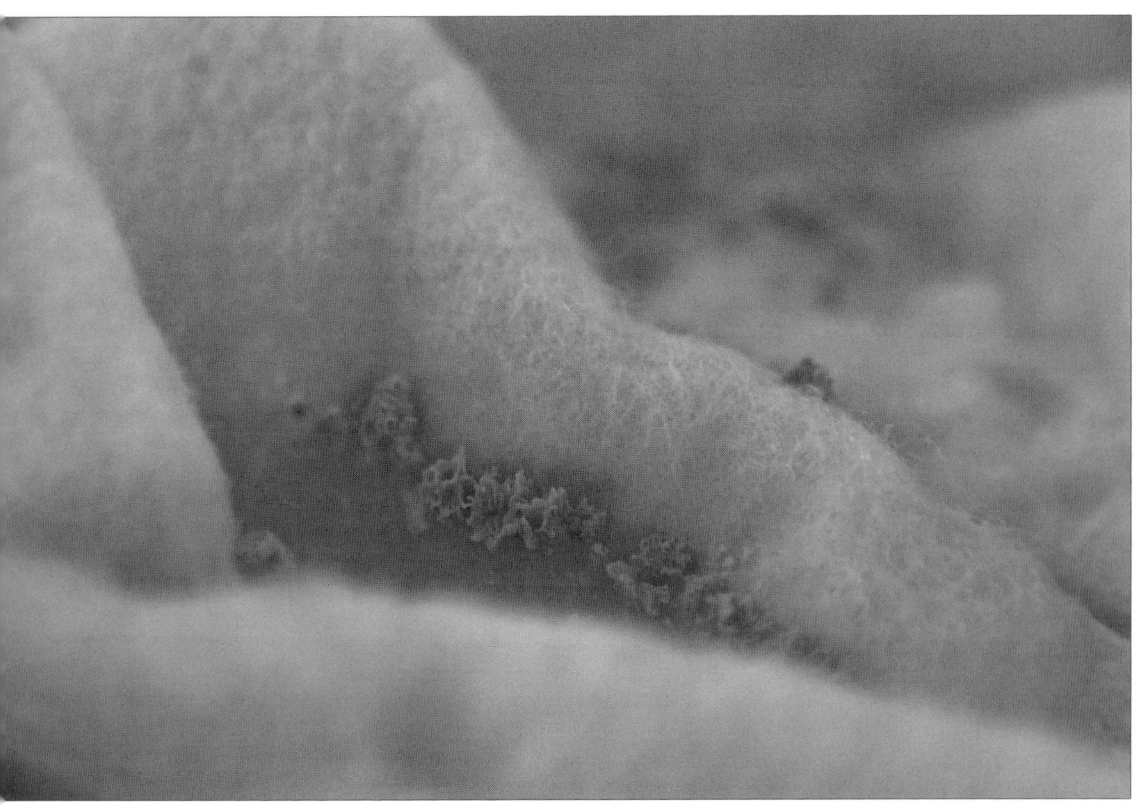

Figure 12b: Maurizio Montalti, *Bodies of Change* detail (2010) from the *Continuous Bodies* series, felted wool and selected fungal cultures, 180 x 80 x 25 cm. Photo by Officina Corpuscoli/Maurizio Montalti. Courtesy of Officina Corpuscoli/Maurizio Montalti.

recyclers of dead matter on the planet. Unlike plants that get their energy from the sun and atmosphere, fungi get their energy by digesting living or dead organic matter. The artist explains, 'The key role of these recyclers is to break down any dead matter and return the nutrients to the soil to become available to plants once again.'[15]

Montalti collaborated with mycologists from Utrecht University and Central Bureau of Fungal Cultures, Fungal Biodiversity Centre to identify and cultivate the fungi species that are most suited to dismantling human remains. He then conducted further experiments to determine the materials and conditions that could optimize the efficiency of their actions. This component of his undertaking explains the full title of his thesis project: 'Continuous Bodies: Cycles of Decomposition Triggering a Symbiotic Partnership Between Humans and Fungi.'

Montalti materialized this symbiotic partnership between human bodies and fungi as an art project by altering the common practice of dressing the deceased in special burial garments. The shroud he created promotes the corpse's cycling back into the ecosystem. It consists of handmade felt that was inoculated with fungal mycelia from the species *Schizophyllum commune*. Montalti engages them as his creative partners in renewing planetary life. He explains:

> *I believe we could benefit from becoming symbiotic with other organisms, by learning how to communicate with them, reciprocally understanding and exploiting each other in a mutual beneficial way.*[16]

The installation of *Bodies of Change* consists of an inoculated felt shroud that is placed around a structural 'body'. In order to minimize contamination by other microorganisms, the materials were sterilized by placing them in glass jars and boiling them in a pressure cooker. When they were removed, the shroud was inoculated with mushroom spores, then grains of wheat were scattered on the damp felt, which provided food for the fungi that would otherwise eat corpses. These items were then sealed in a display case fabricated out of stainless steel and Plexiglas. Temperature and humidity levels were carefully monitored to maximize fungi growth on the shroud. Viewers observed the processes and imagined their occurrence underground upon a decomposing corpse.

The strains of fungi that Montalti selected are not only adept at dismantling human remains and returning them to the elemental state, which enables the formulation of new forms of life; they also perform another vital ecological function: collecting and neutralizing toxins that have accumulated in the deceased person's body as they were breathed and swallowed during his/her lifetime. The toxin that Montalti paid particular attention to was plastic, because it has become a major pollutant in the human body. Plastics are not only absorbed by touching plastic items; they also enter through pesticides, detergents, cosmetics, medications, and shampoo that include plastic as an ingredient.[17] Furthermore, plastic compounds enter the food chain after they are discarded. Montalti notes, 'This toxic pollution lives inside of us and it's quickly accumulating: it has become something "personal," a part of the background chemistry in our bodies.'[18] The fungi Montalti incorporated into the shroud transformed hazards into harmless minerals. These microorganisms were capable of digesting even the most un-biodegradable types of plastic.

Bodies of Change offers a vitalizing alternative to the environmentally harmful and wasteful effects of traditional end-of-life practices for humans. Embalming inoculates the body with toxic fluids such as formaldehyde, methanol, and ethanol. Cremation pumps dioxins, hydrochloric acid, sulfur dioxide, and carbon dioxide together with mercury compounds into the air, water and soil. Even cemetery burials are problematic because a corpse at the bottom of a six-foot deep grave lacks sufficient oxygen to support healthful decomposition. These bodies lay out of reach to most plants, scavengers, and decomposing organisms.

Bodies of Change offers more than a correction for afterlife protocols. By stating, 'I propose that we adopt the fungi as the organizational model for life in the 21st century,'[19] Montalti reveals his vision of a society in which humans emulate fungi. Humans that behave like fungi would conduct beneficial recycling interactions; they would decontaminate substances; furthermore, they would convert wastes into useful compost. Beyond these environmental advantages, social structures that are modeled on fungi would replace exploitation with reciprocity. Montalti refers to this optimistic vision of the future as HFI (Human Fungi Interaction).[20]

Bodies of Change addresses three widespread forms of distress: dread of death, disgust regarding decomposition, and distaste for microorganisms. Nonetheless, he offers his project as an opportunity for transforming these aversions into admiration and respect. He summarizes the uplifting implications of his initiative by stating:

> *The intention of my exploration is to be able to create a shift in the perception of the generally 'unknown' wonders of life and in the beauty of the decay and in how, by observing them, we could amplify the awareness of us, the human beings, as part of a complex system in which the pattern of life repeats as a paradigm on the macro as well as on the micro scale.*[21]

In all these ways, the *Bodies of Change* project provides a vivid example of how Eco Materialism honors the substances that are associated with 'dirty jobs'.

Your opinion: What's next?

Reader Interaction

Before moving on, interrogate your own attitudes regarding the 'abject' and the 'exalted':

- **How would you convince others that discovering extraterrestrial death would be just as significant as discovering extraterrestrial life?**

- **Should the Environmental Protection Agency outlaw products that thwart decay because they interfere with the cycling of matter that renews life?** Repulsion of materials that foster the regenerative cycling of matter is accompanied by appreciation of the materials and strategies that prevent forms of decomposition. Resistance to decay is apparent in the popularity of wood preservatives, food stabilizers, rust retardants, Botox injections, cryogenics, and fluoride treatments.

- **If you were wounded, would you choose to heal yourself by murdering bacteria with antibiotics or by feeding maggots?** Maggots savor rot. Some doctors are beginning to enlist these miniature flesh-eating organisms into the hygienic field of biotherapy. They are revising a practice that was used during the American Civil War whereby maggots were placed on the wounds of soldiers to facilitate healing. The larvae savor dead or infected tissue, leaving healthy tissue alone. Furthermore, they secrete chemicals that stimulate the production of healthy tissue as they munch. It has been suggested that maggots avoid the problems associated with antibiotic-resistant bacteria.[26]

- **What is an environmentally responsible way to prepare corpses that contain the following artificial amendments for burial:** Breast implants, artificial joints, pacemakers, and prostheses do not degrade biologically after death?

- **Should environmental agencies inspect corpses for drugs, growth hormones, preservatives, pesticides, asbestos, and other toxins before issuing permits for burial?**

- **Should free burials be provided to healthy people who are willing to be buried in depleted soils?**

Your comments: What's next?

Endnotes

1   Gu Wenda, 'Gu Wenda: A Divine Comedy of Our Times,' accessed August 7, 2017, http://wendagu.
     com/installation/united_nations/concept.html.

2   Caitlin Albritton, 'Sweat Prints,' accessed August 7, 2017, https://caitlinalbritton.com/portfolio/
     sweat-drawings/.

3   Rose-Lynn Fisher, 'The Topography of Tears,' accessed August 7, 2017, http://www.rose-lynnfisher.
     com/tears.html.

4   Amanda Cotton, 'Ear Wax Necklace,' accessed August 7, 2017, http://www.amandacotton.co.uk/#/
     ear-wax-necklace/.

5   Cotton, 'Ear Wax Necklace.'

6   Cultural Reproducers, 'Interview: Jess Dobkin,' October 13, 2016, accessed July 23, 2017, http://
     www.culturalreproducers.org/2016/10/jess-dobkin.html.

7   Spurse, email correspondence with the author, September 24, 2017.

8   Rain Embuscado, 'Terence Koh Makes a Grand Return to the Art World With New Show: Our
     favorite punk boy is back,' April 11, 2016, accessed July 31, 2018, https://news.artnet.com/market/
     terence-koh-art-world-return-470669.

9   Tate, 'Abject Art,' accessed July 23, 2017, http://www.tate.org.uk/art/art-terms/a/abject-art.

10   Spiral Workshop at University of Illinois at Chicago, 'Department of Decomposition: Painting,'
     accessed August 7, 2017, https://naea.digication.com/Spiral/Decomposition_Painting_working.

11   Spiral Workshop, 'Department of Decomposition: Painting.'

12   Maurizio Montalti, 'Continuous Bodies: Cycles of Decomposition Triggering a Symbiotic Partnership
     Between Humans and Fungi' (MA thesis, Design Academy Eindhoven, 2010), 17, accessed August
     3, 2017, https://www.mediamatic.net/image/2016/12/14/continuous_bodies_thesis_maurizio_
     montalti.pdf.

13   Montalti, 'Continuous Bodies,' 17.

14   Montalti, 9.

15   Montalti, 25.

16   Montalti, 11.

17   Ronald Grisanti, '7 Ways Plastics Damage the Body,' Functional Medicine University (FMU),
     accessed August 7, 2017, https://www.functionalmedicineuniversity.com/public/919.cfm.

18   Montalti, 'Continuous Bodies,' 17.

19   Montalti, 51.

20   Montalti, 51.

21   Montalti, 15.

ARCHETYPES OF MATERIAL USE & DISUSE?

Dirt

ARCHETYPES OF MATERIAL USE & DISUSE

The bold assertion that this chapter is relevant to every living person is easily justified because it addresses the fact that, as corporeal beings, we cannot avoid contact with the planet's substances. Material interactions are inescapable for us and all other living entities. We are all implicated in the fate and disposition of living systems because we consume resources and produce wastes. Nonetheless, humanity possesses the unique ability to consciously alter the manner in which we manage these material interactions. We are endowed with the ability to design these interactions and control whether they result in environmental disturbances or benefits. Human behaviors and their outcomes are determined by cultural traditions, historic precedents, and ethical principles, which compete for our compliance with personal ambitions, desires, and discontents.

This chapter tackles this unwieldy topic by identifying seven archetypes of human-material interaction. Archetypes are patterns of thought and behavior that exist as shared, cultural motifs. Because they can be easily conceptualized, archetypes are useful tools for assessing humanity's past relationships to materials, reflecting upon current modes of material conduct, and crafting material interactions that optimize the well-being of future generations. Each of the seven archetypes represents a particular category of material interface. Objects can be treated like pets, sacred objects, hazards, specimens, aesthetic stimuli, resources, and merchandise.

All but the last emerged early in human history. Merchandise is unique to the current era of mass production and global exchange. Working and playing, observing and acting, consuming and producing, indeed all human behaviors are constituted out of infinitely varied mixtures and proportions of the interactions represented by these archetypes.

Each archetype is presented from three Eco Material vantage points:

Cultural Engagement: The environmental impact of that archetype's human–material interaction.

Cultural Disengagement: The environmental consequence of human disposal of the objects representing that archetype.

Art Engagement: A single artwork exemplifies the environmental engagement and disengagement with each of the seven archetypes.

In order to facilitate comparing and contrasting, all seven artworks utilize dirt as their medium. Furthermore, all the applications of the archetypes' themes are dirt-related. Dirt was selected because it factors into a wide range of environmental appraisals of materiality. Soil cushions our falls, sequesters our toxins, and mediates on our behalf between minerals and organic matter. It also grows our food, although fertile soils are becoming increasingly rare. In truth, this shrinking resource has *never* been abundant on our planet:

- $\frac{3}{4}$ of Earth is water and $\frac{1}{4}$ of Earth is land.
- $\frac{1}{2}$ of land is available for human use.
- $\frac{3}{4}$ of land available for human use is too rocky, too wet or too dry, too hot or too cold for food production.
- $\frac{1}{4}$ of land available for human use is available for food production.[1] It is the source of sustenance and well-being for all forms of life.

The artworks presented in this chapter awaken awareness of the underground universe that typically exists out of sight beneath multistory skyscrapers, paved highways, and parking lots. Dirt manifests the seven archetypes by appearing in these artworks. It appears as compost, hallowed ground, contaminated soil, underground rubble, repository for hidden treasures, indicator of place, and fertile growing medium. Readers are invited to explore the rich vein of imaginative opportunity provided by dirt as they consider which behavioral archetypes are responsible for assaulting soil with nasty concoctions of sewage, pesticides, herbicides, industrial effluents, medical waste, spilled oil, and chemical substances; and which enrich and protect it.

Pet Archetype

 Imagine the state of the physical environment if the relationship humans have with pets was applied to bees, forests, reefs, streams, cell phones, shoes, cars, plastic containers, soils, and all other forms of materiality. Pet owners are protective caregivers. They are attentive to their pets' needs and strive to prolong their pet's physical and mental well-being. Such consideration is granted even though it means routinely conducting tasks most of us would otherwise avoid, such as cleaning up messes, tolerating inconvenience, and sharing personal resources. Furthermore, the time required to perform these duties is factored into each day without complaint. Ultimately, affectionate rapport transforms these tedious tasks into acts that yield deep personal satisfaction. Thus, if pet interactions became the model for all humanity's material relationships, caretaking would replace careless disregard, health would be monitored and maintained, protection would be provided, and grief would accompany loss.

 Pets are attended to with care and consideration when they die as well as when they are alive. Thus, if all our material possessions were treated to pet end-of-life protocols, they would not be discarded as litter, and they would not be dumped in trash destined for a landfill. However, the environmental benefits might be reversed if laying an unwanted object to rest was accompanied by the extravagant 'pet-aftercare' paraphernalia available in the marketplace, such as luxury cremation urns, coffins, and grave markers.

Compost

Amy M. Youngs

Art Engagement:

Amy M. Youngs (b. 1968, Chico, California, United States) is an artist currently based in Columbus, Ohio. She creates biological art, interactive sculptures, and digital media works that explore interdependencies between technology, plants, and animals. Youngs has exhibited her works nationally and internationally at venues such as the Te Papa Museum in New Zealand, the Trondheim Electronic Arts Centre in Norway, the Biennale of Electronic Arts in Australia, Centro Andaluz de Arte Contemporáneo in Spain, and the Peabody Essex Museum in Salem, Massachusetts. In 1999, she received an MFA from the School of the Art Institute of Chicago. In 2001, she joined the faculty at the Ohio State University, where she is currently working as an associate professor of art.

Amy Youngs treats soil as a pet by nurturing the worms that break down organic materials and transform them into an enhanced fertile growing medium. She created an artwork that is aptly titled *Digestive Table* because it includes a place for her to dine with her worms in the midst of the soil they co-created. Youngs applies to worms the cross-species conviviality that humans normally cultivate with cats, dogs, and parakeets, despite the fact that worms offer little in the way of companionship. They don't greet Youngs at the door; or take walks with her in the park; or sleep at the foot of her bed. Nonetheless, they are amiable pets. Worms require little caregiving; they are inexpensive to maintain; they never need grooming; they don't need veterinary attention; they don't require behavioral training; and they make no noise so they don't bother the neighbors. Youngs highlights an additional benefit offered by few other pets: worms are service providers. They transform rotting organic matter into rich productive soil by processing leftovers from their owners' meals, transforming pizza crusts and apple cores and used tea bags into sweet-smelling compost that is so valuable as a growing medium it is often described as 'black gold'.

Youngs acknowledges that expanding caregiving to worms is complicated by 'their power to repulse us.'[2] She comments, 'There is a lot of cultural work to do if we are to develop symbiotic relationships with them.'[3] There is also a lot of practical work to do to overcome the fact that worms require underground habitats that are not congenial to humans. Mutually alien environments make it difficult for humans to bond with their pet worms. Youngs developed several strategies to enable these divergent species to dine together. For example, she attached an opaque bag to the underside of the table to provide the dark, moist environment that worms prefer for dining. At the same time, she provides humans with a well-lit dining environment, an attractive table to place food upon, and a stool to sit on while eating. Since feeding is an essential pet-care pleasure, she inserted a cavity into the top of the table for the immediate dispatch of leftover foods to the worm bag as ready-made ingredients for their next meal. This built-in receptacle converts the chore of waste management into the pleasure act of feeding a pet.

Next, Youngs installed an infrared camera inside the worm bag to provide a communication channel between these divergent diners in their contrasting territories. The images are transmitted to an LCD screen that is inserted into the top of the table, enabling Youngs to observe the worms' munching on her scraps, a virtuoso waste-disposal performance, as they dine together.

Youngs then made this symbiotic relationship reciprocal. At the same time that the worms are benefiting from human waste, humans are supplementing their diet with worm waste. This nourishing exchange occurred at the table's base

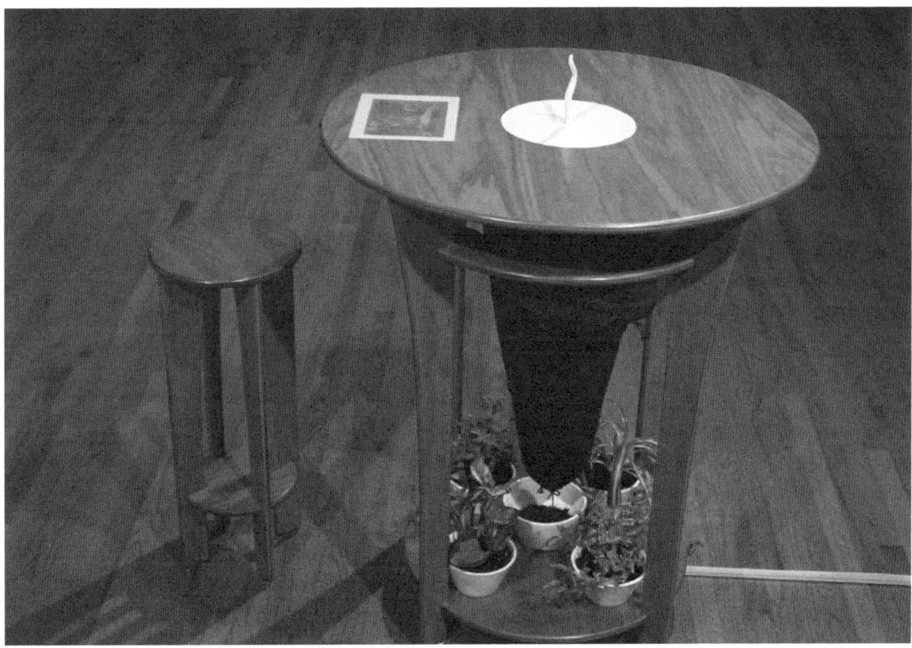

Figure 1a: Amy Youngs, *Digestive Table* (2006), live red wiggler composting worms, sowbugs, food scraps, shredded paper, landscaping fabric, polyethylene, security camera, LCD screen, infrared filters, live houseplants, and Forest Stewardship Council certified oak plywood, stained with a mixture of boiled red cabbage and worm compost tea, 48 x 36 x 48 in. Photo by Amy M. Youngs. Courtesy of the artist.

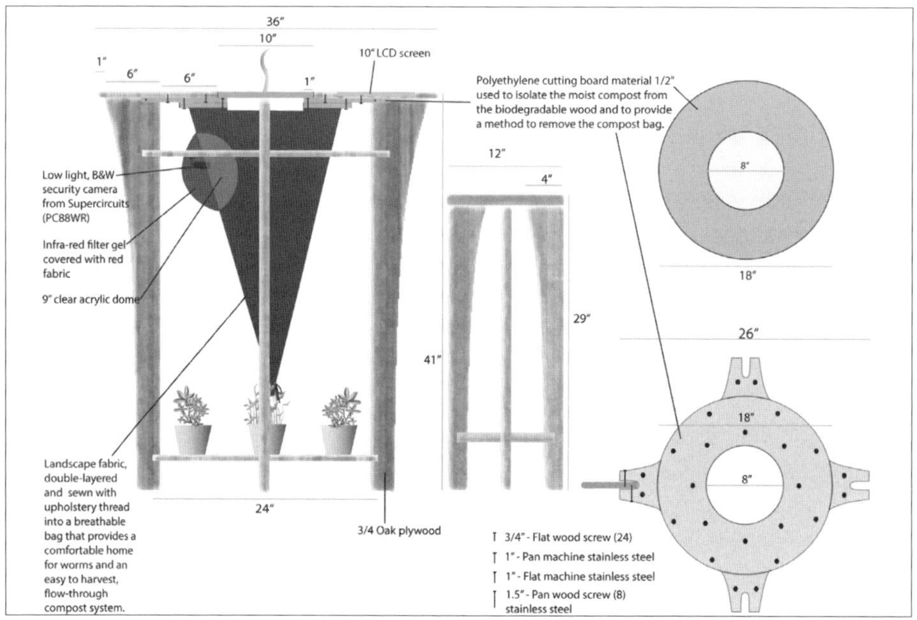

Figure 1b: Amy Youngs, *Digestive Table* (2006), vector drawing made and shared online to open source this artwork. http://creativecommons.org/licenses/by-nc/2.5/. Courtesy of the artist.

that was outfitted to receive the nutrient-rich soil that the worms crafted as they digested their meals. This special soil became the growing medium for the ingredients of a tasty salad for the next human meal. Finally, by welcoming the worms into her home, Youngs satisfied the ultimate condition for establishing a pet-like relationship by welcoming co-habitation, a privilege only granted to animals that are pets.

The *Digestive Table* provides the worms with food, bedding, warmth, and moisture. In exchange, the worms manage human food scraps and produce the growing medium for salad greens. The advantages of this reciprocal material interaction accrue to the worms, to their human caregivers, and to the environment. But Youngs also promotes her scheme as an 'archetype' that offers bonuses missing from many material interactions. She explains:

> I live and work with domesticated technological and biological beings that sustain me aesthetically, gastronomically, intellectually and emotionally. Emitting waste, I seek co-habitation with creatures that make use of it. I like to be reminded that my human existence depends on interdependence.[4]

She concludes by issuing an invitation to readers to apply pet protocols to all environmental interactions, 'I hope you will join me in prototyping a future that appreciates our shared world with other living things. We start with the worm, and we work together to create new culture and new ecosystems.'[5]

Sacred Archetype

 Imagine the state of the physical environment if the relationship humans have with sacred objects was applied to bees, forests, reefs, streams, cell phones, shoes, cars, plastic containers, soils, and all other forms of materiality. As embodiments of un-Earthly perfection, sacred objects and places are reminders of the inadequacy of human accomplishment. We humans are humbled by their divine glory. Routine protocols cannot embody the overwhelming emotions they stir. Sacred objects demand the honor bestowed upon them by ceremony and the heightened wakefulness of ritual. The awe such objects instill is tinged with fear and expressed as veneration. As such, altering, consuming, or improving something sacred is an arrogant presumption and deeply disrespectful.

Many indigenous cultures bestow sacred status upon every component of the material environment. Within these cultures, animals, plants, rocks, rivers, mountains, thunder, wind, sun, rain, and shadows are all imbued with supernatural powers. As such, even the routine material interactions carry consequences of a godly blessing or punishment. Anthropologists refer to the consecration of all material entities as 'animism'. Environmentalists call it 'deep ecology'. In art, this attitude is known as 'the sublime'. In all three contexts, material encounters are conducted with the solemnity and grace of a sacrament. By endowing such contemporary commodities as plastic bags, cell phones, and sneakers with the attributes of divinity, careless material interactions would be replaced with reverential attentiveness.

 Extraordinary care does not end when a holy person dies or a hallowed object is decommissioned. End-of-life procedures that honor a saint's consecrated remains, for example, are ritually laid in a shrine, mausoleum, or reliquary. Equal veneration is evident in the end-of-life procedures that honor sacred objects. For example, Navajo masks imbued with the power they absorbed during religious ceremonies are lavished with formal burial rites. If banal materials were retired with the solemnity reserved for sacred objects, expediency would be replaced with heightened regard for their continued impact on the lives of the living.

A landmark installation by Walter De Maria, entitled *The New York Earth Room* (1977),[6] is still on display despite the fact that it was first installed in 1977. That is because it was deemed so significant from its inception that the Dia Art Foundation not only funded its production, it also agreed to fund its perpetual exhibition, maintenance, and protection. Visiting the site of this ongoing presentation resembles a religious pilgrimage. Like a shrine, the *Earth Room* is elevated above the bustle of the urban street on the second floor of a loft building in Manhattan. One enters, climbs stairs, navigates a dark hallway, and only then beholds a 3,600-square-foot space filled to its outer walls with soil. It is quintessential soil – dark brown, crumbly, moist, dense, and sweetly aromatic. Within the paved expanse of a modern city, beholding any expanse of soil is amazing. The special qualities of this soil contribute to this amazement. It seems too perfect to be associated with such an Earthly function as planting. It belongs to consecrated realms where humans dare not tread. In fact, entry is blocked by a knee-high Plexiglas barrier, the same height as the soil. The view is unimpeded, but access is forbidden. Reverential distance is mandated.

The New York Earth Room reverses the notion that dirt exists down low – the territories associated with depression and despair. It sheds all its connotations with filth and moral degradation. De Maria created a shrine to soil's glory. It induces veneration, not distaste. This soil exists to be worshipped, like the ashes of a saint or the relics of a martyr.

The backstory that is often told about this artwork was not confirmed by De Maria before his death in 2013, but it is instructive nonetheless. It is widely reported that he rejected all 280,000 pounds of dirt that he ordered from a Pennsylvania farm for this art installation because it sprouted! The evidence

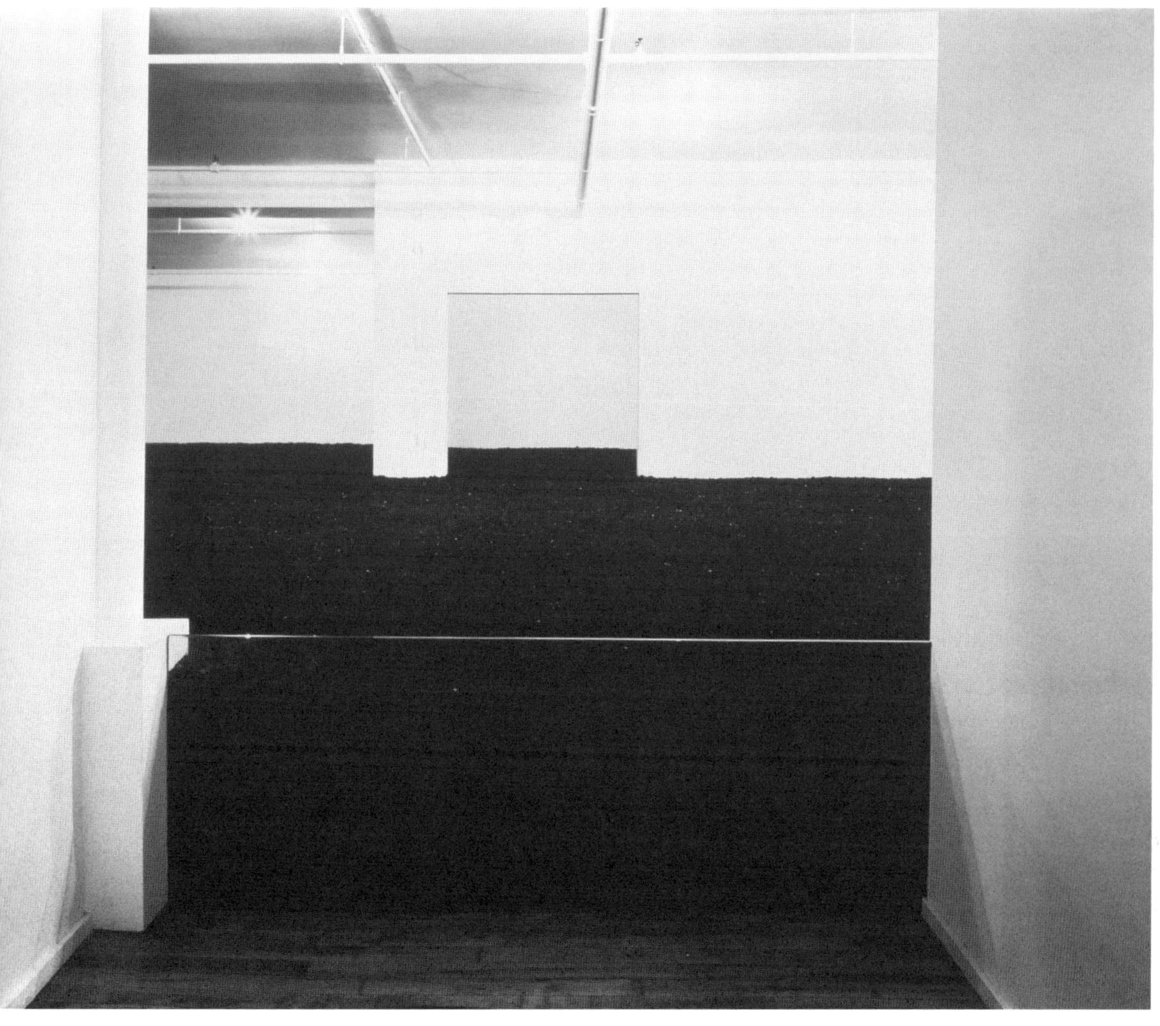

Figure 2: Walter De Maria, *The New York Earth Room* (1977). The Estate of Walter De Maria. Photo by John Cliett. Courtesy of the Dia Art Foundation, New York.

of organic life belongs to Earthly functions. Thus, the entire load was removed, fumigated, sterilized, and then reinstalled so that not weed, mold spore, or mushroom could germinate, disrupting the pristine purity of a sacred substance. Only barren soil could transcend mundane realities and emulate divinity.

What are the ecological implications of the *Earth Room*? Is this a constructive model for human interactions with the material world? The answer is complicated. On the one hand, environmental abuse is avoided by the 'don't touch' policy that accompanies sacred items. Likewise, material neglect is countered by the reverence bestowed upon a sacred substance. Nonetheless, worshipful demeanors inhibit stewardship, remediation, and restoration.

Hazard Archetype

 Imagine the state of the physical environment if the relationship humans have with hazards was applied to bees, forests, reefs, streams, cell phones, shoes, cars, plastic containers, soils, and all other forms of materiality. This category already applies to many substances. While poisons occur naturally among some plants, animals, minerals, metals, and gases, complex concoctions of chemicals that have recently been contrived have swelled the list of endangering substances. According to the United Nations Environmental Program's report, *Global Chemicals Outlook*,[7] poisonings from industrial and agricultural chemicals are among the five leading causes of death globally. They also lurk in the home, at work, and in public places in the form of disinfectants, antifreeze, batteries, bug sprays, epoxy, fluorescent bulbs, gasoline, hair dye, paint, and shampoo.[8] This list does not exhaust current dangers. Acid rain, PCB pollution, cyber attacks, airplane collisions, space and ocean debris, radioactive waste, dam collapse, and noise pollution did not concern our ancestors. Indeed, an extensive vocabulary has developed to distinguish between the severity and type of hazard encountered in everyday situations: poison, warning, danger, caution, ignitable, corrosive, explosive, and toxic, for example.[9]

Plants, because they are rooted in place, depend upon chemical defenses against herbivores by acting as repellents or toxins. Animals have the advantage of supplementing such 'fight' responses with 'flight' that allows them to escape from a site of danger. Humans supplement these instinctual defense mechanisms by developing offensive tactics to guard against dangers. Some examples include inoculations that protect against infectious diseases, radar installations that detect enemy invasions, water treatment plants that decontaminate sewage, and weather satellites that anticipate storms.

 Government regulations govern the handling and disposal of hazardous sludge, solvents, dioxins, radioactive uranium tailings, refrigerants, glues, pharmaceuticals, pesticides, paints, batteries, chemicals, etc. Indeed, so many toxic substances now encumber the material environment in the United States that a Resource Conservation and Recovery Act was passed to regulate their disposal; the US Department of Transportation, Pipeline and Hazardous Materials Safety Administration created a Code of Federal Regulations; the Environmental Protection Agency imposed Hazardous Waste Recycling Regulations; and the US Department of Transportation requires hazardous-materials training for all 'hazmat employees'.

There are several destinations for discarded dangerous substances, and none resemble the disposition of a pet or a sacred object. **Landfills** are engineered sites where non-liquid hazardous waste is deposited and covered. **Surface impoundments** are natural and man-made depressions where liquid hazardous wastes are stored or treated. **Waste piles** are non-containerized piles of solid waste. **Land treatment units** apply hazardous waste directly on the soil surface where microbes and plants transform hazardous constituents. **Injection wells** place hazardous fluid deep underground in porous rock formations. **Underground caves** are natural geologic repositories. A less common tactic involves detoxifying the substances so they can be reused. An Eco Material response to the extreme measures designed to manage the ill-effects of material interactions is to avoid the problems in the first place. Eliminating hazardous materials and decommissioning hazardous processes avoids these problems.

Contaminated Soil

Art Engagement:

Hilda Hellström

Hilda Hellström (b. 1984, Gothenburg, Sweden) is an artist based in Copenhagen, Denmark who works primarily in sculpture. In 2012, she graduated with an MA at the Royal College of Art in London. Her work has been exhibited at the Gothenburg Museum of Art; the British Craft Council; MAK Vienna; the Victoria & Albert Museum; the London Design Museum; MUDAC Lausanne; Tramway Glasgow; and the Shanghai Power Station Art Museum. In 2017, she presented a solo exhibition of new works, titled I Do Believe in Nature, *at the Berg Gallery in Stockholm, Sweden.*

The Swedish artist, Hilda Hellström, confronted a material horror when she traveled to Fukushima, Japan in 2012, one year after a rare double earthquake devastated the region. Damage from the earthquake skyrocketed when a huge tsunami followed in its wake. The tsunami inundated over two hundred square miles. Its human death toll topped nineteen thousand. Over a million buildings were destroyed. This horror escalated because the Fukushima Daiichi Nuclear Power Plant lay in the path of destruction. All six external power-supply sources were lost due to earthquake damage. Then the tsunami wave slammed into it, releasing radioactive isotopes into the air, water, and soil. Populations in the vicinity evacuated. Hilda Hellström's modest earthen vessels tell this terrible tale. They were created with soil from the Daiichi evacuation zone. She explains:

A chronological definition to our present geological age is Anthropocene, which defines human impact as the most significant recent geological development. This might sound obscene, but each time a natural disaster occurs I imagine that it is nature's revenge on that same upper hand.[10]

Hellström purchased a Geiger counter on eBay and took it with her to Japan to meet Naoto Matsumura, the last man still living in the evacuated zone surrounding the damaged Daiichi power facility. Matsumura had planned to evacuate with his neighbors, but discovered he had nowhere to go because the recovery centers were full and his relatives rejected him. They feared he was radioactive. In the end, he stayed to devote himself to caring for the pets and farm animals abandoned when people fled. Hellström documented Matsumura's new life as a shepherd and survivalist. Then, side-by-side, she and Matsumura put on masks and gloves, knelt in an abandoned rice field within the 'exclusion zone', and dug the soil she later used to create a serving dish, bowl, vase, goblet, and covered container. Hellström explains how she proceeded:

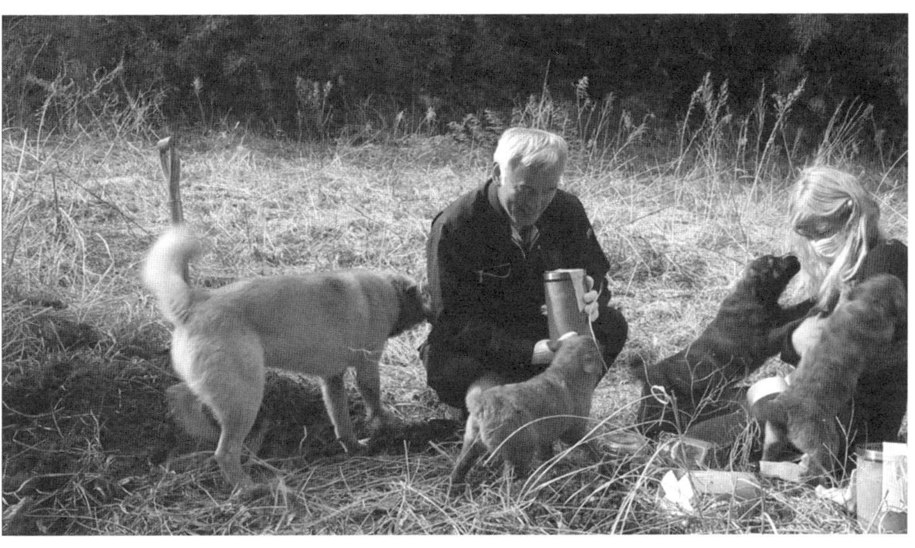

Figure 3a: Hilda Hellström, *The Materiality of a Natural Disaster* (2012), still from *The Materiality of a Natural Disaster*. The Royal College of Art, London. Courtesy of Hilda Hellström.

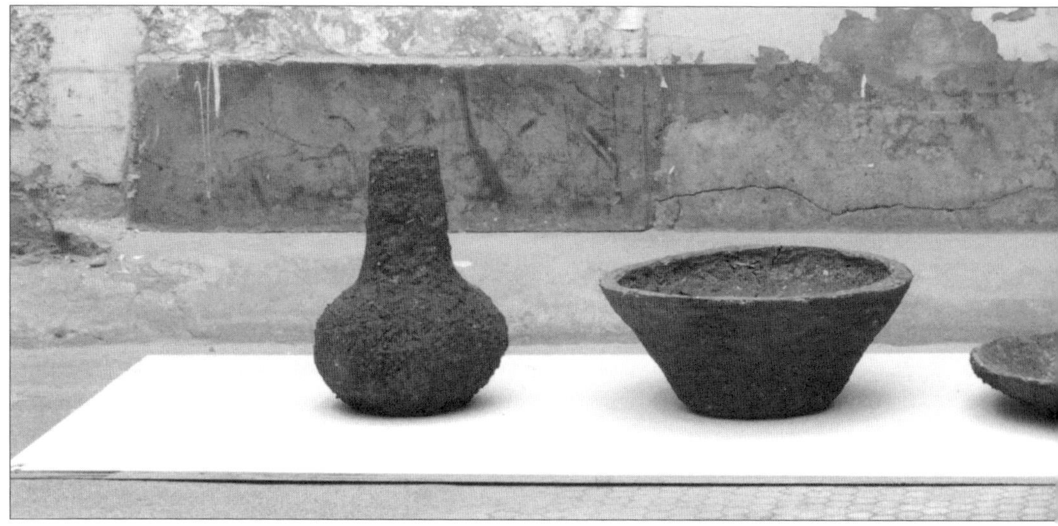

Figure 3b: Hilda Hellström, *The Materiality of a Natural Disaster* (2012), hand-packed soil with PVA binder. Food vessels from soil and pva. From left to right: water jug, salad bowl, fish plate, flower vase, rice bowl. Photo by Hilda Hellstrom. Courtesy of Hilda Hellstrom.

I wish I could tell you that my way of sourcing the soil was more analytical. [...] I think 'ad-hoc' is the best way to describe it. I had read up about it, mostly on the internet, and even though the valley where I was working was inside the exclusion zone, it wasn't very affected in comparison to other parts. The wind blew in the opposite direction during the breakdown so the radiation I was exposed to there was supposedly no worse than a few scans in the hospital. I met up with a Japanese scientist before the project began who guided me. I collected soil from about 30 cm below the surface where very little radiation had gone down. It was unnecessary that the soil should be dangerous as long as it was loaded with the narrative of the devastating situation.[11]

While the danger is inferred, the loss of utility due to the devastation is actual. These vessels are useless for serving food because there is real danger in ingesting radiation, just as the soil from which they are made is useless for growing rice, and the region is useless for habitation.

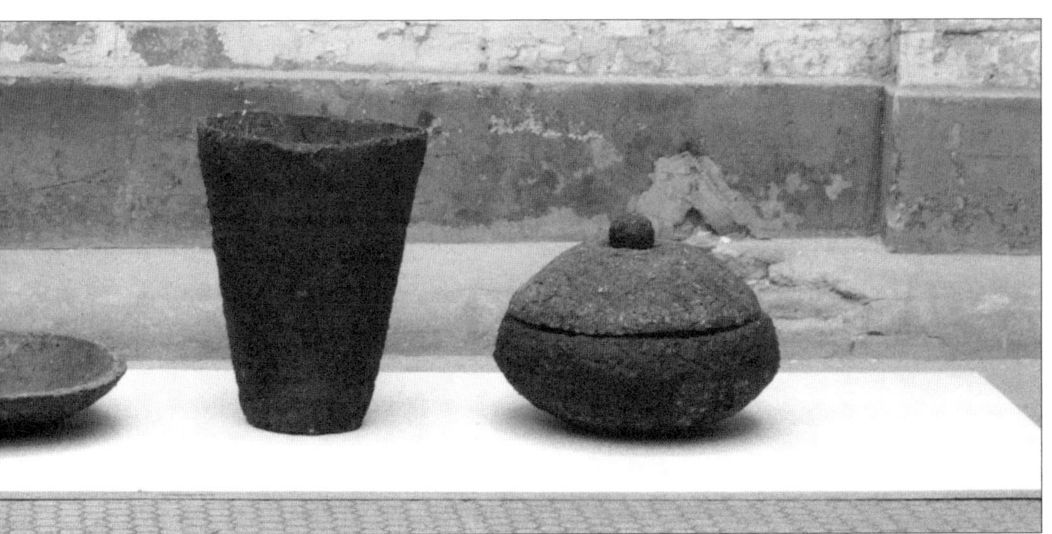

Gray, rough, thick, cracked, and unembellished, the materiality of these vessels evokes their sordid history. Hellström explains that the specific theme she intended to evoke is dread of radiation, not radiation itself.

> I started out the project with the intention to create objects that would materialize the events in Fukushima, like iconized or mythological objects. But the further the project went I realized that the radiation itself was the real mythological phenomenon that evoked fear in people. I think the fear of radiation is worse than the fear of a natural catastrophe or pandemics. Partly because it's not natural so we don't have the same acceptance to it as we have to those events evoked by Mother Earth. Partly also because it's so intangible.[12]

Specimen Archetype

Cultural Engagement: Imagine the state of the physical environment if the relationship humans have with specimens was applied to bees, forests, reefs, streams, cell phones, shoes, cars, plastic containers, soils, and all other forms of materiality. Unlike the pets that are appreciated for their unique characteristics, specimens represent the standard characteristics of an entire set of entities. The more 'typical' they are the better because their role is to represent a general category. Specimens are useful in scientific testing, research, display, diagnostic examination, and evaluation. Rocks, plants, insects, worms, embryos, blood, bones, tissues, cells, and viruses have all been enlisted as specimens.

Those who interact with specimens are receptive to all possibilities because they are concerned exclusively with the concrete accumulation of verifiable facts. Unlike pets, emotional attachment is omitted from human interactions with specimens. Such material exchanges are characterized by objective collection and identification. These results are typically assembled and indexed into systematic categories. Such rigor enables specimens to contribute to scientific knowledge. The field collection of specimens remains today, as it has in the past, an essential component of documentation and analysis in the scientific disciplines, despite the development of technologies such as high-resolution photography and tissue sampling for DNA analysis. Specimens play a particularly important role in ecological research because specimens register changes in complex biological, geological, climatic, and social conditions. Such research facilitates informed decisions about species management and conservation.

Because subjective desires and aversions are the twin drivers of many consumer decisions, the clinically detached scrutiny that typifies interactions with specimens might foster consumer restraint. Furthermore, specimen-based attitudes might inhibit the emotional impulses that lead to wasteful or redundant purchases regarding wardrobes, diets, electronic devices, and modes of transportation.

Cultural Disengagement: Because specimens are vehicles for future study, two complementary material engagements are required. The first is collection, which can occur on the surface of the moon, in deep-sea vents, within the cells of tissues, and in backyards. The second is fixation, which refers to the processes undertaken to maintain the specimen for future research. While a stable environment usually suffices for non-biological specimens, special procedures are required to thwart the natural forces that would decompose biological specimens. Such entities are either 'pickled' in a liquid preservative; dehydrated by removing the body's moisture; or frozen to stop enzymatic decay. All biological objects treated like specimens would depend upon technology to provide a stable, non-reactive state that would enable them to persist into the future. Besides producing clutter in the short run, widespread application of specimen treatments would interfere with the renewal of life in the long run.

Art Engagement:

Rubble

Alexandra Regan Toland

Alexandra Regan Toland (b. 1975, Boston, Massachusetts, United States) is a visual artist and environmental planner currently living and working in Germany. In 2015, she received a doctorate degree from the TU-Berlin Institute for Ecology. Toland has held lectures and teaching positions at the Technische Universität Berlin, Universitat der Kunste Berline, and Leuphana University. From 2011 to 2015, she co-chaired the German Soil Science Societies (DBG) Commission on Soils in Education and Society. Currently, she is editing a book that brings together fifty international artists and scientists to discuss the challenges and possibilities of soil protection. Toland worked as a freelance photographer for the environmental visualization firm Lenne 3D from 2006 to 2008 and participated as a design researcher at the Wriezener Park Open Space Lab in Berlin-Friedrichshain from 2005 to 2009.

The 'specimens' that the artist Alexandra Regan Toland collected to create *Rubble Mapping* (2010)[13] consisted of dirt that provided a tangible repository of Berlin's history. Working with Prof. Dr. Gerd Wessolek of the Department of Soil Protection in Berlin, Toland collected sand, stones, roots, worms, and microbes from beneath the city's pavement to discover its biological and geological history. Cultural history was on their agenda as well. Thus, the specimen collection included artifacts and substances that reveal human settlement, urban development, agriculture, and war. Toland comments, 'Peel back the pavement and what do we find? Dirt, sand, stones, debris, roots, worms, turds – in short urban soil.'[14]

Modern Berlin provided an ideal site to conduct such an exploration because it was rebuilt on top of the remains of the seventy-five million cubic meters of rubble and debris left by the allied bombings during World War II in the 1940s.[15] Beyond curiosity, discerning this geological history became an ecological necessity because sulfates from buried mortar were leaching into the city's groundwater system and contaminating the water. Toland explains that 'a steady release of sulfur trickles from the crumbling mortar, ashes, and coal buried in mountainous memorials

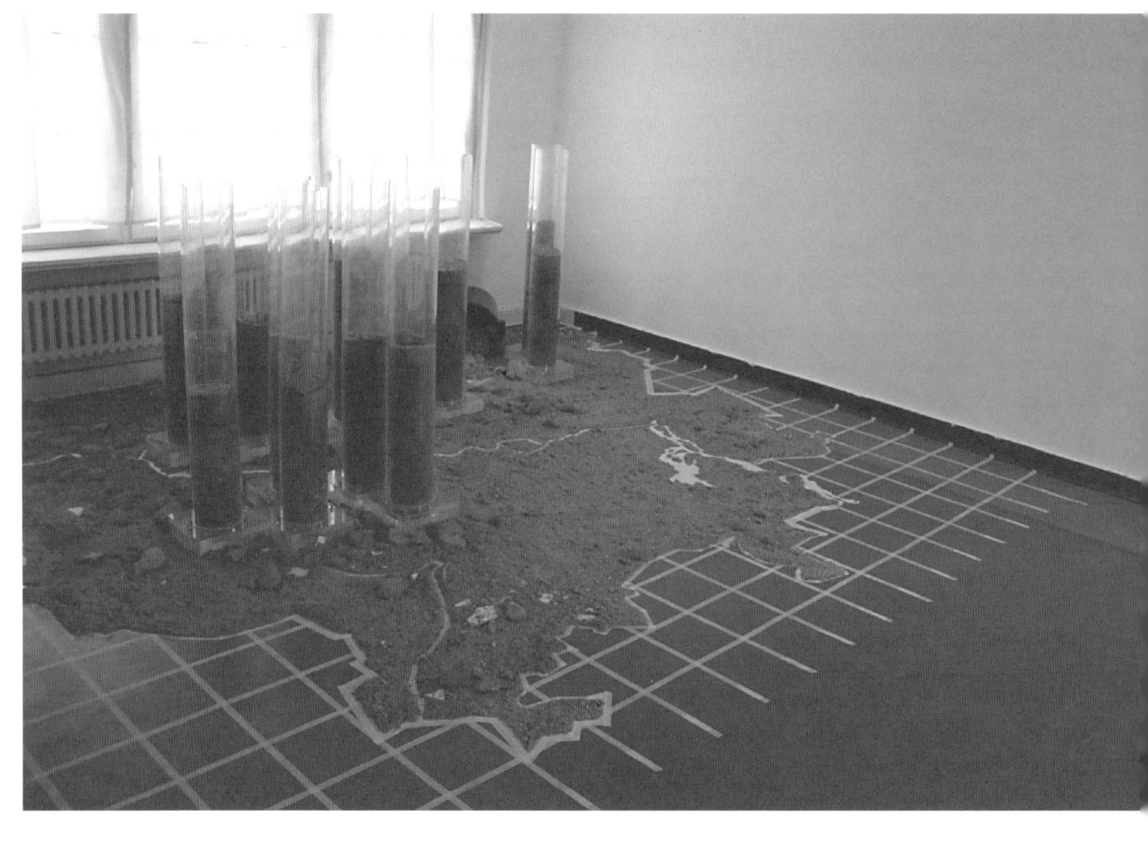

Figure 4a: Alexandra Regan Toland, *Rubble Mapping* (2010), rubble map installation showing concentrations of WWII rubble deposition in Berlin, dimensions variable. Photo by Alexandra Toland. Courtesy of the artist.

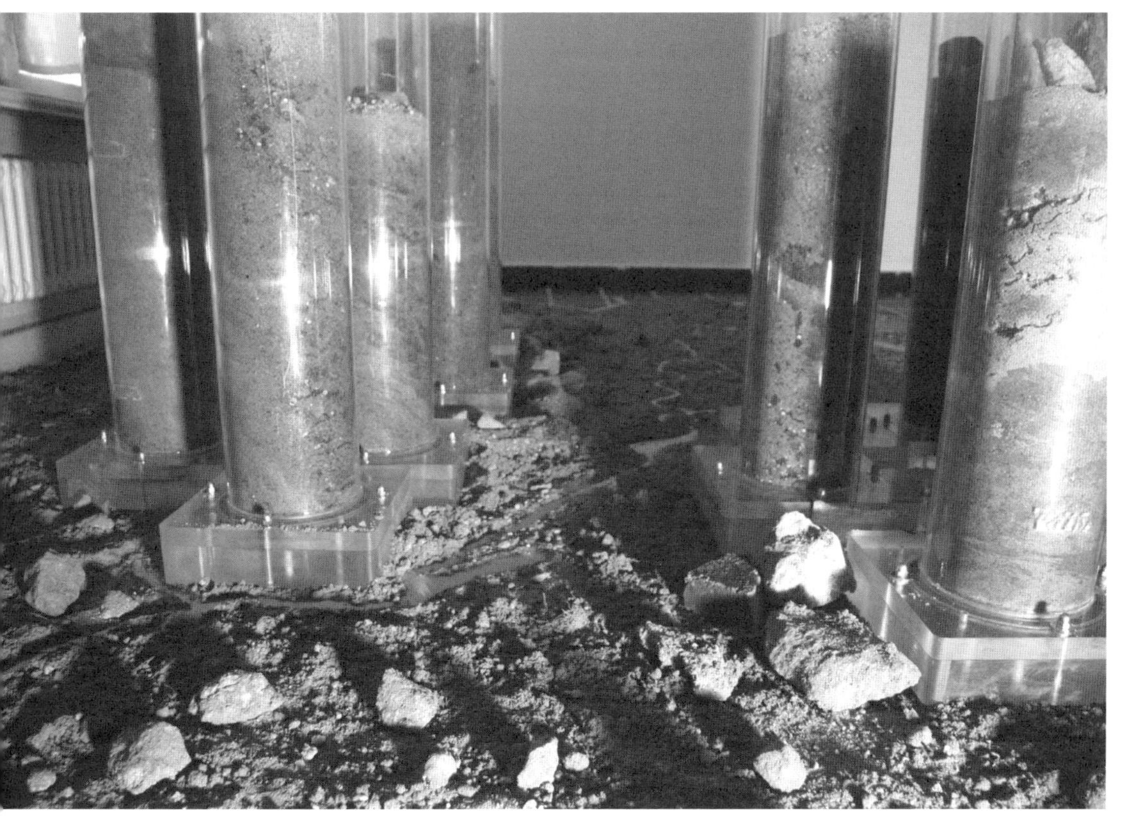

Figure 4b: Alexandra Regan Toland, *Rubble Mapping* (2010), rubble map installation showing concentrations of WWII rubble deposition in Berlin, dimensions variable. Photo by Alexandra Toland. Courtesy of the artist.

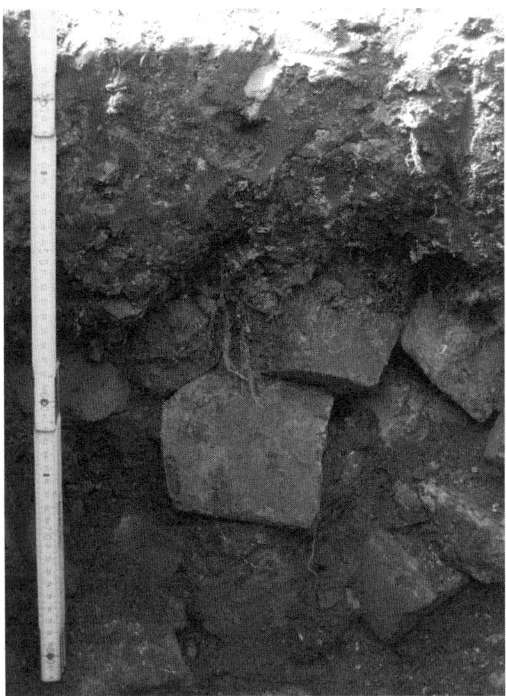

Figure 4c: Alexandra Regan Toland, *Rubble Mapping* (2010), rubble map installation showing concentrations of WWII rubble deposition in Berlin, dimensions variable. Photo by Alexandra Toland. Courtesy of the artist.

that still dot the landscape of Berlin and cities across Europe.'[16] She goes on to explain:

> *Historical tragedy and political amnesia aside, the nature of rubble has a fundamental effect on the physical, chemical, and biological properties of the soil, contributing to the amount and availability of nutrients and heavy metals as well as the water quality.*[17]

This lingering postwar legacy escalated into a political platform in 2014 when the German government passed a Federal Soil Protection Act to preserve the soil's natural functions regarding fertility and filtration, as well as its cultural function as an archive of the city's history.

Alexandra Toland collected specimen soils for an art installation that conveyed this rich history. It consisted of a map of the city that covered the floor of a large gallery at the Altes Museum. Rubble that had been dug up was scattered over the entire surface. It included bits of brick, mortar, metals, ceramics, glass, slate, marble, etc. The areas on the map that depicted the city's waterways were left exposed, but they too contributed to the artwork's environmental theme because, instead of a placid blue, the artist painted these waters a lurid yellow, the color of sulfur in its elemental form. This acrid element was augmented by long yellow curtains that covered the windows. The installation not only documented the insidious seepage of sulfur throughout the city's underground territories, it actually evoked sulfur's acidic composition and foul smell.

The rigorous methodology of specimen collection that the artist undertook was indicated in the exhibition by nine tall glass test tubes. Each was placed on the map at a location where the greatest rubble deposition had accumulated. Some sites were in popular city parks. Others were clustered in surrounding forests that had been used as dumping sites. Each column was filled with proportional quantities of the specific rubble discovered at these sites: decomposed leaf litter, rotting food waste, excrement, spilled motor oil, cigarette stubs, and plastic packaging, as well as rubble from bricks, beams, mortar, and glass that had broken down, decomposed, or become pulverized by pressure from pedestrians and motor vehicles. These specimens represented all twenty-five million cubic feet of rubble[18] that comprised an underground monument to Berlin's past. They also testified to the extensive analysis that accompanied the collection of these specimens. Toland published the results of her studies in collaboration with Dr. Gerd Wessolek, Department of Soil Protection, TU-Berlin. The data they collected considers rubble sediments from different types of building materials; when each building was demolished; how its debris was recycled; and how the materials were sorted. In addition, they account for spatial deposition, chemical composition, size, density, history, and climate. Toland inserts these specific facts into a broader cultural context when she states:

> *Rubble is one of the indicators of the Anthropocene, or the current epoch characterized by traces of human activity in the geologic record of the Earth [...] Teufelsberg Mountain is part of the cultural landscape of the Anthropocene; its parent materials of rubble and ash are studied objectively, as if they were natural bedrock.*[19]

By presenting the urban soil samples as specimens, Toland removed them from the zone of speculation, and imbued them with the certainty of fact. In this manner, she offered visually persuasive evidence that the destructive legacy of war continues long after peace treaties are signed. She not only conveyed this continuing history visually, she installed a desk in the exhibition space and stacked it with books, notes, maps, and calculations. These documents provided data regarding the relationship between rubble soils, sulfate leaching, post-WWII reconstruction, rubble management, and other research examining rubble soils. The specific material interactions of this archetype are most powerfully revealed by the aspects of this narrative with which Toland did not engage. She did not attempt to remediate the contaminated soil, and she did not express alarm or any other subjective reaction to it. Instead, she presented the urban dirt without comment, in the objective manner of a specimen.

Resource Archetype

 Imagine the state of the physical environment if the relationship humans have with resources was applied to bees, forests, reefs, streams, cell phones, shoes, cars, plastic containers, soils, and all other forms of materiality. A resource is any object, any substance, any person, or any quality, including human services and knowledge that is perceived to be useful. All resources are positive assets whether they are comprised of matter that is 'natural' or engineered, renewable or nonrenewable. Conditions of ecosystems are increasingly affected by the way humans produce, distribute, utilize, and dispose of resources.

There is nothing inherently immoral or wasteful in the consumption of resources. Life depends upon acquiring inputs for growing, surviving, and reproducing. Animals require food, water, oxygen, and territory. Plants depend upon sunshine, nutrients, water, and a place to grow. Humans, however, diverge from dependence upon need-based, locally satisfied resources. Many of the resources that are consumed culturally satisfy desires that are contrived by entrepreneurs, promoted by marketing, and facilitated by finance. These economic structures not only provide necessities; they arouse desire for non-essential comfort, leisure, excitement, and convenience. While the consumerist version of the resource archetype tends to be associated with waste, depletion, and pollution, the Eco Material version strives for prudence, restraint, and stewardship.

 By definition, a resource is the opposite of waste. Of course, there are wasted resources, as when oil spills from tankers, or foods spoil, or trains wreck, or chickens contract the avian flu. In some of these instances, items that had been a 'resource' become a 'hazard'. Material Recovery Facilities apply the designation 'resource' to disposed materials. They selectively extract reusable materials, recycle non-biological waste, compost biological waste, and combust waste for energy generation. These strategies to extend the usefulness of the by-products of production, discarded items, leftovers, and obsolete items involve delaying their classification as 'waste' by extending their status as 'resource'.[20]

Jon Cohrs (b. Cleveland, Ohio, 1977) is a recording engineer and visual/sound artist who lives in Brooklyn, New York. His work explores technology and its connection with wilderness. He created a documentary, The Door to Red Hook: Backpacking through Brooklyn *in 2008. In 2009, he won the Futuresonic Art Award for* The Urban Prospector. *In 2015, he premiered his feature film* Back Water *about the New Jersey Meadowlands and the Anthropocene at the Vision Du Reel Film Festival in Nyon, Switzerland. He has had residencies, installations, and performances at W2, I-Park, Banff New Media Institute, Futuresonic, and Eyebeam. He is currently a Fellow at the Eyebeam Atelier working on creating 'designer salt' with selective serotonin reuptake inhibitors (SSRIs); writing a book on Urban Wilderness; and researching sonic weapons in collaboration with Audint.*

In the project entitled *The Urban Prospector* (2009),[21] the artist Jon Cohrs searched for valuable resources in urban soils. His treasure hunt bypasses lost coins and jewelry. The valuable item he seeks is residue of past oil spills! This form of treasure-hunting is not as bizarre as it may seem. Cohrs's ludicrous proposition coincided with skyrocketing crude oil prices. Oil was so valuable at the time that its availability determined the economic well-being of entire nations. He notes, 'Given the current high cost of oil, these urban spills or potential gold mines are waiting to be tapped.'[22]

Cohrs's quest for the dark, tarry pollutant can be explained by his reworking of the popular narrative relating to oil spills that focuses on dependency and jeopardy. Cohrs introduced 'independency' and then added 'resource' to indicate the possibility of individuals tapping the value of contaminating deposits. He conveys this message by referring to the harmful leakages of oil as 'reserves', not oil 'spills'.[23] Furthermore, he uses the term 'urban prospector', not 'hazardous waste workers', to identify those who seek sites of oil accumulations. These strategies evoke the frenzied American Gold Rush of the 1800s when thousands set out across the country driven by the dream of exclaiming 'Pay Dirt!' when they discovered gold. Cohrs's reference to this euphoric era in US history offers an ironic commentary on the comparable contemporary obsession with oil.

These exploratory expeditions are conducted with a handheld metal detector that he modified to detect spilled oil lurking in city soils. First he outfits the detector with a combustible gas sensor; then he removes the metal sensor and replaces it with a 'petro' sensor designed to detect petroleum hydrocarbons (TPH).[24] Analog data emitted by the sensor is then analyzed by a microchip connected to a computer. Cohrs boasts that it costs less than one hundred dollars to build a

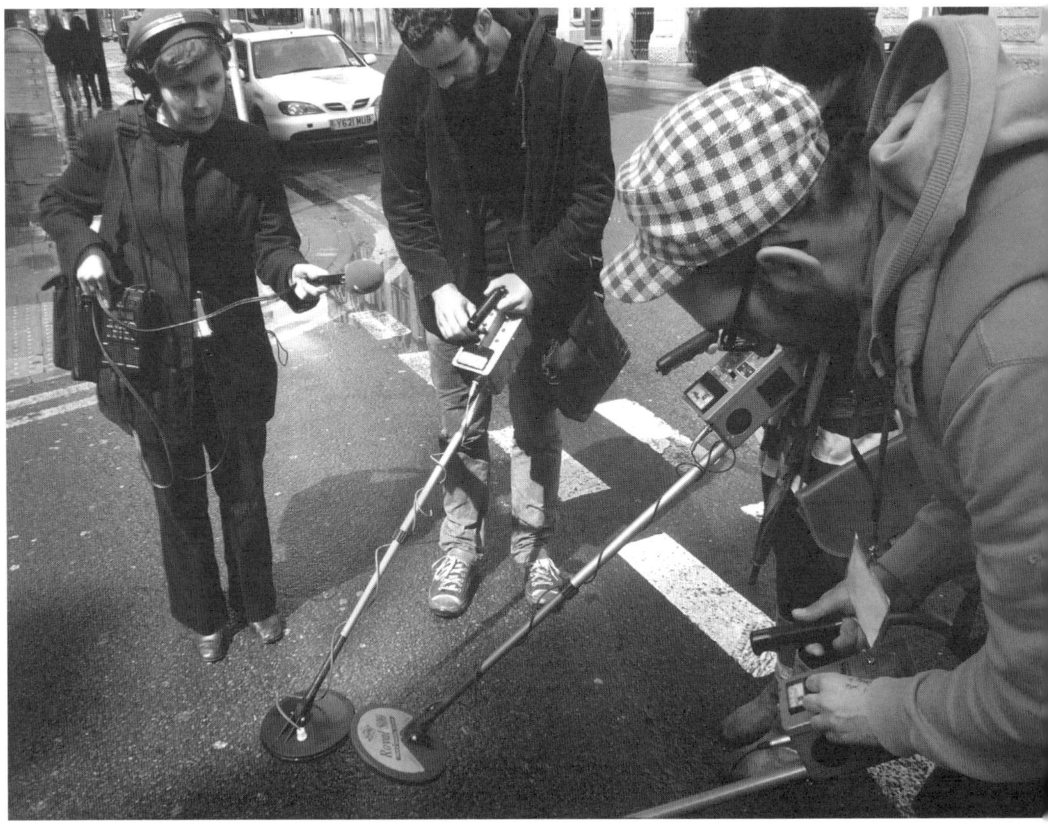

Figure 5a: Jon Cohrs, *Urban Prospector* (2009). Futuresonic Festival Manchester, United Kingdom. Photo by Jonah Brucker-Cohen. Courtesy of the artist.

device to detect this valuable and plentiful resource. He delights in its potential to disrupt the oil-prospecting industry. Currently it is monopolized by professionals and corporations that can afford the expensive tools needed for detection and can assemble people with expertise to utilize them. The prospecting tool that Cohrs devised scales these efforts down to the level of the nineteenth-century gold prospector's saddle bag. Furthermore, it invests oil-drenched urban soils with an allure that is comparable to nuggets of gold glistening in stream beds. The artist's film of the process of sweeping the terrain with his detector is not the least tedious because each instant carries the anticipation that he will exclaim, 'Pay Dirt!' Instead of signaling danger by wearing a protective slicker suit, safety

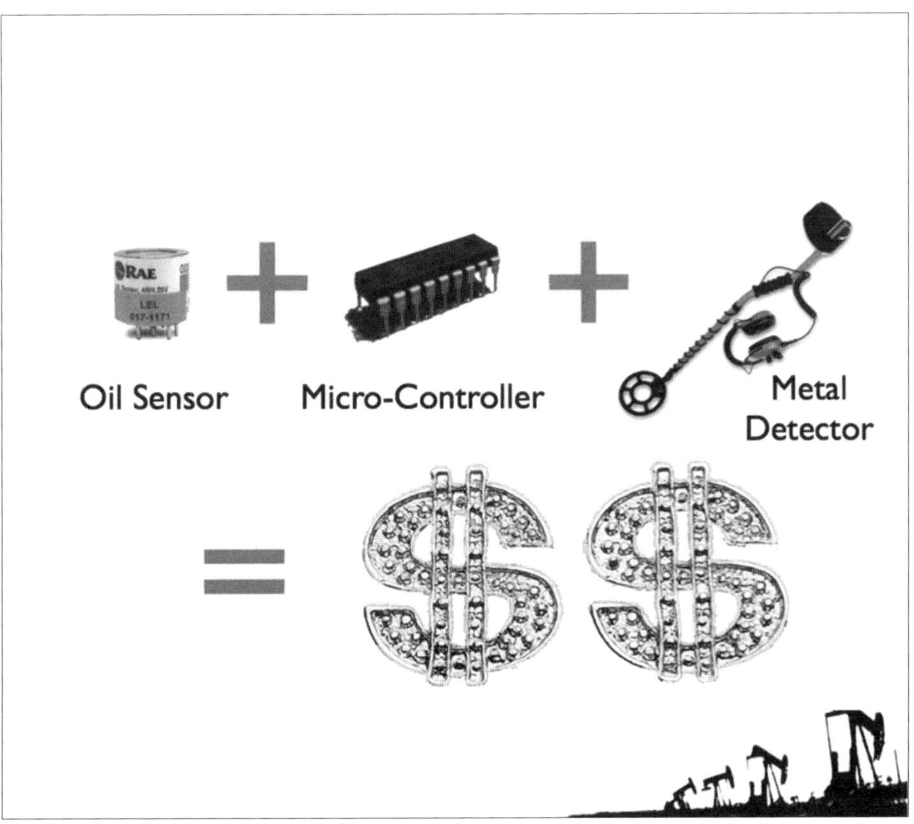

Figure 5b: Jon Cohrs, *Urban Prospector* (2009), video still Greenpoint, New York. Photo by Jon Cohrs. Courtesy of the artist.

goggles, hard hat, and respirator, Cohrs appears in jeans, cotton shirt, and sunglasses, an attire that transforms the unsavory job of toxic waste clean-up into a casual pastime.

One might dispute the lust for oil that Cohrs both ridicules and exploits; but the magnitude of the spilled-oil 'reserve' that he tapped in this project is not debatable. His prospecting scheme was carried out in the neighborhood where he lives in Brooklyn, New York – the site of the Greenpoint Oil Spill that has been leaking crude oil from processing facilities in the area for over fifty years. Estimates of the quantity of spilled crude oil range from seventeen million gallons to thirty million gallons, twice the size of the Exxon Valdez disaster.[25] While this environmental offense lies on-site, it exists out of sight. The spill essentially destroyed the local aquifer, because the environmental hazard it poses extends the entire length of the food chain, from microorganisms to apex predator species. Previous attempts to clean it, contain it, or deal with it have been rudimentary. Cohrs introduced a novel approach to remediation by presenting the contaminating goo as a potentially lucrative resource.

By posting instructions for his do-it-yourself urban prospecting device online,[26] Cohrs gives victims of global oil corporations opportunities to forage for their own oil resources. He even publicizes the best prospecting locations online, referring to these dangerous hot spots as 'rich deposits'. Cohrs also plans to expand outreach by developing web-based mapping software so amateur prospectors can use their mobile devices to post their findings in real time. These generous acts of information-sharing do not include extraction. For now, the located oil will remain in the ground. However, by presenting an invisible and insidious menace as a resource, *Urban Prospector* offers a parody of cultural values and individual greed. Perhaps its witty incentives for citizens to participate in soil remediation will ignite a frenzy of activity, like that of the Gold Rushes in the 1800s when new discoveries of gold convinced hordes of miners, stirred by the hope of attaining instant wealth, to seek their fortunes in Australia, New Zealand, Brazil, Canada, South Africa, and the United States.

Aesthetic Archetype

 Imagine the state of the physical environment if aesthetic considerations were applied to bees, forests, reefs, streams, cell phones, shoes, cars, plastic containers, soils, and all other forms of materiality. Every item would be viewed as a composite of sensory components. Visually, aesthetic attention involves shape, hue, tonality, size, space, symmetry, surface, transparency, pattern, geometry, and dozens of additional qualities. Kinetically and temporally, aesthetics introduces awareness of tempo, sequence, duration, rhythm, meter, cadence, and interval. The aesthetics of touch tap into the rich vein of sensory interactions that include weight, tensile strength, flexibility, absorption, temperature, solidity, liquidity, softness, hardness, elasticity, brittleness, viscosity, durability, fragility, reactivity, and so forth. These sense opportunities continue to expand when the qualities of taste, sound, and smell are added. Eco-psychologist, Dr. Michael J. Cohen, claims that humans actually have fifty-three senses, not just five.[27] He adds, for example, the sense of light, temperature, season, radiation, pressure, gravity, weight, balance, proximity, motion, humidity, fear, playtime, design, etc. While beauty is constructed out of these sensory perceptions, Cohen asserts that this load of anatomical, neurological, and perceptual connectors evolved because they were essential for human survival. Until sensory perceptions were taken over by industrial tools and electronic technologies, humans depended upon conscious cultivation of aesthetic awareness to predict the weather (changes in the atmosphere were observed and smelled); analyze the balance of the soil (it was tasted); read the temperature of an ill patient (the forehead was felt); discern the time (shadows were observed); monitor the heat in an oven (the color of the flame was examined).

Aesthetic perceptions are direct links between external stimuli and visceral emotions. Personal experience and cultural conventions converge to please or irritate. While artists and art viewers cultivate aesthetic sensitivity, it is possible to apply such active and inquisitive forms of experience to every substance and every object.

 Aesthetic experiences originate in materials and objects, and are received by the body's sensory apparatus. These receptors deliver these stimuli to various regions of the brain where they evoke mental states. Some aesthetic stimuli produce pleasure. Others irritate. This outcome influences disengagement from the materials and objects that produced them.

Generators of aesthetically pleasing experiences tend to be savored and preserved, while disagreeable sources are more likely to be hastily discarded. But aesthetics is not the exclusive determinant of delight and disgust. Culture intervenes when, for example, bruises, shattered glass, and muddy shoes are not appreciated for their aesthetic arrays of hues and patterns, but are viewed as disagreeable. Any increase in the range of aesthetic appreciation could augment human caregiving efforts.

Sensory Stimulation

Laura Parker

Art Engagement:

Laura Parker (b. San Francisco, California, United States) focuses on agriculture, the environment, and social structures. In 2001, Parker began her installation work with a project called LandScape: The Farmer as Artist. *Her work from 2001,* How Far Are You from the Farm? A Mile or a Generation?, *was exhibited at the Jewett Gallery, San Francisco Public Library. In 2006, Parker created the work* A Taste of Place, *which is an investigation into the relationship between soil, food, and labor. Parker's artwork has been exhibited at Rautenstrauch-Joest Museum, Cologne; MR Gallerie; and the Chateau de Nieuil, France.*

Some contemporary artists create works that literally 'ground' aesthetic experiences by culling the spectrum of visual and textual sensations that different soils provide. They draw or paint with earth pigments to capitalize on the diverse tones and textures that result from the interactions between soil's infinitely varied combinations of organic and mineral components. Instead of using earth as an art medium to create a representational or expressive artwork, Laura Parker presents soil as her art subject. For example, in a work entitled *Palette* (2001), she displayed twenty samples of soil in clear containers. The earth colors ranged from black to golden. Likewise, *Clod* (2000) is a sculpture that takes the form of a chunk of earth 6 x 8 x 12 inch. The dirt was not altered by the artist. It represented itself.

In 2006, Parker extended her aesthetic exploration of dirt beyond its visual qualities to include smell and taste. *A Taste of Place* was an art installation that resembled a wine tasting. To assemble the components, Parker travelled throughout Sonoma County in California, a region known for its organic farms and ranches, gathering samples of soils along with samples of the produce the farms grew. The soils were placed in Mason jars and installed on shelves behind a counter and stools in the gallery. Like a wine steward, Parker invited visitors to take a seat so she could offer them a sniff of a few soil samples that she had transferred into elegant stemmed wine glasses. She then provided tastes of the collard greens, squash, radishes, eggs, or cheese these soils produced and urged them to seek a correspondence. The correlation between aroma and taste is well-documented. Approximately seventy percent of what we perceive as taste is derived from smell because our taste buds can detect only five basic flavors: sweet, salty, sour, bitter, and umami.[28]

In this manner, the structured formality of a wine tasting was applied to soil sniffing. Parker explained to each participant:

> *First the scent of the soil will be stimulated by adding a small amount of water and stirring to release the earth's aromas as if from a fresh rain. Then you will smell, identify the scents you recognize, and note their properties. You will then be served food grown in the same soil you have just smelled. See if you can taste in the food the same properties you smelled in the soil.*[29]

Thus Parker extended the model of wine tasting by applying the refined appreciation of a wine connoisseur to dirt. Wine connoisseurs can detect the aesthetic qualities of the wine being tasted and sniff to discern its specific growing conditions. The correlation between place and taste, known as 'terroir', accounts for soil's role in shaping the uniqueness of wines. This is because plants cannot pick themselves up and move to drier, warmer, or more protected areas as animals can. Thus, they adapt internally to the particular combination of inert minerals and living microbes in their soil, along with weather and elevation. These adjustments are the source of the aesthetic qualities of the plant's chemistry and structure. As a result, the unique tastes and textures of the collard greens, squash, and radishes she served are products of the unique aesthetic qualities of the farm soil.

Parker has held tastings of meat from pigs; cheese from goats; and eggs from chickens. Was there a correlation between them and the soils they grew upon? Parker reports, 'It was harmony.'[30] She describes the aroma of soil from Philo Apple Farm as 'a bit less exotic in aroma, but more varietal, with olive and mineral notes, and a weightier finish.'[31] In contrast, she describes an 'underlying presence of cream opens up to hints of citrus and spice'[32] in the soil from T & D Willey Farms. These adjectives also describe the tastes of the foods each farm produces.

Aesthetic stimulation of the senses provides the vehicle by which *Taste of Place* connects landscape and food, a relationship that is severed when family farms are replaced by globalized and industrialized agriculture. Such modern forms of food production obliterate local characteristics because land is treated with carefully calculated chemical infusions to ensure uniformity of the products. These technologies boost productivity, but reduce the variety of flavors offered by locally grown foods.

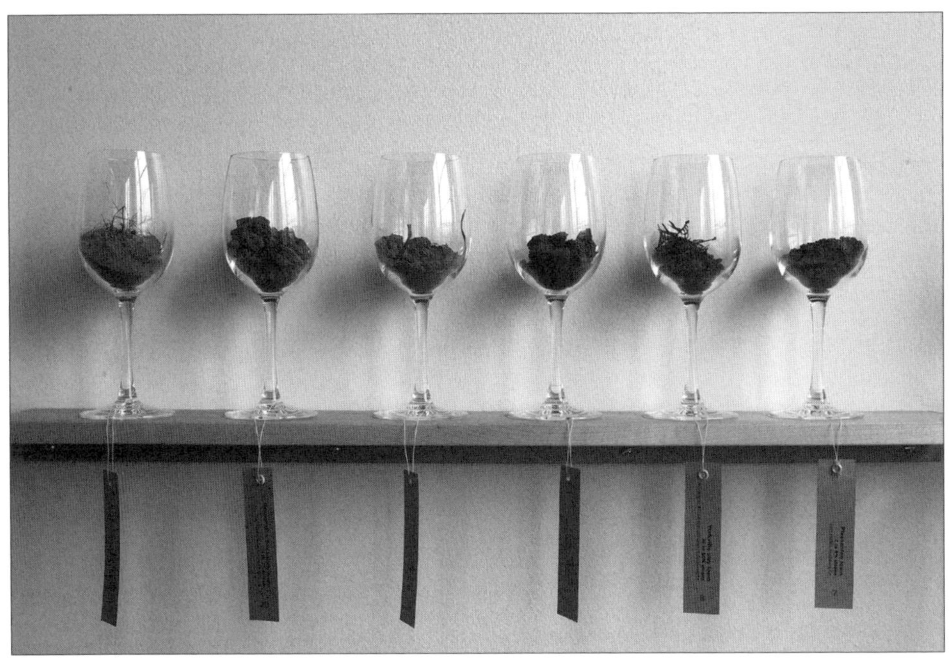

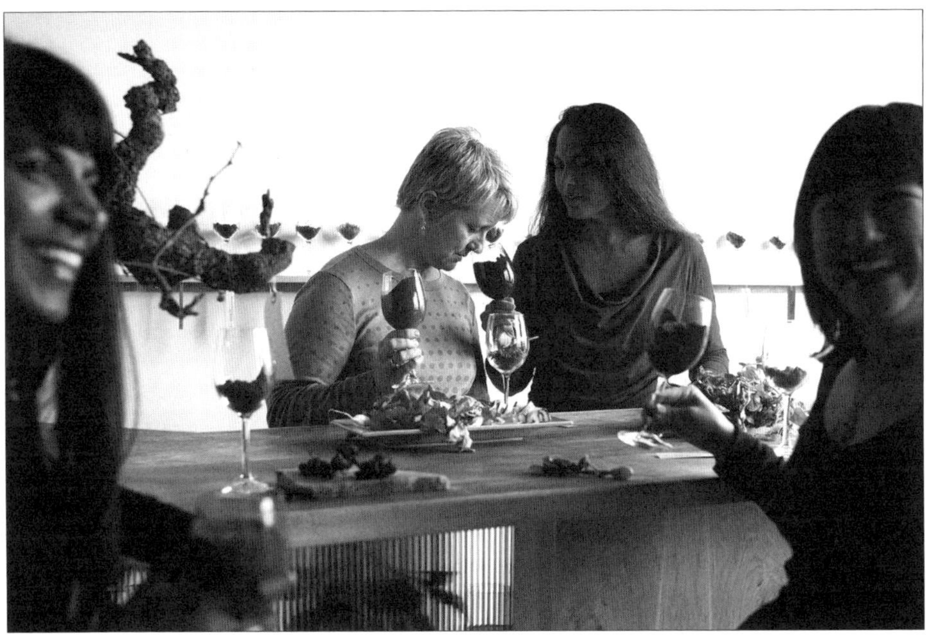

Figure 6a: Laura Parker, *A Taste of Place* (2006–ongoing), soil samples in wine glasses. Photo by Eszter and David Matheson. Courtesy of Soil Tasting, A Taste of Place.

Figure 6b: Laura Parker, *A Taste of Place* (2006–ongoing), soil tasting. Photo by Eszter and David Matheson. Courtesy of Soil Tasting, A Taste of Place.

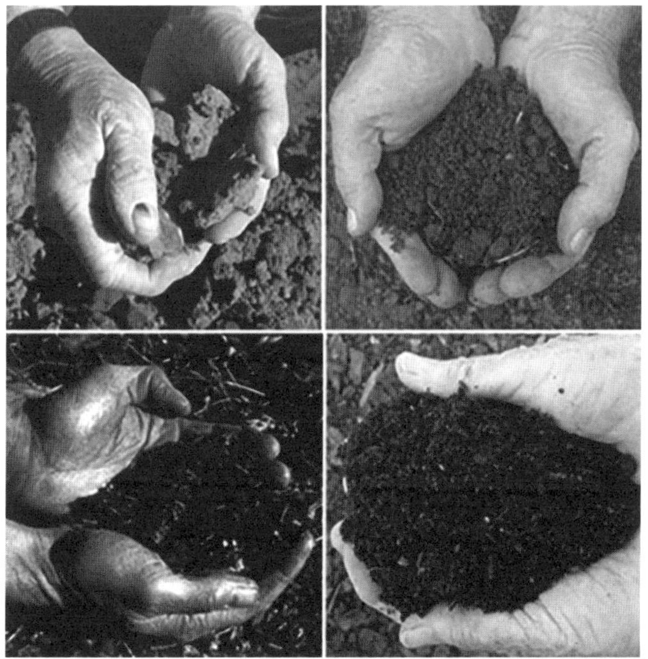

Figure 6c. Laura Parker. *A Taste of Place* (2006–ongoing), soil samples. Courtesy of Soil Tasting, A Taste of Place.

By applying the practice of 'terroir' to dirt, *Taste of Place* reverses two dismissive attitudes regarding soil. On the one hand, it restores dignity to a material that currently serves as a metaphor for the lowest of low values, as indicated by such phrases as 'dirty words', 'dirt cheap', and 'dirt poor'. On the other hand, this art project transforms dirt, and its demeaning association with squalor, into Earth's honored life-sustaining substance. *Taste of Place* gives museumgoers an opportunity to experience dirt within the context of aesthetic appeal and environmental necessity.

In addition, restoring respect to soil's status is the prerequisite for developing a connoisseur's ability to distinguish between soil's excellence and inferiority. There are fourteen thousand varieties.[33] Each type of soil provides a particular combination of aesthetic and functional value. Such connoisseurship could endow fine dirt with the respect bestowed upon fine wines and fine art. Because we humans tend to lavish care upon things we perceive as beautiful and pleasurable, such appreciation might result in soil conservation and sustainable land-use.

Parker discovered soil's ability to evoke beauty and pleasure when she was just a child. She explains,

> *As a city kid, I was always the one sent to the garden to harvest for dinner and for supper. I learned to love what it means to pick things when they're ripe. There had to be one for me and one for the basket so that I really knew what I was taking into the house. I think it was during these times when I was in the garden by myself, picking and looking and feeling the soil between my toes, that I became totally enamored with the farm. I would take books out and read them, laying in the dirt. It was heaven to go and lay in the beans between the rows and have the sun on my back, reading my book....* [34]

Merchandise Archetype

 The seventh archetype differs from the six previously discussed because it is the only one that did not originate when civilization's impact on the environment was limited to handheld tools controlled by the human body's motor and sensory apparatus. While those six archetypes persist in our own era, it is the recent addition that seems to be dominating current human interactions with materiality. This archetype is merchandise, defined as mass-manufactured products that are identified by brands, intended for purchase by individuals known as 'consumers', and sold through retail outlets.

Because merchandise is the dominant mode of material interactions for most people today, little imagination is needed to envision the state of the physical environment if the relationship humans have with merchandise was applied to bees, forests, reefs, streams, cell phones, shoes, cars, plastic containers, soils, and all other forms of materiality. Contemporary lives depend, almost exclusively, upon the purchase of goods that are mass-produced, packaged, marketed, and sold. This sequence accounts for provisions of food, shelter, and fuel, as well as transportation, defense, hygiene, health, and entertainment.

It was only about two hundred years ago that the production of multiple identical units began to usurp the uniqueness of handcrafted goods. The takeover was propelled by the engines of industry, the efficiencies of mass production, the availability of fossil fuels, and the development of new tools. Over time, these advantages were bolstered by the communication capacities of electronics and the organizational capacities of international corporations. The prevalence of merchandise has made material extravagance available, for the first time, to people who are not royalty, wealthy gentry, or heads of state. The unrealistic dreams of past generations have become commonplace realities.

The myriad diverse examples of 'merchandise' presented to consumers tend to share three primary characteristics. They are predictable, cheap, and replaceable. The accounting of these unprecedented assets runs in two directions. One column tabulates the ease of acquiring, using, and discarding the commodities we purchase. The entries on the other column, however, offer a sobering reckoning of the toll that merchandise-dependent lifestyles inflict upon ecosystem functions, social stability, and wildlife diversity. Environmentalists cite evidence that the short-term advantages that merchandise affords humans are endangering the long-term prospects of the planet.

 Uncommon strategies for dispensing with unwanted merchandise include gifting, bartering, refurbishing, repairing, recycling, and upcycling. It is far more common for unwanted merchandise to come to rest, without ceremony or environmental consideration, in storage bins, landfills, incinerators, gutters, and alleyways. The end-phase of merchandise ownership may occur when an item is still functional. Such premature discards are indications that a consumer has yielded to the temptation of upgrades, has succumbed to boredom, or been lured by the appeals of advertisers. According to the Environmental Protection Agency, Americans generated about 258 million tons of trash in 2014. Only eighty-nine million tons of these materials were recycled or composted.[35] The trash statistics account for durable goods (e.g. tires, furniture), nondurable goods (e.g. newspapers, plastic plates/cups), containers and packaging (e.g. milk cartons, plastic wrap), and other wastes (e.g. yard waste, food).[36] Totals have increased by about sixty-five percent since 1980. These worrisome statistics would increase if non-manufactured and non-commercial entities were treated like merchandise.

Art Engagement:

Fertile Medium

Joe Scanlan

Joe Scanlan (b. 1961, Columbus, Ohio, United States) is a New York-based artist and educator. Scanlan holds a BFA in sculpture from the Columbus College of Art and Design. He served as an assistant professor and, later, an associate professor in the Sculpture Department at Yale University (2001–09). He previously was an assistant director of the Renaissance Society at the University of Chicago (1987–94). Scanlan's work is in the public collections of K21 Westphalia, Germany; Tate Modern, London; Centre Georges Pompidou, Paris; the Van Abbemuseum, Eindhoven; the Stedelijk Museum voor Actuele Kunst, Ghent; and the Museum of Contemporary Art, Chicago.

While the components of merchandising typically emphasize the role of the consumer: shopping, buying, using, and disposing, the artist Joe Scanlan activated the role of the producers: sourcing, processing, packaging, pricing, and distributing. By applying the complex sequence of material interactions that precede the sale of the merchandise, to the creation of an artwork, Scanlan satirized the entire system that guides the economic distribution of industrially produced goods. This artwork is entitled *Pay Dirt* (2003).[37]

Scanlan denounces the prevailing market system and commerce, declaring 'capitalism's ruthless success' and 'the cruddy workings of commerce.'[38] Nonetheless, he incorporates these structures into a project titled *Pay Dirt*. However, he enters the market on his own terms, defying the common lament regarding the infiltration of market practices into the refined arena of fine art. Scanlan explains, 'The conventional view of artistic change (the avant-garde) is: What can I say that is unique? My view… is: What can I do that is profitable?'[39]

This market-driven goal was achieved by developing an exceptional soil – an unlikely example of merchandise. The product was crafted, refined, and cultivated with all the attention worthy of an artistic masterpiece. Accomplishing this feat required six years of experimentation, rigorous research, and product development with agricultural chemists at bio-tech labs. Scanlan's claim that his concoction exceeded previous measures of soil excellence was not a dubious marketing campaign. Patent No. 6,488,732 is proof of its uniqueness – 99.85 percent of its ingredients originate in post-consumer waste. The official text on the patent describes this special soil as a material composed primarily of dirt,

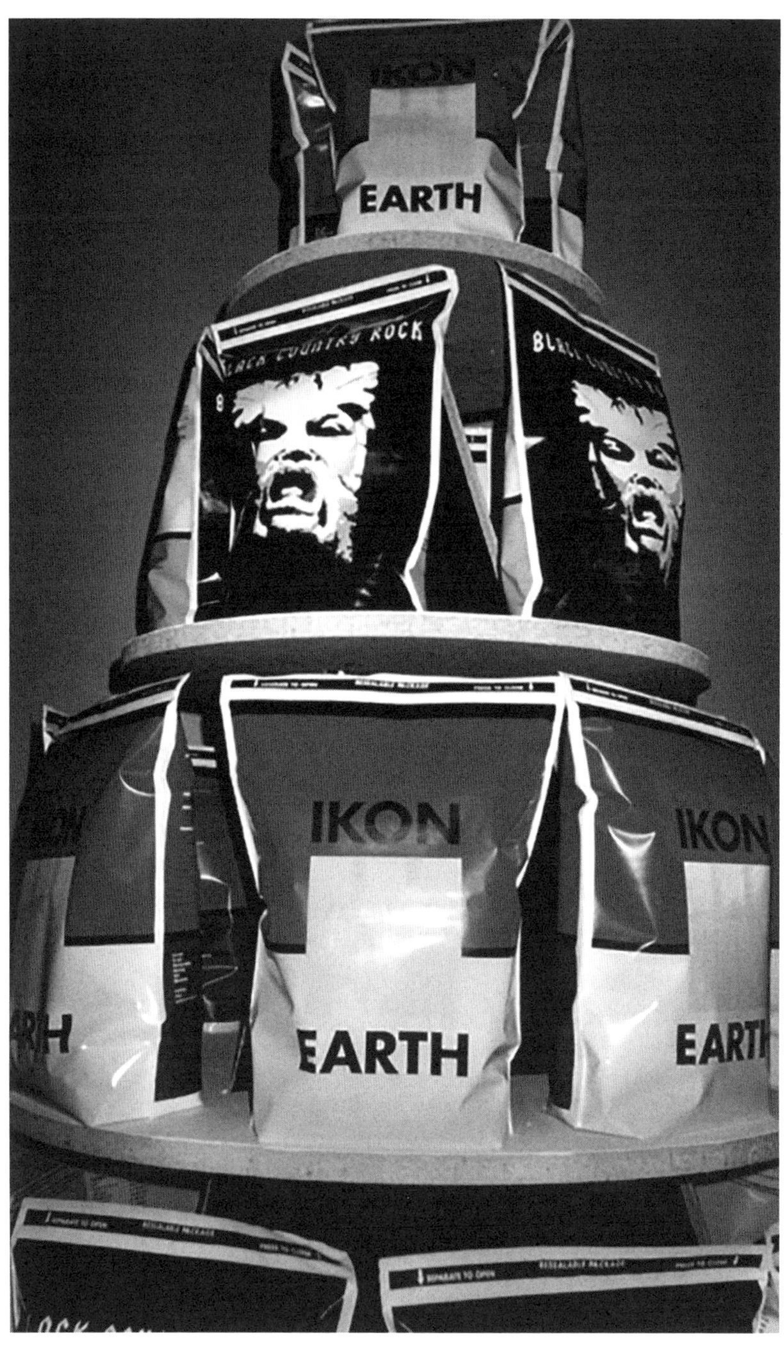

Figure 7a: Joe Scanlan, *Pay Dirt* (2003), installation view. Ikon Earth and Black Country Rock soil, synthetic potting soil packaged in one liter bags. Co-production with Ikon Gallery, Birmingham, United Kingdom. Courtesy of the artist.

along with other commercial and industrial dirts, in which the materials are uniformly pulverized, skillfully measured and combined to form homogeneous particulate dirt. The composition is alternately turned and rested in a windrow or like apparatus for several months until natural aerobic thermophilic fermentation causes the dirt to have an overall pH level of 5.0 to 3.0 for the purpose of making the minerals latent in the dirt composition soluble.

Scanlan disagrees with the common opinion that the term 'saleable commodities'[40] is demeaning when it is used to describe artworks. Such critiques, he maintains, do not serve the well-being of artists living within a 'capitalist democracy' where 'name recognition is essential to the reception and purchase of their work.'[41] He goes on to explain:

> *Unfortunately, from 1968 until your reading of this sentence it has been very, very hard to change the subject from an irrelevant class struggle that condemns artists to a state of purity or poverty or both, to an appreciation of agile, realist, freelance artists plying their skills in a project-based, network economy.*[42]

Yielding to the economic reality of surviving as an artist does not necessarily compromise the integrity of the work of art. Indeed, Scanlan integrates the mandates of the market into his creative contemplations. *Pay Dirt*, for example, fulfills the following assertion: 'Many critical artists (myself included) would agree. They understand that they could never exist outside or above the market but that their only viable option is to try to shape the kind of market they want to inhabit.'[43]

The artistic shaping that created this artwork took two forms. It was applied to both the merchandise (soil) and its marketing (sale). Scanlan created two versions of his dirt to maximize sales: an inexpensive version for the budget-minded, and a luxury version for the wealthier clients. Ikon Earth is the popular brand. It is offered in six-liter bags, each priced at $20.00/bag. The promotional material claims that Ikon Earth contains optimum amounts of all the nutrients your garden and potted plants need to thrive.

> *Nitrogen for robust growth. Potassium for water uptake. Phosphorus for bountiful fruiting and flowering. Calcium for root development. Magnesium for photosynthesis. Sulfur for promoting new growth. A high cation exchange rate for maximum absorption of nutrients. All released in the slow, organic way plants like best, without the use of synthetic polymers or nutrient-retardants.*[44]

Black Country Rock is the luxury soil. It costs $49.00. It, too, is packaged in six-liter containers. Furthermore, it is made from the same formula as Ikon Earth. Indeed, although one was affordable and the other was expensive, Scanlan announced that the two versions of his synthetic potting soils were equally 'conducive to healthy plant life and therefore usable as dirt.'[45] Scanlan explains that the expensive soil is reserved for an extremely exclusive demographic. It consists of people who can afford to spend a premium price for common merchandise; they are 'Bowie fans who garden.'[46]

This absurd mockery of normal market constructions is actually rooted in a convoluted version of rock music history. 'Black Country Rock' is a song written in 1970 by the rock superstar David Bowie. His lead guitarist at the time was Mick Ronson. Ronson had a previous career as a municipal gardener before joining Bowie's band as the lead guitarist. His signature song, 'Black Country Rock', was inspired by the industrial region north of Birmingham, England, otherwise known as 'the Black Country' because of its soot-laden air. Very few people could decipher this complicated narrative explaining the title, and even fewer are likely to appreciate that the name of Scanlan's fertile soil was inspired by contaminated soil. Through these antics, Scanlan pokes fun at the pretensions of luxury goods buyers. Furthermore, his mischievous distortion of the market applies 'conspicuous consumption' to a material that has rarely been coveted as merchandise: dirt.

Scanlan admits that his product offers little aesthetic appeal. He describes his soil as 'quite inert and unspectacular to the naked eye. It is dirt, like and unlike every other kind of dirt, whether commercially or naturally formed.'[47] Perhaps that is why he designed attractive three-color packaging, a tactic many manufacturers employ to attract customers. Those willing to pay extra to acquire the luxury version of the soil might pride themselves in having a deeper respect for the 'artistic' components of Scanlan's design. Or perhaps, purchasers of the expensive version value this soil as an artwork. As such, they would honor the artist's virtuosity in producing it. Furthermore, they would recognize the aesthetic properties it shares with historic masterpieces: proportion (the relationship between the size, shape, and position of the soil's particles) and harmony (the congruity between its ingredients). Or perhaps they perceive the beauty in Scanlan's soil, not because of how it appears, but because of how it functions in terms of water-retention, pH, electrical conductivity, etc. As in other market contexts, high pricing soil is a strategy for elevating the stature of the owner.

Manufacturing protocols were included in this parody of economic norms in an exhibition at the Ikon Gallery in Birmingham, England in 2003, the region that inspired Scanlan to name his luxury soil, Black Country Rock. Scanlan transformed the Ikon Gallery into a mini-processing plant that produced the special dirt. Scanlan explains the ironic twist he bestowed upon a familiar production strategy:

> *The idea was to collect precise categories of post-consumer data – i.e., garbage – from the shopkeepers of Birmingham, pass it through the Ikon Gallery as if the museum were a kind of refinery, and then sell it back to the people of Birmingham as an exclusive local brand. The price was £7.99 for a six-liter bag, pretty expensive for potting soil but pretty cheap for a work of art – especially one that, if properly exhibited, could produce gladiolus or Brussels sprouts.*[48]

In other words, the residents of Birmingham could buy back their own garbage at an inflated price!

Thus, Scanlan's multifaceted project presents soil as merchandise in order to assert its desirability; exposes the manipulative tactics of merchandising; and applies these influences on peoples' material interactions to the creation of fine art. The artist explains:

> *Pay Dirt is unique to me and therefore useful as a kind of soapbox on which to stand and proclaim my proud participation in – but distinction within – the global economy. I'm very proud of this dirt and I enjoy making it. It is a vital, healthy product that promotes nourishment and beauty wherever it goes. So, as both a product and a philosophy, I want Pay Dirt to be infectious. I want it to get under people's nails.*[49]

Of course, he also wants to make a profit.

Your opinion: What's next? ➔

Reader Interaction

Reader Interaction #1 - Archetypes: A Personal Survey

The preceding discussion is neither theoretical nor metaphorical. It is a tool for scrutinizing the impact of one's own material behaviors and their environmental impact from an Eco Material perspective. The following chart has been prepared to facilitate this process. Material archetypes are presented on one axis; material possessions are listed on the other. Please mark the archetype that most closely resembles your relationship with that material item. Tabulating these responses discloses behavioral and emotional relationships between these variables. A single object may elicit more than one archetype. It is hoped that self-interrogation will lead to self-reflection, and will ultimately encourage you to apply Eco Material considerations to your professional and personal behaviors.

	PET ARCHETYPE	SACRED ARCHETYPE	HAZARD ARCHETYPE	AESTHETIC ARCHETYPE	SPECIMEN ARCHETYPE	RESOURCE ARCHETYPE	MERCHANDISE ARCHETYPE
CELL PHONE							
PLASTIC BAG							
SHAMPOO							
COAT							
CREDIT CARD							
COMB							
CHAIR							
NAIL CLIPPINGS							
PAPER NAPKINS							
KEYS							
MIRROR							
URINE							
MONEY							
SCISSORS							
APPLE							

Reader Interaction #2 – Soil: A Personal Survey

Please check the choices provided on this survey that describe your relationship with soil. What do the results reveal? Does your current relationship confirm or contradict Eco Material principles and attitudes?

The locations where I most often encounter soil are:

Parks ___, zoos ___, botanical gardens ___, golf courses ___, sports fields ___

Residential yards ___, lawns ___, gardens ___, pathways ___

Atriums and plazas ___, window boxes ___, planters___

Vacant lots___, alleys ___, curbs___

Farms ___, meadows ___, forests ___

Other?____

The aspects of soil that I personally engage are:

Genetically modified seeds, fertilizers, pesticides, herbicides, tractors, columbines
____

Worms, bacteria, viruses, fungi, protozoa, algae, nematodes, arthropods _________

Carbon, hydrogen, oxygen, nitrogen, phosphorus, potassium, sulfur, calcium, magnesium, iron, manganese, boron, zinc, molybdenum, nickel, copper, cobalt, and chlorine ____

Agricultural run-off, oil spills, litter, clear-cut forests, strip mining, landfills, toxic waste dumps, radioactive waste, overgrazing, leaching, salinization ____________
Other ?____

My psychological relationship to the soil in my watershed is characterized as:

Responsibility

Liability

Resource

Pride

Shame

Devotion

Security

Jeopardy

Other

The following groups have significant impact on the condition of the soil in the watershed where I live:

Corporate managers

Government officials

Religious leaders

Financial authorities

Marketers

Artists

Educators

Farmers

Engineers

Consumers

Other

Your comments: What's next?

Endnotes

1   Utah Agriculture in the Classroom, 'How Much Is Dirt Worth?,' accessed March 21, 2016, http://extension.usu.edu/AITC/teachers/pdf/dirt/worth10.pdf.

2   Amy Youngs, curatorial statement for *Vermiculture Makers Club* exhibition, 849 Gallery, Kentucky School of Art, Louisville, Kentucky, March 6–April 17, 2015.

3   Youngs, curatorial statement.

4   Youngs, curatorial statement.

5   Youngs, curatorial statement.

6   *The New York Earth Room* specs are 250 cubic yards of earth; 3,600 square feet of floor space; twenty-two-inch depth of material; total weight of sculpture: 280,000 lbs. *The New York Earth Room* is the third *Earth Room* sculpture executed by the artist, the first being in Munich, Germany in 1968. The second was installed at the Hessisches Landesmuseum in Darmstadt, Germany in 1974. The first two works no longer exist.

7   'Global Chemicals Outlook – Towards Sound Management of Chemicals,' UNEP 2013, accessed August 13, 2017, https://sustainabledevelopment.un.org/index.php?page=view&type=400&nr=1966&menu=35.

8   ChemSec, 'Hazardous Chemicals Found in Many Household Products,' accessed August 13, 2017, http://chemsec.org/hazardous-chemicals/.

9   *Inventory of Hazardous Substances in the Home*, accessed September 13, 2017, https://keec.ky.gov/Documents/Home%20Inventories/Hazardous%20Chemicals%20Home%20Inventory.pdf.

10   'The Materiality of a Natural Disaster,' *DisegnoDaily*, accessed August 13, 2017, https://www.disegnodaily.com/article/the-materiality-of-a-natural-disaster.

11   Hilda Hellström, email correspondence with the author, August 20, 2017.

12   Hellström, email correspondence.

13   Alexandra Regan Toland, 'Rubble Mapping,' February 19, 2012, accessed August 13, 2017, https://artoland.wordpress.com/2010/09/10/spatial-analysis-of-wwii-rubble-deposition-in-berlin-2/.

14   Alexandra Regan Toland in Amy Franceschini and Myriel Milicevic, eds, *Beneath the Pavement: A Garden Paperback* (Loughborough, UK: Loughborough University, 2010), 80–105.

15   Amy Franceschini and Myriel Milicevic, *Beneath the Pavement*, 84.

16   Amy Franceschini and Myriel Milicevic, 87.

17   Amy Franceschini and Myriel Milicevic, 89.

18   Amy Franceschini and Myriel Milicevic, 92.

19   Alexandra Toland and Gerd Wessolek, *Devil in the Sand – the Case of Teufelsberg Berlin and Cultural Ecosystem Services Provided by Urban Soils*, ResearchGate, April 2017, accessed March 21, 2016, https://www.researchgate.net/publication/316739086_Devil_in_the_Sand_-_the_Case_of_the_Teufelsberg_Berlin_and_Urban_Soil_Ecosystem_Services.

20   Matt Bradbury, 'A Brief Timeline of the History of Recycling,' May 20, 2017, accessed December 4, 2017, https://www.buschsystems.com/resource-center/page/a-brief-timeline-of-the-history-of-recycling.

21   *The Urban Prospector* has been presented at Ars Electronica Linz, Austria, September 2010; Amplified, Swedish Embassy – Berlin DE, January 2010; RIXC, Riga, Latvia, October 2009; Futuresonic Festival Manchester, United Kingdom, May 2009.

22   'Urban Prospecting Detector,' *Instructables*, May 12, 2016, accessed September 26, 2017, http://www.instructables.com/id/Urban-Prospecting-Detector/.

23   Interview with author in Kingston, NY, November 12, 2017.

24   TPH means the detectors operate in air, water, and soil. They can detect any semi-volatile hydrocarbons and related compounds.

25   Riverkeeper, 'Greenpoint Oil Spill on Newtown Creek,' accessed August 13, 2017, https://www.riverkeeper.org/campaigns/stop-polluters/newtown/.

26   'Urban Prospecting Detector,' *Instructables*.

27   Michael Cohen, *Reconnecting with Nature: Finding Wellness through Restoring Your Bond with the Earth* (Corvalis, OR: Ecopress, 1997), 48–50.

28   Newton's Apple, 'Taste and Smell,' accessed August 13, 2017, http://www.reachoutmichigan.org/funexperiments/agesubject/lessons/newton/tstesmll.html.

29   Nicola Twilley, 'Sweet and Sour Soils,' *Edible Geography*, December 9, 2009, accessed August 13, 2017, http://www.ediblegeography.com/sweet-and-sour-soils/.

30   Twilley, 'Sweet and Sour Soils.'

31   Twilley, 'Sweet and Sour Soils.'

32   Twilley, 'Sweet and Sour Soils.'

33   Twilley, 'Sweet and Sour Soils.'

34   Quoted in Terri Cohn, 'How Far Are You from the Farm: A Mile or a Generation? The Agricultural Art of Laura Parker,' *Places Journal*, 2004, accessed June 26, 2018, https://placesjournal.org/article/how-far-are-you-from-the-farmlaura-parker/.

35   Facts and Figures about Materials, Waste and Recycling, United States Environmental Protection Agency, accessed July 31, 2018, https://www.epa.gov/facts-and-figures-about-materials-waste-and-recycling/advancing-sustainable-materials-management-0

36   Municipal Solid Waste Factsheet, 2014, accessed October 3, 2016, http://css.umich.edu/factsheets/municipal-solid-waste-factsheet.

37   Joe Scanlan, *Joe Scanlan: Object Lessons* (Ostende: Mu. ZEE, 2012), 31.

38   Joe Scanlan, 'Things That Fall,' *Things That Fall*, accessed August 13, 2017, http://www.thingsthatfall.com.

39   'Joe Scanlan,' accessed August 13, 2017, http://www.undo.net/it/mostra/68445.

40   Joe Scanlan, 'Modest Proposals,' Things That Fall, accessed August 13, 2017, http://www.thingsthatfall.com.

41   Scanlan, 'Modest Proposals.'

42   Scanlan, 'Modest Proposals.'

43   Scanlan, 'Modest Proposals.'

44   Joe Scanlan, 'Ikon Earth,' *Things That Fall*, accessed August 13, 2017. http://www.thingsthatfall.com/dirt-ikon-earth.php.

45   The Serving Library, Exhibition Research Lab, School of Art & Design, Liverpool John Moores University, accessed October 5, 2016. http://servinglibrary.org/journal/8/23-thoughts-about-dirt.

46   Scanlan, 'Ikon Earth.'

47   Joe Scanlan, 'Commodify Your Dissent: François Piron Interviews Joe Scanlan,' accessed August 13, 2017, http://www.thingsthatfall.com/interviews-piron.php.

48   Scanlan collected precise categories of post-consumer garbage from shopkeepers in Birmingham, England. They deposited this garbage in the Ikon Gallery, where it was 'refined' into a fertile medium.

49   Scanlan, 'Commodify Your Dissent.'

ECO MATERIAL OPERATIVES ?

People

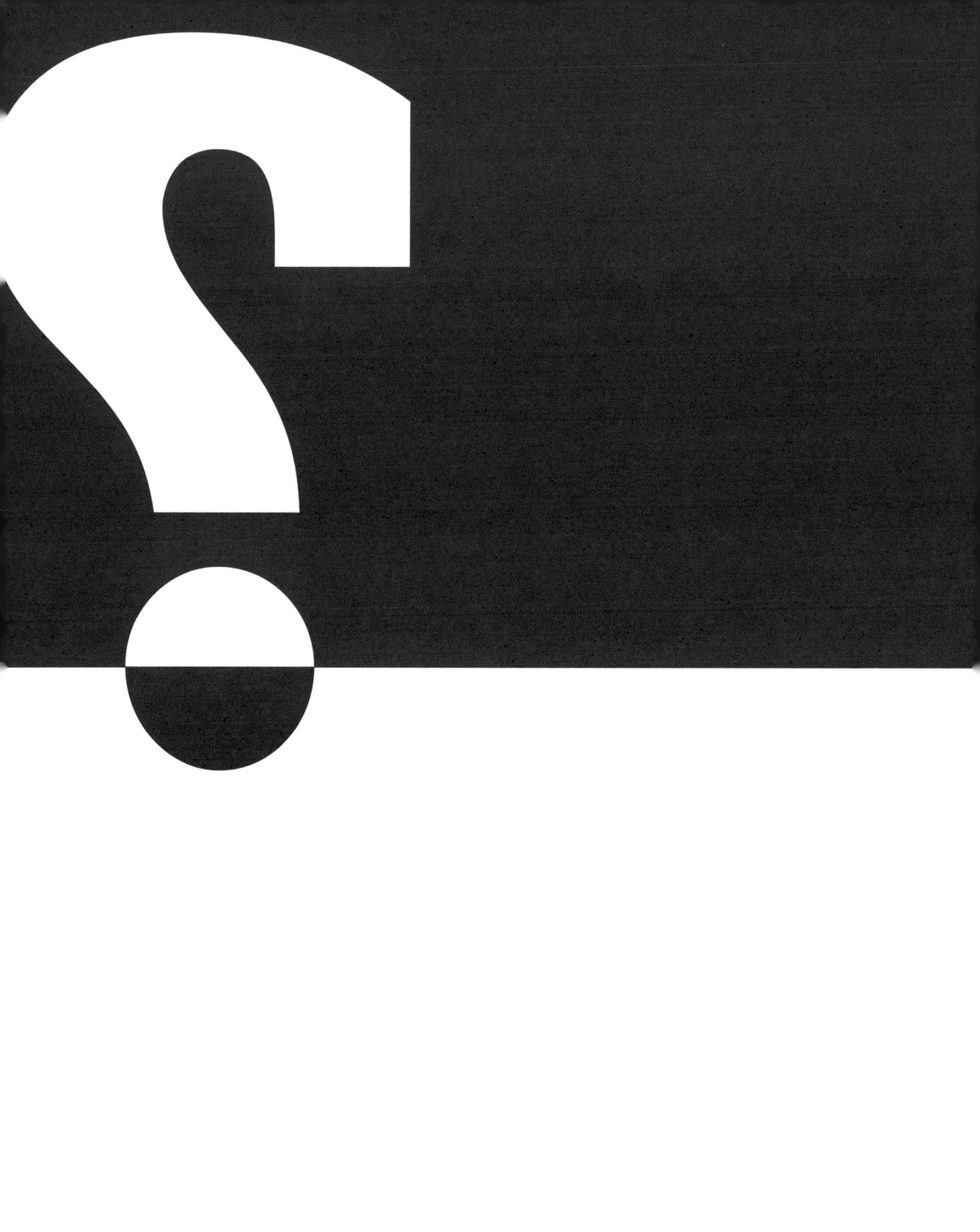

ECO MATERIAL OPERATIVES

In this chapter, you are challenged to suspend the cultural habit of thinking <u>about</u> the material components of the world. Instead, you are challenged to think *as* a material component <u>of</u> this world. This requires suspending your preferences, aversions, memories, hopes, and disappointments, and focusing on yourself as a physical entity that inhales, exhales, makes noise, emits smells, compresses soil, casts shadows, and consumes resources. This text pays particular attention to what the physical bodies of humans produce. It offers an alternative to the anticipated answer – biological waste. Besides waste, human bodies produce useful outcomes. Muscles, bones, tendons, sensory receptors, appendages, and brains comprise an efficient apparatus capable of conducting many kinds of work. The artists in this chapter test the limits of the human body's functional capacities. Each challenges the unaided physical organism to accomplish something remarkable through its interactions with the material environment. In the process, they offer a means for contemporary humans to reunite with the elemental substances and conditions of Earthly systems.

Do-It-Without-People

James Bridle

James Bridle (b. 1980, London, United Kingdom) is an artist, writer, and technologist currently living in Athens, Greece. He has a Master's degree in computer and cognitive science from University College, London. He is currently an adjunct professor on the Interactive Telecommunications Program at New York University. Bridle has exhibited in Europe, North and South America, Asia, and Australia. He has had works commissioned by organizations including the Victoria & Albert Museum; the Barbican; Artangel; the Oslo Architecture Triennale; the Istanbul Design Biennial; and has been honored by Ars Electronica, the Japan Media Arts Festival, and the Design Museum, London.

James Bridle provides an artistic embodiment of extreme separation from planetary substances and conditions. Bridle was named one of the '100 Most Influential People in Europe' by *Wired* magazine[1] the year he completed *Dronestagram* (2012–15). This technologically savvy art project replaces physical immersion in actual substances and conditions, with immaterial transmissions via communication technologies. *Dronestagram* is an Instagram artwork that presents aerial photographs of drone-strike locations. Bridle began this project by searching international news media for unacknowledged drone strikes around the world. Using open-source satellite imagery, he consults news articles, Wikipedia pages, Google Maps, Google Earth and other publicly available satellite maps to locate the drones and snap photos. Then he captures Google satellite images of these locations as they appeared prior to the attack. These images, therefore, present the quaint villages and orderly suburbs as they existed before they were bombed. While the images do not reveal demolished buildings, human casualties, and terrorized survivors, text accompanying each image conveys these devastating conditions. Access to this artwork is transmitted via social media. People experience *Dronestagram* on Twitter, Tumblr, and Instagram.

The processes that produced this artwork replace the functions of the physical body with those performed by sophisticated technologies. Bridle methodically eliminated human perceptions from the production of imagery, and human material interactions from its formal embodiment:

- Instead of hand-to-hand combat, bombs are delivered via remote telecommunications between an unmanned predator drone and an operator thousands of miles away from the battlefield. Bootcamp training is irrelevant to such contemporary military actions.

- Instead of the artist observing actual conditions in his vicinity, satellites in space capture the images and transmit them, via remote communications, to his terminal on land.

- Instead of manipulating a medium to create a tangible artwork, Bridle presents captured, dematerialized visual information, eliminating the need for manual skill or personal expression.

- Instead of producing a unique artwork, this artwork consists of pixels that circulate globally to unlimited viewers simultaneously. There is no authentic version.

By summoning images from around the globe via remote-sensing technologies, *Dronestagram* presents locations the artist never visited and events he never witnessed. Bridle reveals the significance he assigns to the distances that intervene between him and this artwork when he states, 'The political and practical possibilities of drone strikes are the consequence of invisible, distancing technologies, and a technologically-disengaged media and society.'[2] He then expresses doubts about the benefits of such advanced tools: The technology that was supposed to bring us closer together is used to obscure and obfuscate. We use military technologies like GPS and Kinect for work and play; they continue to be used militarily to maim and kill, ever further away and ever less visibly.[3]

While Bridle describes drones as 'among the most efficient, the most distancing, the most invisible'[4] military technologies, he also observes that the extreme separation of his body from his process and his product conveys cultural insights that extend beyond art and war. The retirement of the human apparatus from active duty also prevails among basic strategies of contemporary sustenance, protection, investigation, and recreation. In each instance, physical distancing has been associated with the escalation of human power. Bridle explains, 'The advent of ballooning was the first time we got an aerial perspective of the earth, and from then on it has always been a perspective that has been held by military power. It is the perspective of power. That fact is incredibly revealing about the nature of technology vis à vis the nature of power. The elevated point of view gives the viewer the illusion of a kind of control or power....Throughout history, he who commands the high ground has the power.'[5] He concludes, 'History, like space, is coproduced by us and our technologies.'[6]

The jurors who included Bridle among the '100 Most Influential People in Europe' rewarded him for encapsulating, within a single art work, diverse strands of sophisticated technologies that are capable of replacing human mental and physical exertions. By withdrawing from active engagements with materiality, Bridle represents the extreme opposite of Eco Material operatives.

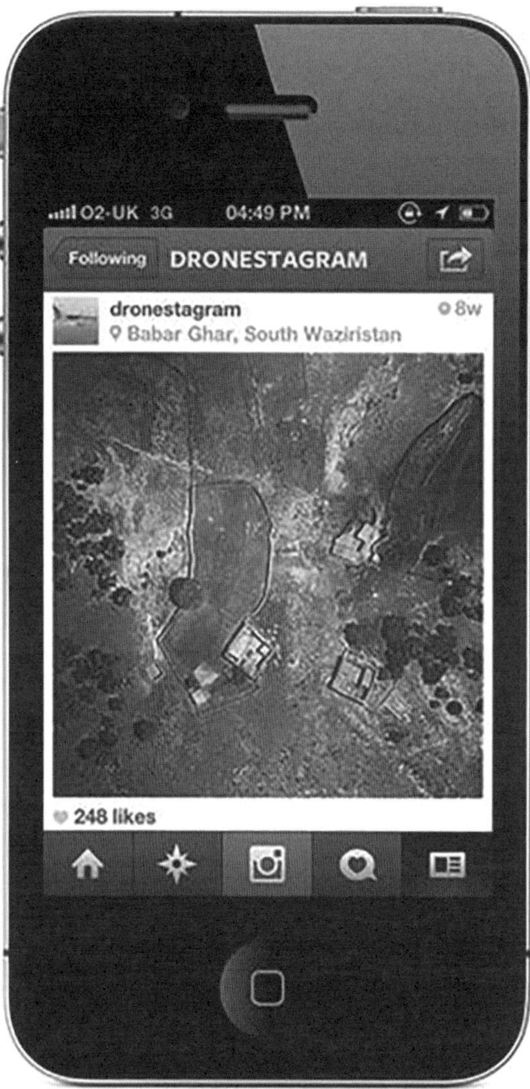

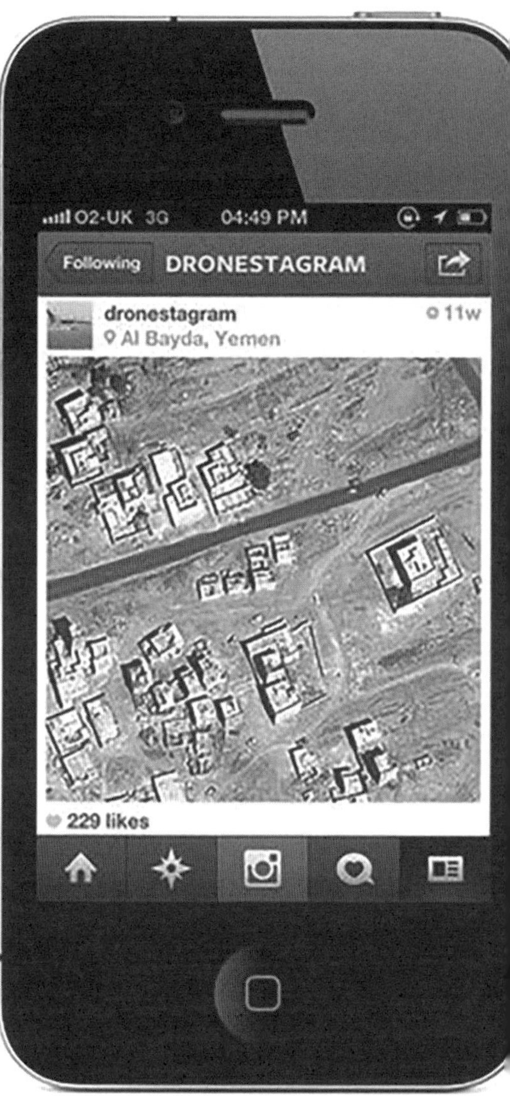

Figure 1a: James Bridle, *Dronestagram, an artwork*, (2012–2015). Screenshot of satellite images of drone strike locations from Google Maps posted on Instagram, Tumblr, and Twitter. 'February 8: Up to 9 killed and 6 injured in a strike on two separate mud-built houses on the North-South Waziristan border. Local sources reported that "Six drones were hovering in the sky at the time of the attack" and "Fear prevailed in the area as more drones were flying in the air halting the rescuers to launch an operation to take out bodies from the debris of the destroyed house." #drone #drones #pakistan. Courtesy James Bridle via Flickr.

Figure 1b: James Bridle, *Dronestagram, an artwork*, (2012–2015). Screenshot of satellite images of drone strike locations from Google Maps posted on Instagram, Tumblr, and Twitter. 'January 19th: A house destroyed in the Beida valley, killing 8. Three more were killed in Maarib, in a separate strike. Earlier this month, tribespeople blocked roads through the region in protest at the death of innocent civilians which fed anger against the United States.' #drone #drones #yemen. Courtesy James Bridle via Flickr.

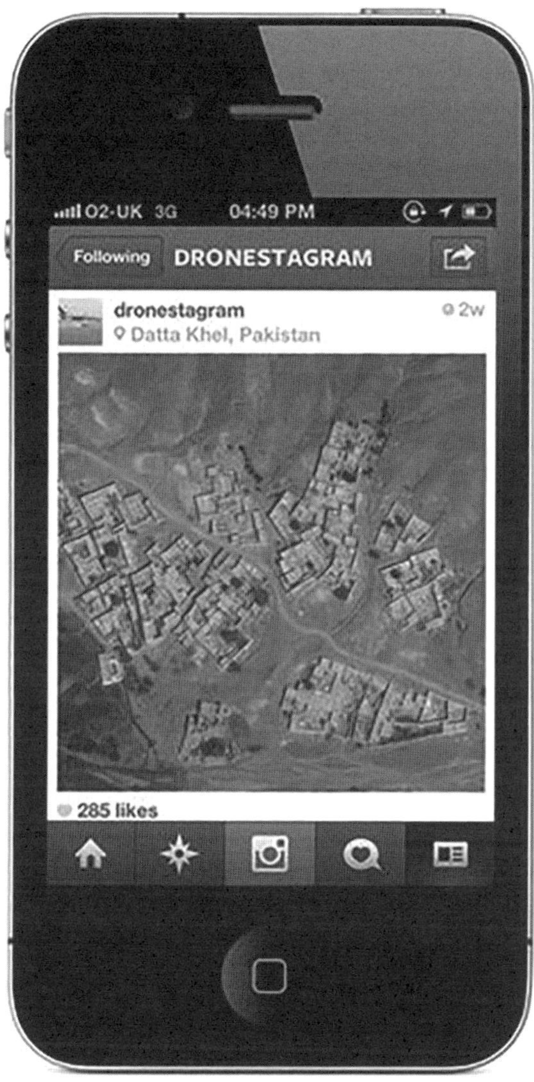

Figure 1c: James Bridle, *Dronestagram, an artwork*, (2012–2015). Screenshot of satellite images of drone strike locations from Google Maps posted on Instagram, Tumblr, and Twitter. 'March 21, 2013: A drone strike near the Afghan border kills at least three people. Possible houses or vehicles hit on the road to Datta Khel, with conflicting reports of the target. (Datta Khel, Pakistan).' Courtesy James Bridle via Flickr.

People

The Eco Material artists presented in this chapter diverge from the dependence upon technologies that Bridle's art epitomizes. Instead, they recognize the capacity of humans to think, plan, lift, scrape, bend, grasp, and perform myriad other actions that produce beneficial outcomes. The artworks presented in this chapter exemplify four operative categories of productive work performed by humans:

Do-it-yourself refers to the capacity of the human organism to conduct the tasks required for sustaining itself; it liberates humans from dependence upon the mass-produced goods and services that spew from anonymous, multinational corporations.

Do-it-as-community involves assembling and orchestrating the varied skills and capacities of contributors.

Do-it-as-troop demonstrates the effectiveness of organized efforts in which individuality is suppressed in favor of synchronized and coordinated actions.

Do-it-with-site refers to the ability of the human body to form productive affiliations with proximate materials and processes.

'What's next?' will look very different from 'What's now?' if this repertoire of Eco Material modes of interaction becomes a cultural norm. James Bridles's drone project epitomizes the physical detachment that results from the widespread dependence upon technical intermediaries. In contrast, all four Eco Material artworks in this chapter explore the potential of unmediated engagements. The first two of the projects reported are so ambitious they exceed the ability of the artists to fulfill them. The third documents a massive effort that succeeds, but only produces a miniscule result. The fourth is modestly scaled, and therefore successful. Despite their varied outcomes, the physical engagements undertaken by these artists are significant because they reinstate sensual experience that is omitted from interactions with machined goods and digital technologies.

Thomas
Thwaites

Thomas Thwaites (b. 1980, London UK) is a designer interested in the way technology, science, and economics are shaping the future. Thwaites studied economics and biology at University College London and in 2009 earned an MA in design interactions at the Royal College of Art. He has authored two books about his artwworks; The Toaster Project and GoatMan, published by Princeton Architectural Press. Thwaites's work has been acquired by the Victoria & Albert Museum. He has exhibited at the National Museum of China; the Museum of Modern and Contemporary Art in Seoul; and the Science Museum in London. In 2009, Thwaites gave a TED Talk on The Toaster Project. He is currently a visiting assistant professor in industrial design at Rhode Island School of Design.

Prior to the Industrial Revolution, entire cultures were fed, housed, clothed, and entertained by do-it-yourself crafting. People also fabricated the tools and fashioned the supplies required to accomplish these tasks. Thus, when the British artist, Thomas Thwaites, decided to test the limits of the DIY processes, he dismissed obvious DIY successes such as dipping candles, weaving cloth, carving bowls, and forging knives. Instead, he produced, single-handedly, from scratch, a device that is a product of the modern era. He created a toaster. By attempting to create a handmade version of the cheapest toaster he could find (it cost less than $7.00), Thwaites approached DIY feasibility in the spirit of an extreme sport.

The feats of technology that typically fuel contemporary imaginations are likely to be accomplished through micro-scaled manipulations of particles in three dimensions as small as a few tens of nanometers,[7] or mighty macro-scaled endeavors such as drilling seven thousand feet beneath the ocean floor.[8] Thwaites's efforts attracted a great deal of media attention despite the fact that the tasks he accomplished relied upon some simple tools and the physical apparatus he acquired at birth.[9] The following description of the outcome of his efforts that appeared in the *Boston Globe* suggests he was not likely to be praised by the DIY community of creative amateurs and hobbyists. The reporter writes, '(Thwaites) single-handedly built a pop-up toaster that has steadfastly refused to toast, is ghastly looking, and would likely rank among the most dangerous kitchen appliances ever created – if, that is, its maker dared to plug it in.'[10]

Thwaites began his DIY initiative by reverse-engineering the toaster he had purchased. He disassembled it and was dismayed to discover that this cheap, banal device contained 404 individual parts! Using these parts as patterns, he then proceeded to gather the raw materials to produce the substances that had been used to construct the manufactured version. The list of requirements included copper to make pins for the electric plug, the cord, and internal wires; iron to make

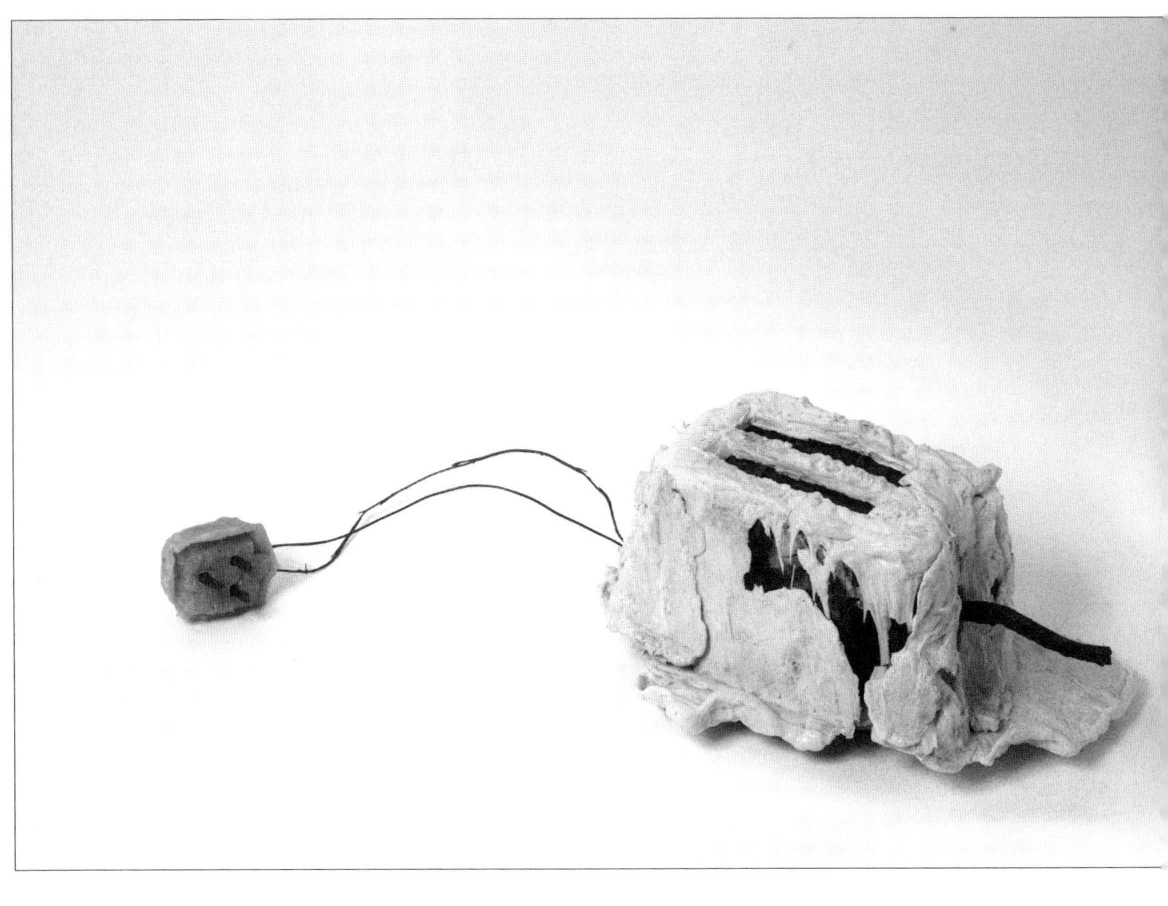

Figure 2a: Thomas Thwaites, *Toaster Project* (2010), iron, copper, mica, nickel, and crude oil (for the plastic case). A completed DIY toaster. Objects used in its production: two microwaves, a melted ceramic furnace powered by hairdryers, a two-part compression mold carved from a wooden tree trunk, water-cooler bottles filled with dirty water from a mine, a camping stove, and buckets. Photo by Daniel Alexander. Courtesy of the artist.

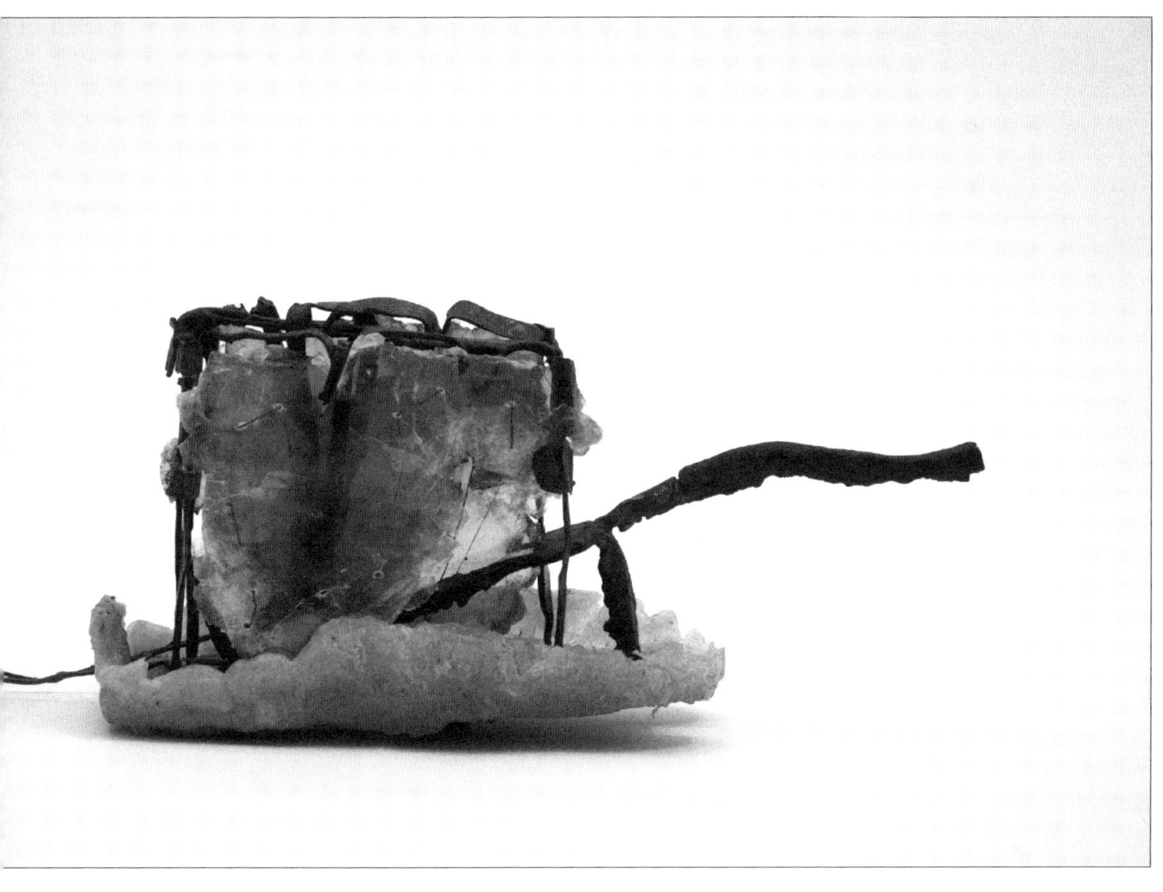

Figure 2b: Thomas Thwaites, *Toaster Project* (2010), iron, copper, mica, nickel, and crude oil (for the plastic case). A DIY toaster after plug-in. Objects used in its production: two microwaves, a melted ceramic furnace powered by hairdryers, a two-part compression mold carved from a wooden tree trunk, water-cooler bottles filled with dirty water from a mine, a camping stove, and buckets. Photo by Daniel Alexander. Courtesy of the artist.

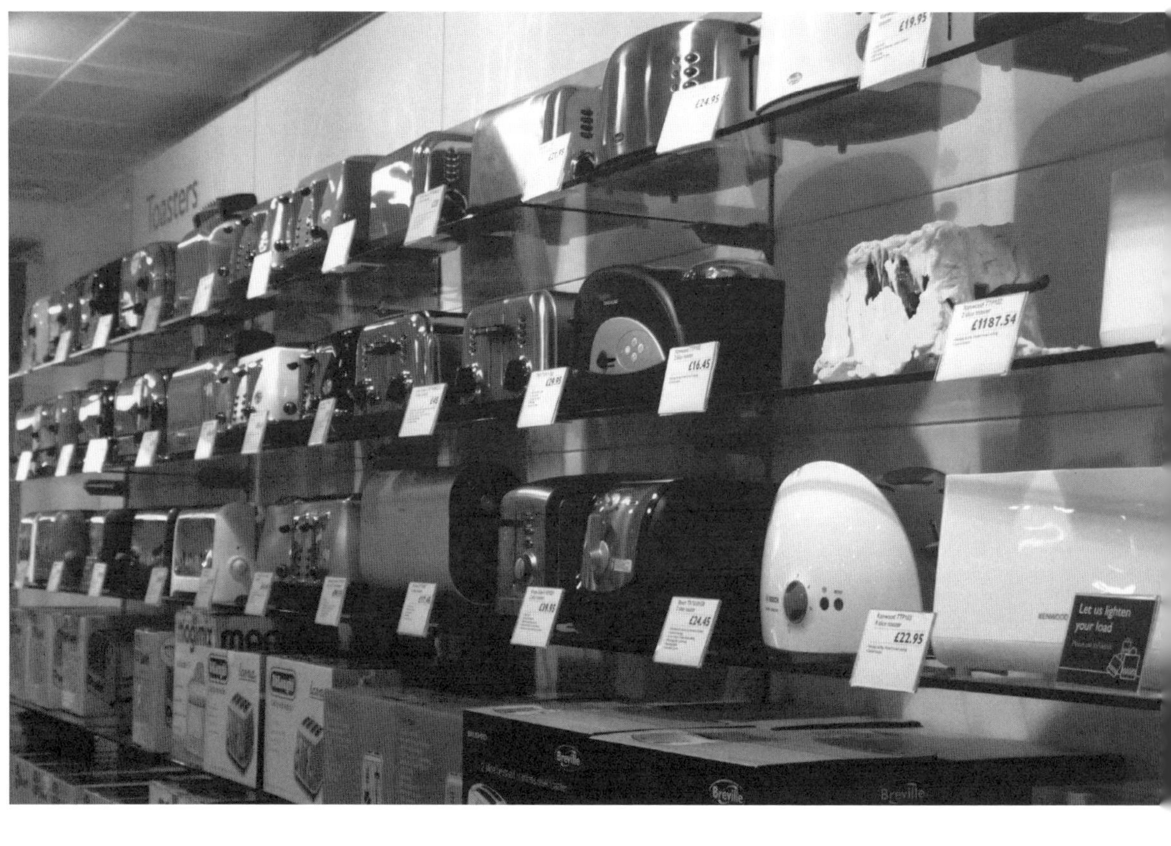

Figure 2c: Thomas Thwaites. View of the DIY toaster displayed in a store beside commercial toasters. Photo by Thomas Thwaites. Courtesy of the artist.

the steel grilling apparatus and the spring to pop-up the toast; nickel to make the heating element; mica to insert inside the heating element; and plastic for the plug, the cord insulation, and the casing. Thwaites's artistic process, therefore, entailed journeys to the sites where the materials originated; interactions with materials he had never touched; and the design and manufacture of tools he had never used before. The book, *The Toaster Project*, documents his ambitious exploits. It is part travelogue, part 'how-not-to' manual, and part confession. As opposed to the grand claims made by advertisers, Thwaites used this opportunity to recount his comically bungled efforts and to describe in amusing detail his DIY toaster's flaws.

None of the required materials spring from the ground ready to be assembled into toaster parts. Thus, Thwaites began by researching where these elements and minerals still existed as components of the planet, not items on a shopping list. The quest for iron to make steel exemplifies the ludicrous complexity of this undertaking. Thwaites discovered that an iron mine that was operating as a tourist attraction was accessible from his location in London. It was attended by a retired miner who mistakenly thought the chunks of iron ore he placed in Thwaites's suitcase were intended to make a 'poster', not a 'toaster'.

Thwaites then set about figuring how to transform the chunks of ore in his cache into toaster parts. Transforming rock into metal required constructing a furnace to smelt the ore. Current methodologies were not helpful because their operations were geared to industrial scales. His DIY ambitions were scaled to artisan skills that originated in the Iron Age and Bronze Age, and prevailed until industrialization usurped their functions. Thwaites discovered a description of this defunct process in metallurgy text published in the 1500s. The author of this text would be dismayed by Thwaites's approximation of the Iron Age technique. The artist admits, 'The first attempt to smelt ore occurred in a car-park in London. I built this furnace out of a dustbin and an antique chimney pot I'd taken from my mom's garden and a leaf blower. Only it didn't work.'[11] Then he discovered that you could melt a lump of iron oxide in a microwave, if you keep pumping energy into it for thirty to forty-five minutes, and add some carbon. This transforms iron oxide into iron. Where did Thwaites acquire this carbon? He confesses, 'I used my mum's ramekins.'[12] This triumphant moment led to the next challenge: to shape the metal into the toaster's frame. To accomplish this, Thwaites borrowed an anvil, heated the bits of iron he had smelted, beat them into long strips, and bent them into a rough approximation of a toaster frame.

Creating plastic proved to be equally challenging. Since plastic is made from oil, Thwaites's first attempt involved trying to convince BP to fly him out to an oil rig to acquire a few ounces of oil for the toaster's plastic casing. When this failed, he tried to create plastic from vegetable starches. This also failed because snails pillaged the concoction. His third attempt involved recycling plastic trash. Thwaites explains:

> *It was also a good opportunity to start thinking about the other end of the spectrum. Here I'd been thinking and working with mining stuff and taking it out of the ground, but then it will go back into the ground as landfill, so I wanted to focus on the other end of the lifecycle. I mean this stuff is sucked out as oil, turned into plastic, then sits on your desk for a couple years, and then where does it go? It goes back into landfill or recycling centers, which is just down-cycling.*[13]

This logic convinced Thwaites to conceive of the slow-degrading plastic products that currently overflow landfills as the geological 'rocks' that future generations will mine. He anticipated this future by digging in the landfill to procure his plastic, a practical solution that compromised his commitment to work 'from scratch'.

Thwaites plugged his version in once, but because he wasn't able to make insulation for the wires, the toaster started melting itself in about five seconds. Nonetheless, he displayed his impractical, ungainly toaster alongside rows of its slick and glossy prototypes. Like the others, its price was noted on its tag. Unlike the others, his cost 250 times more than a store-bought version! Ironically, this vivid disclosure of his toaster's deficiencies may testify to the success of this undertaking. Toasting bread was not this artwork's primary function. Instead, Thwaites's amusingly clumsy appliance is an elegant transmitter of a contemporary Eco Material quandary. He asks, 'What if the world came to an end? Would I be able to survive?'[14] Devoting nine months to create a toaster that couldn't toast is a material embodiment of the complexity of producing the commodities that are currently taken for granted. Thwaites's floundering testifies to the material complexity of dropping two slices of bread into a device and retrieving them a minute later all warm and crisp.

Eco Materialists might scrutinize the homemade toaster and the mass-produced toaster to evaluate which is more desirable environmentally. From this perspective, judgment of excellence may favor Thwaites even though his courageous attempt to achieve DIY self-sufficiency failed. Nonetheless, it succeeded by exposing the expenditures of energy and resources associated with mass production, consumerism, packaging, chain stores, advertising, marketing, and global trade. Thwaites offers a thought-provoking defense of his undertaking when he states:

My quest is perhaps absurd, but the contrast in scale between the products we use and the industry that produces them also seems absurd. Massive industrial activity in the pursuit of additional modicums of comfort at lower prices – small trifles, like an evenly crispy piece of toast, that we quickly become accustomed to.[15]

The folly of attempting to handcraft products that are spewed off assembly lines in factories also dissolves when he states, 'The provenance and the fate of the things we buy is too important to ignore.'[16] Thwaites emerges as the Eco Material winner with regard to the contested fates of his DIY toaster and the mass-produced version. On the one hand, dismantling a $7.00 toaster for recycling would entail separating all four hundred components according to their substance, an effort as ludicrous as Thwaites's effort to create his own. In addition, useful toasters are likely to be discarded in the dismissive manner of cheap, mass-produced versions, while his useless toaster will probably be retained because of the immaterial values it embodies: the effort invested in it, the personality conveyed by it, and the tale of its ludicrous creation.

Ultimately, the DIY toaster affirms its status as an artwork by conveying essential cultural insights. First, it draws attention to the impoverished homogeneity of industrially produced commodities, as well as the disregard that characterizes interactions with them. Second, it manifests the creative opportunities offered by doing, as opposed to purchasing. Thwaites proposes an environmental defense of his bungled exertions by concluding:

The laboriousness of producing even the most basic material from the ground up exposes the fallacy in a return to some romantic ideal of a pre-industrialized time. But at a moment in time when the effects of industry are no longer trivial in relation to the wider environment, the throwaway toasters of today seem unreasonable.[17]

Do-It-As-Community

Mary Mattingly

Mary Mattingly (b. 1979, Rockford, Connecticut, United States) is a visual artist based in New York. She explores issues of sustainability, climate change, and displacement. Mattingly combines photography, performance, portable architecture, and sculptural ecosystems into strategies for adaptation and survival. In 2015, she completed a two-part sculpture, titled Pull, *for the International Havana Biennial with the Museo Nacional de Bellas Artes de la Habana and the Bronx Museum of the Arts. Mattingly's work has been exhibited at the International Center of Photography; the Seoul Art Center; the Brooklyn Museum; the New York Public Library; deCordova Museum and Sculpture Park; and the Palais de Tokyo. Her most recent project,* Swale *(2016), is a floating food forest.*

Throughout her career, Mary Mattingly has actualized the system-wide reforms she believes will be required to address threats associated with global warming by activating the capacity of humans to perform the fundamental tasks of survival. Early in her career, she designed wearable gear outfitted for survival after habitable Earth conditions collapsed. More recently, she conducted a bold experiment introducing a survival-optimizing lifestyle. She explains her motivation by stating:

> *In preparation for our coming world with an increase in population, a decrease in usable land, new challenges with water, and a greater flux in environmental conditions, people will need to rely closely on immediate communities and look for alternative living models.*[18]

Efforts to provide the necessities of daily life include independent food production, waste management, recycling, energy production, and so forth. These functional self-sustaining operations were common until they were usurped by power tools and sophisticated technologies. Then they became remarkable. Mattingly tackles them all, but unlike DIY advocates, she doesn't work alone. She builds communities to ensure survival after some disaster has occurred and conventional supply chains have been disrupted, production has dwindled, and infrastructures have collapsed.

Mattingly conducted this cooperative venture with a small team of participants upon a 3,300-square-foot barge named *Waterpod* that circumnavigated New York City, docking at all five boroughs between spring and fall 2009. By relying on available resources that included simple technologies, the sun, the river ways, plants, and a few chickens, this prototype for community self-sufficiency repeals and replaces current dependence upon multinational corporations to supply human needs. Thus, the *Waterpod*'s codes of conduct did not depend upon advanced technology. They avoided mass production and a global market. They did not utilize wage labor. Nor did they deplete resources, pollute waterways, guzzle fossil fuels, diminish soil fertility, or contaminate the air. Instead, *Waterpod* tested the feasibility of humans forming small communities to provide for their own needs in a manner that offers personal gratification and avoids the harmful outcomes of lifestyles that depend upon industry. Their ambitious agenda included collecting water from rain and rivers and producing energy onboard. It also involved generating sufficient food to feed the five residents and an average of three guests three meals a day for five months by relying on the four hundred square feet of growing space on the barge, catching fish from the city's waterways, and generating produce from hanging gardens, window farms, hydroponics, and sub-irrigation planters that were installed throughout the barge.

These efforts reveal three survival strategies that are embodied by the do-it-as-community initiative. 'Do' refers to utilizing the biological operating equipment of the human body that has been undergoing continuous upgrades over the entire course of its two million years of evolutionary tune-ups and adjustments. 'It' consists of strategies that ensure survivability. 'Community' is comprised of communal relationships formed to accomplish survival skills. By combining these operative capacities, Mattingly hoped to satisfy the physical and psychological needs of the participants.

Mattingly explains:

> At first, I designed [Waterpod] as a personal space, but as the idea evolved, it became clear that it needed community to be sustainable and to benefit from multiple inputs and interpretations. I became more interested in the benefits that could be gained from a diverse community living on and interacting with the pod. I started to form a group of people who were interested in the project, either from an artistic, infrastructural, or technological point of view. Eventually, we formed a democratic group, a meritocracy, and developed a set of guidelines. Right now, there are five people who will be living on the Waterpod. Everyone will have to help out with repairs, gardening, cooking, and composting. Basically, everyone will learn how to take care of everything. I think this is really important – as the first industrial and technological age in the developed world is drawing to a close, people need to relearn how to do a lot of things.[19]

The *Waterpod* community combined the talents of artists, designers, builders, civic activists, scientists, environmentalists, and marine engineers. Their cross-disciplinary collaboration was bolstered by thirty volunteer growers that had been crowdsourced from Facebook and enlisted from classrooms around New York City. The contributors constructed four spherical pods on the barge using debris wood from abandoned water towers, old art crates, and a disused Parks Department pier. They also acquired metal railings from the Broadway play *Equus*; wallpaper

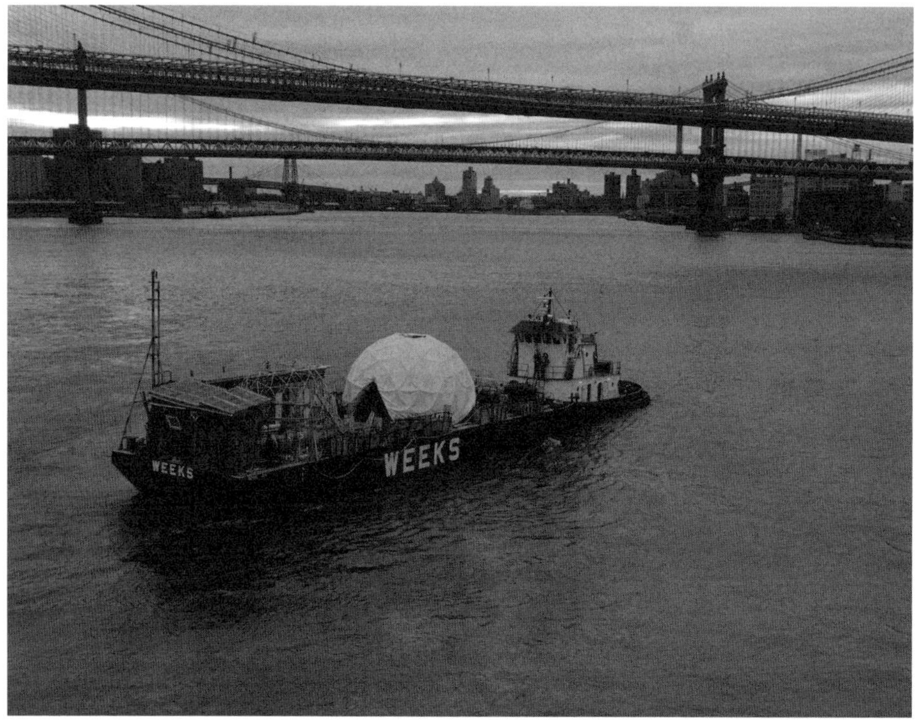

Figure 3a: Mary Mattingly, *Waterpod* (2009). Floating sculptural living structure. Repurposed wood, metal, plastic, fabric, and other materials on top of and adjacent to an industrial barge outfitted with a wind turbine, solar PV panels, bicycle power and a picohydro system. 3,300 square feet. Photo by Mary Mattingly. Courtesy of Mary Mattingly and the Waterpod Project.

from the television show *As the World Turns*; and billboard advertisements for the luxury goods that, ironically, share the blame for the impending disasters this project is attempting to avert.

One spherical pod was designated for community engagements in the arts. The second sphere housed a vertical garden and gray water recycling system in which rain was collected in a large bucket. This water was filtered for bathing, dishwashing, and drinking. This sphere also provided on-site power generation from a wind turbine, solar panels, bicycle power, and pico hydro systems. The third sphere served as a kitchen and shower room. The fourth contained four 8 x 8 ft. bedrooms for the crew and guests. All food, water, and human waste were cycled through the onboard living systems that produced fertilized soil and purified water. Many of these technologies were designed and built by students from the Environmental Resources Engineering Department at Humboldt State University, California.

The public was invited onboard to observe, participate in survival tasks, and suggest improvements. Thus the onboard community's efforts to explore resource sharing and creative reuse of local materials from the New York waste stream expanded through community outreach efforts. Daily events added inducements for the public to visit. Music was performed; art was created; vegetables were

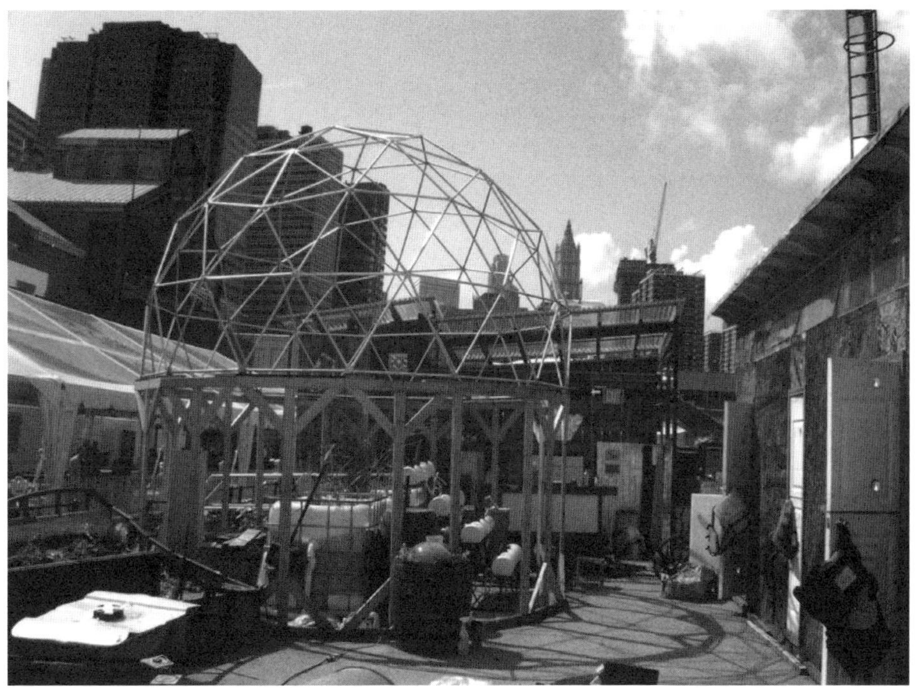

Figure 3b: Mary Mattingly, *Waterpod* (2009). On board floating sculptural living structure. Repurposed wood, metal, plastic, fabric, and other materials on top of and adjacent to an industrial barge outfitted with a wind turbine, solar PV panels, bicycle power and a picohydro system. 3,300 square feet. Photo by Mary Mattingly. Courtesy of Mary Mattingly and the Waterpod Project.

picked; wine was brewed. In addition, lectures and tutorials were offered on global warming, sea-level change, interactive art, Manhattan culture, futuristic architecture, water purification, hydroponics, air quality, and more. Workshops were conducted to teach members of the public how to build a boat with urban debris, construct a portable garden, preserve food, tie knots, and fabricate pulleys. In addition, discussions were held that anticipated the consequences of rising sea levels on population, migration, and homelessness. These wide-ranging activities all resolved back to the originating premise for this expansive project: the relevance of communal cooperation to changing paradigms of sustainability and domestic shelter due to global warming. As Mattingly notes, *Waterpod* demonstrated how to 'live like we will probably all need to live sooner or later.'[20]

Does the *Waterpod* experiment suggest that humanity's creative resilience will replace the industrial era? Conducting life onboard a floating vessel in a waterlogged environment offers several delightful alternatives to land-based habitation. Consider the pleasure of living in the midst of sea breezes, sleeping to the murmur of rippling water, and enjoying effortless mobility. For many, this setting is currently restricted to a two-week vacation each year. It might offer a welcome relief from confronting an aggravating commute each workday, and working for wages in a stuffy office or noisy factory.

Pragmatic advantages can be added to these allures:

- When buildings relinquish fixed footings and foundations onboard a barge, architecture and transportation merge. The home life of barge occupants is therefore uninterrupted even while they must move to escape turbulent weather, or climate disturbances, or political turmoil.

- Efficiencies abound when sites of consumption (living spaces) and sites for production (work environments) coincide. Homes also function as farm, factory, community center, and studio.

- Residents are not dependent upon the delivery of outsourced goods and services to survive because the production of essentials is self-managed.

Mattingly tested these assertions by keeping meticulous records of conditions on the barge. The research component was essential to this experiment. She explains, 'Everything was more concentrated and more comprehensible. We could observe the success of compost, the effects of insects, the quantity of rainwater. We could see the causes and effects very easily. Mostly, we could see lack of balance.'[21]

What did this experiment in communal effort reveal? *Waterpod* residents quickly realized the difficulty of surviving according to the ambitious agenda they had devised. Despite their elaborate preparations, Mattingly candidly admits that their survival strategies included bartering with a local green market, and accepting gifts from land-based visitors, and she enumerates the discomforts they endured:

- The laborious process of washing dishes by pumping water from a big bucket.

- The awkwardness of cooking meals in a fire-fed rocket stove or plug-in hot plate powered by the solar panels.

- A diet consisting mostly of vegetables and eggs.

- Exposure to the elements in all kinds of weather.

- Cramped living quarters.

- Needing to pedal an electricity-generating bicycle to use a laptop.

- Limiting showers to the three minutes of hot water only delivered by the solar array on sunny days.

- Being surrounded by space that you could not walk on.

She tops this list by noting that the barge rocked whenever another boat passed. 'Imagine if the earth under your feet swayed every time a car passed.'[22]

Mattingly is still seeking a definitive answer to the question she asked at the outset regarding communal efforts to sustain life: 'Was it possible?'[23] While the *Waterpod* did not attain the ultimate standard of self-sufficiency, it accumulated successes as an art project because, as Mattingly notes, 'Art is an agitation to the status-quo space and artists are agitators.'[24] As an agitation to the status-quo, the project succeeded before it was actualized. This is because acquiring official permission to establish this floating experiment in sustainable living/working required three years spent negotiating state and local permits, and guaranteeing compliance with four different agencies governing state and local codes. At each juncture, Mattingly explained her visionary urban survival experiment to municipal authorities. She explains, 'We wanted to show the authorities that people could be good stewards. We wanted to convince them of the benefits when they were only concerned about the risks.'[25]

The project also succeeded during the months of the venture by introducing – to all two hundred thousand people that climbed onboard – strategies designed for survival if water should inundate the land. Media coverage offers another measure of the project's effectiveness. More than three hundred global newspapers and magazines, and twelve television networks in the United States and abroad featured stories on *Waterpod*.[26] Nonetheless, when asked about the most positive part of the experiment, Mattingly provides a noteworthy answer. It was 'developing a holistic system. The community was functional. We depended on each other.'[27] She explains, '*Waterpod* is an expression of a collective intent, based on available resources, trial and error, as well as an object and a space that continues to be negotiated through democratic participation and implementation.'[28] Ultimately, *Waterpod* was operant because it capitalized on cooperation and collaboration.

Both Thwaites and Mattingly undertook daring attempts to replace dependence upon automated production lines and elaborate distribution channels. The limited success of their local efforts suggests that including the entire world within the network of specialized producers and global exchanges optimizes productivity. However, from an Eco Material perspective, the consumer profusion that is generated by worldwide networks appears doubly tainted because generic exchanges among anonymous producers diminish social cohesion, and because mountains are sliced, rivers are dammed, wetlands are paved, and the seas are mined to conduct large-scale industrial production.

What size of human community is optimal for autonomous sufficiency while preserving ecosystems? Thwaites admits that 'self' does not accurately describe his DIY experiment because he relied upon the contributions of advisers from the Clearwell Ancient Iron Mine, the Royal School of Mines at Imperial College, the Royal College of Art Foundry, Mull Geology, Axion Recycling, the National Non-Food Crops Centre, and so forth. Likewise, *Waterpod*'s futuristic experiment assembled fifteen specialists, numerous volunteers, private donors, and fiscal sponsors.[29] Both artists relied upon inputs from partners and collaborators. The experiments they conducted suggest that communal efforts among people willing to share resources, knowledge, and labor might not produce as much output, but they do produce greater satisfaction. By replacing DIY (do-it-yourself) strategies and DII (do-it-industrially) dependency both artists offer DIWO (do-it-with-others) as a means to avoid the loneliness of self-sufficiency and the anonymity of mass production. Such personalized networks are effective (functionally practical) because they eliminate the need for costly tools, protective packaging, and lengthy channels of distribution. Simultaneously, they are affective (emotionally satisfying) because work that satisfies personal needs combines the skills of the artisan, the pragmatism of the designer, the values of the environmentalist, and the creativity of the artist.

*Francis Alÿs (b. 1959, Antwerp, Belgium) is a conceptual artist based in Mexico City. His work emerges in the interdisciplinary space of art, architecture, and social practice. In Alÿs's wide range of projects and interventions, he has taken a roundabout trip threading the United States–Mexico border to California to highlight the increasing obstacles imposed along the border (*The Loop, *1997); filmed an attempt to enter the center of a tornado (*Tornado, *2000–10); trailed a leaking can of paint along the Israel–Palestine border (*The Green Line (Sometimes doing something poetic can become political and sometimes doing something political can become poetic), *2005). Recent exhibitions include* Francis Alÿs: A Story of Depiction *at MoMA (Museum of Modern Art) and MoMA PS1, both in 2011.*

Move a mountain? Throughout history, mountains have provided the ultimate symbol of 'constancy, eternity, firmness and stillness.'[30] These qualities are embedded in the phrase, 'To move a mountain', which is only summoned when someone proposes a preposterously unrealistic ambition. Accomplishing any aspiration that merits the phrase would be remarkable. The absurdly of this proposition is precisely why the Belgian artist, Francis Alÿs, attempted to actually move a mountain.

Alÿs enlisted five hundred volunteers. He provided the white shirts they wore and distributed the shovels they were to use to move the mountain. Then he instructed them to line up along the base of a massive 1,600-foot sand dune located on the desolate landscape surrounding Lima, Peru. On signal, each volunteer removed a shovelful of sand from the bottom and deposited it on the slope above them. Then they repeated the action again and again. With each step, their shovelfuls of sand moved closer to the top of the mountain. Once they reached the peak, they transferred the sand that was subtracted from one side of the mountain, and added it to the opposite side. In this manner, an entire mountain was moved!

How far was it relocated? Exactly the distance of one shovelful of sand.

How enduring was this action? It lasted until a gust of wind undid the action.

Was *When Faith Moves Mountains* (*Cuando la fe mueve montañas*) (2002), a folly or a triumph? Photographic and filmed documentation records the volunteers' arduous efforts to navigate the incline exposed to the searing heat and blowing sand. But it also records the exhilaration that infused the group as they approached the top, and later arrived at the base on the opposite side. Unlike the oppressed slaves building the pyramids in ancient Egypt, they were energized because they had achieved an unthinkable goal. Yet Alÿs's contributors shared

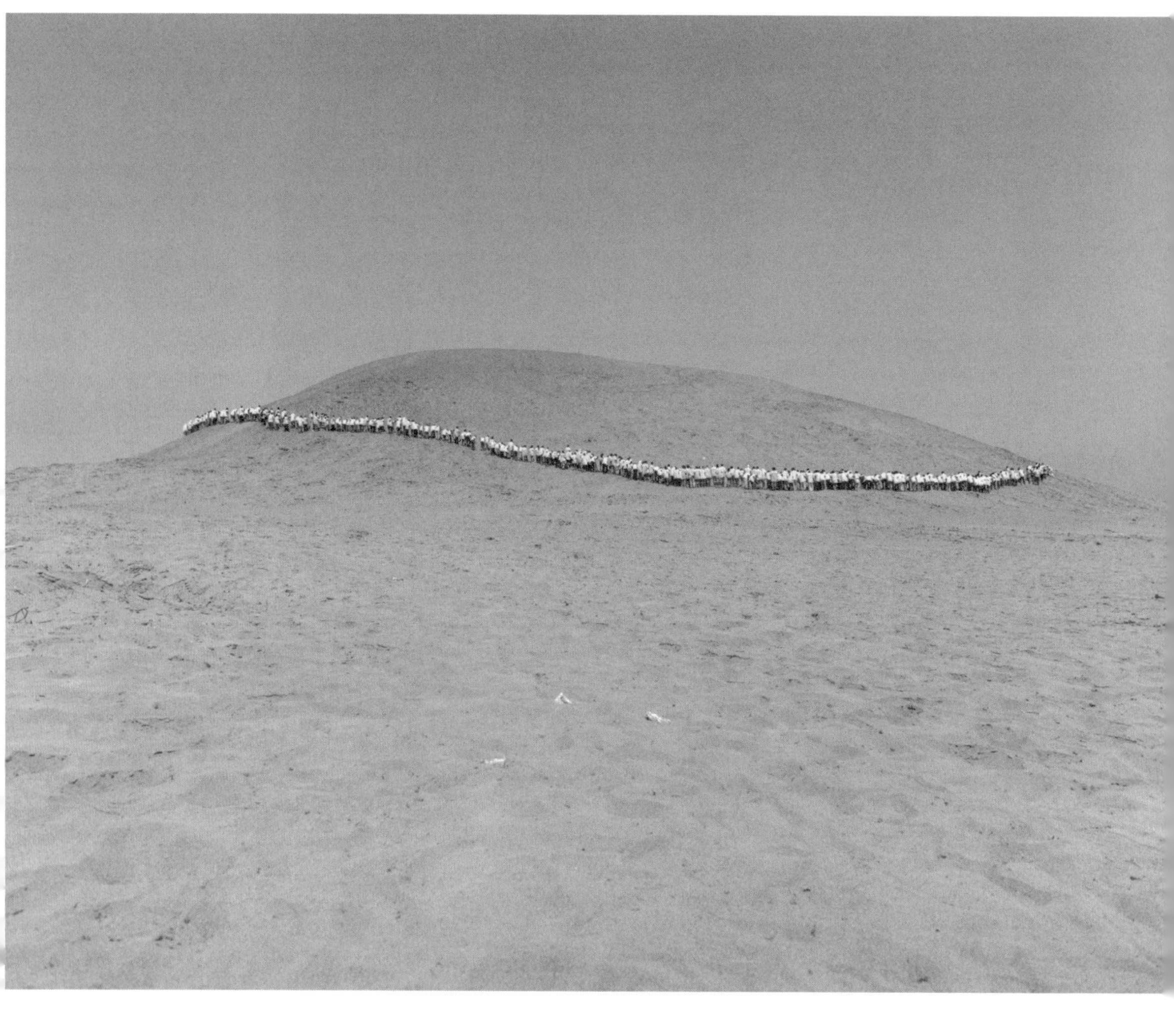

Figure 4a: Francis Alÿs in collaboration with Cuauhtemoc Medina and Rafael Ortega, *When Faith Moves Mountains* (*Cuando la fe mueve montañas*) (2002), video and photographic documentation, Lima, Peru. © by Francis Alÿs. Courtesy of David Zwirner, New York/London.[34]

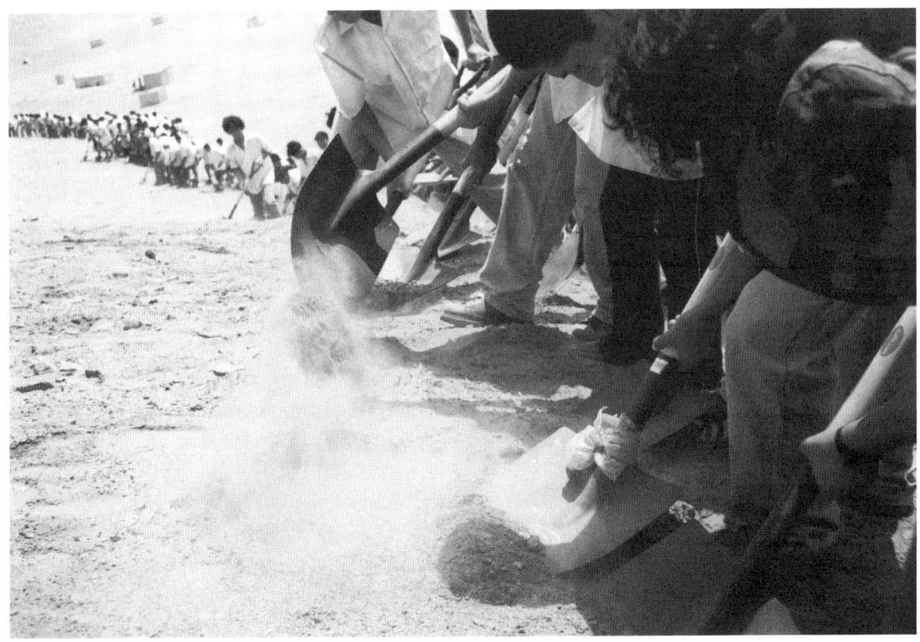

Figure 4b: Francis Alÿs in collaboration with Cuauhtemoc Medina and Rafael Ortega, *When Faith Moves Mountains* (2002), video and photographic documentation, Lima, Peru. © by Francis Alÿs. Courtesy of David Zwirner, New York/London.

with the slaves the obligation to obey an authority. Duress and free choice supply two ways that troops are created, whether they are comprised of soldiers, police officers, boy and girl scouts, or gang members. Unlike teammates that contribute individual skills to a shared goal, troops suppress their individual wills and merge into a composite unit.

Alÿs explains the social significance of this minor change in the landscape by recalling his first visit to Lima in 2000:

> *There were clashes on the street and the resistance movement strengthened. It was a desperate situation, and I felt that it called for an epic response, a 'beau geste' at once futile and heroic, absurd and urgent. Insinuating a social allegory into those circumstances seemed to me more fitting than engaging in some sculptural exercise.*[31]

Lima is surrounded by enormous sand dunes on which shanty towns have sprung up. They are populated by economic immigrants and political refugees who escaped the civil war fought there during the 1980s and 1990s by the military and guerrilla groups like Shining Path. Alÿs recalls, 'After a week of scouting, we chose the Ventanilla dunes, where more than seventy thousand people live with no electricity or running water.'[32]

Although variations on the phrase 'faith can move mountains' appear in Christian, Judaic, and Islamic sacred texts, the 'faith' component of *When Faith Moves Mountains* is not related to religion. In this instance, 'faith' refers to the operative potential of well-organized and directed humans to accomplish great deeds. The five hundred local residents who volunteered shared faith in the possibility of accomplishing a super-human feat by coordinating the efforts of ordinary people, even if the only tools they used were handheld shovels. The mountain that was moved by this artwork was, therefore, more psychological than geological. It reshaped despair into confidence, not only for the five hundred participants, but for the friends and relatives that occupied the vast, barren territory where the action occurred. They all learned that a banal task could accomplish deeds of epic proportions, if this task was performed as a unified effort. Thus, as each participant described the effort to those who hadn't experienced it, the event accumulated the stature of an epic tale or mythic action, instilling hope in the despondent residents of this forsaken territory. Alÿs explains:

> *The dune moved: This wasn't a literary fiction; it really happened. It doesn't matter how far it moved, and in truth only an infinitesimal displacement occurred – but it would have taken the wind years to move an equivalent amount of sand. So it's a tiny miracle. The story starts there.*[33]

Ultimately, what Alÿs created was a compelling symbol of the operative potential of human beings working together toward a common goal.

Rachael Mellors (b. 1954, Yorkshire, United Kingdom) creates artworks that emerge from a deep experience of the natural world, which explains why she primarily works with clay, a material of the Earth. In 2015, she began an ongoing project called Illuminations – *a series of photographs showing the artist's shadow on the cliff during sunrise. Mellors received her MA from the Royal College of Art and her BA from Portsmouth Polytechnic. She was awarded the Royal College of Art Prize for Drawing; the Crafts Council New Craftsman Grant; the Digswell Arts Trust Fellowship; Artist in Residence Harwich School; and the Crafts Council Index of Selected Makers.*

When humans engage their bodies in the flow of energy and matter that sustains life within their local ecosystem, they journey as far from contemporary norms of experience as Marco Polo sitting in front of a computer console. This is because standardized mass-produced commodities, and reliance upon electronic data-gathering and manipulation, are pervasive. As a result, experiences that are strikingly unusual are sensory interactions with the substances and conditions that account for the uniqueness of planet Earth. Direct contact with geological, biological, historical, and meteorological components of the surroundings rarely factors into the lives of citizens in industrialized societies. That is precisely why such interactions constitute an unwavering concern of Eco Materialists who trace disruptions of life-sustaining functions to human behaviors that ignore mutual, interspecies dependencies. Rachael Mellors integrates these fundamental tenets of ecology and environmentalism by demonstrating the operational capacities of her physical organism when it is conjoined with the geological and biological elements in her midst.

Mellors creates clay sculptures that are small enough to nestle in the palm of her hand. These modest artworks originate in the creative partnership she forges with the vast eco/bio/geo/hydro forces that abound in Gargarou, Greece, where she spends part of each year. Olive groves perched on cliffs high above the sea provide the material, inspire the process, and determine the working conditions of her art practice. This Eco Material pursuit could only be achieved by discarding five codes of studio art practice that are enshrined in the recent history of western art, art curriculum, art criticism, and art commerce. One such practice involves the use of industrially manufactured art materials: Mellors replaces them with found materials that embody the geological and cultural history of her site. Dependence on tools is the second practice she dispenses with in order to cultivate tactile connections with her locality through her medium by scraping, shaping, and smoothing her medium by hand. Third, she abandons the art studio because its steady-state conditions isolate her from the dynamic conditions occurring outdoors. Fourth, her creative impulse originates in the growth patterns of trees,

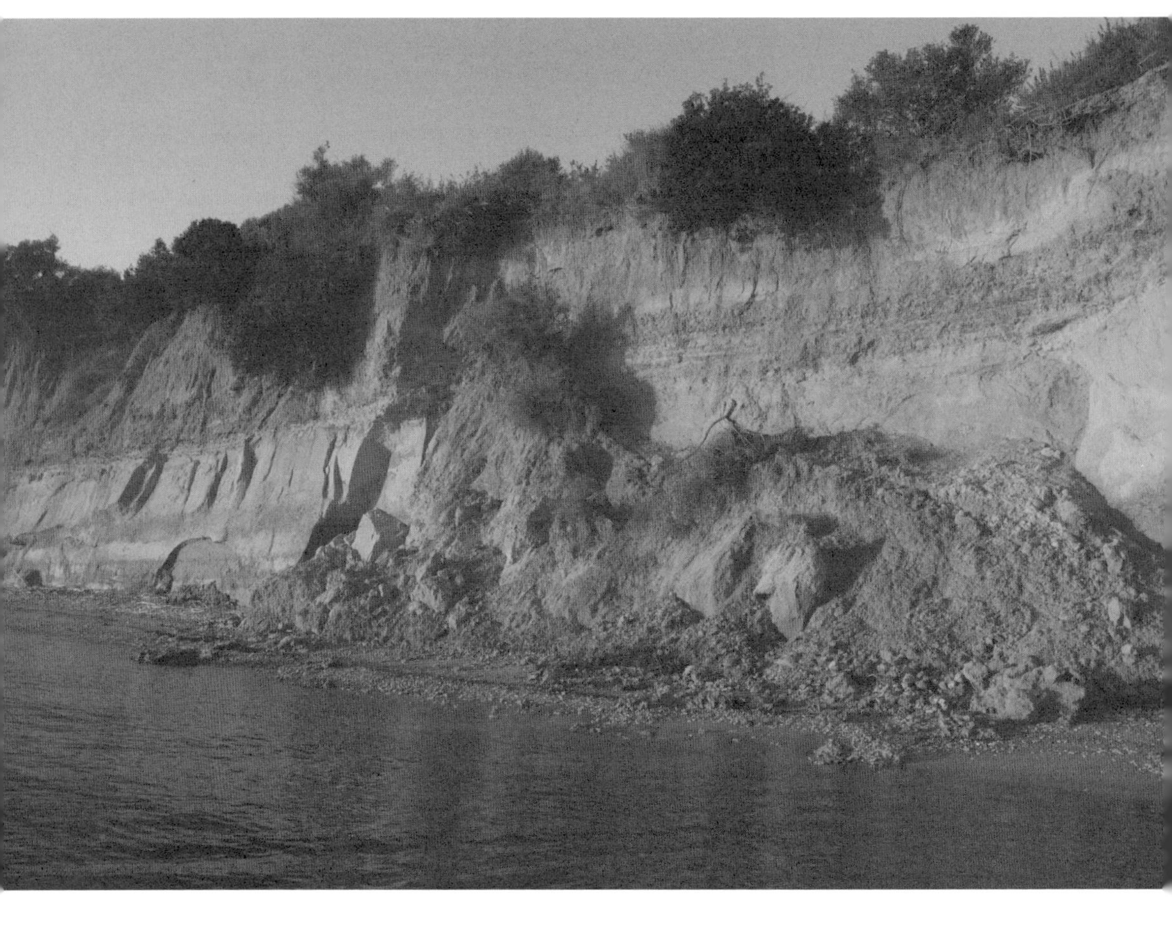

Figure 5a: Rachael Mellors, fallen clay eroding cliff at sunrise, 13 November, 2015. Gargarou, Greece. Photo by Rachael Mellors. Courtesy of Mother Earth.

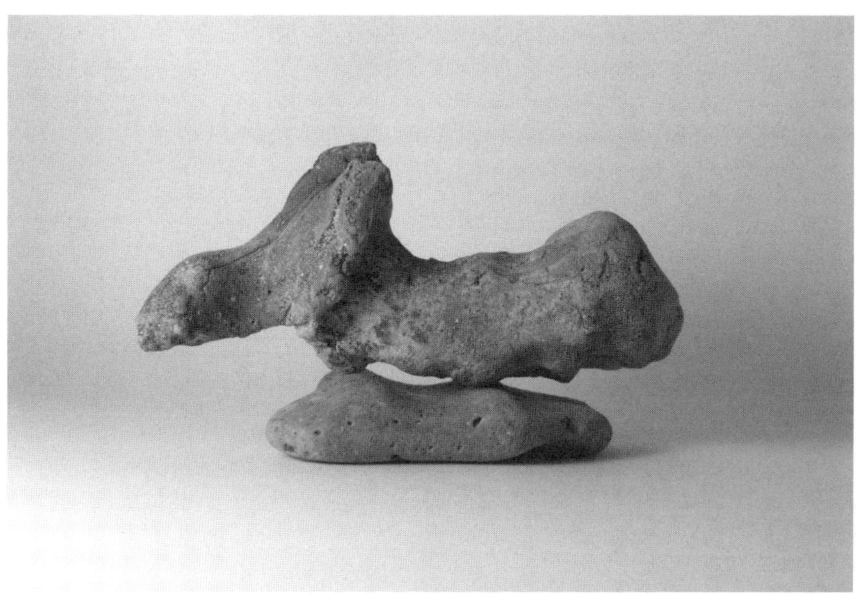

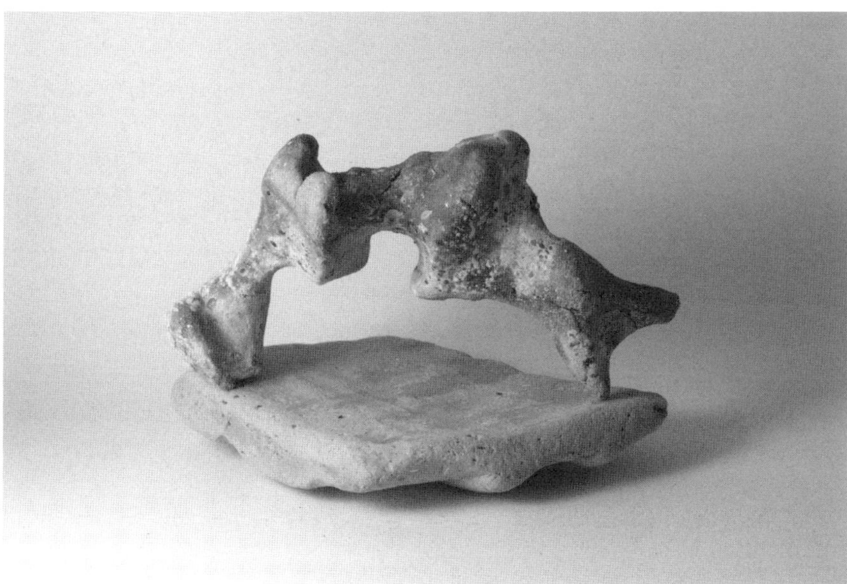

Figure 5b: Rachael Mellors, *Spring 2017 olive grove dusk April fire* (2017), cliff clay, 17 x 10 x 8 cm. Photo by Rachael Mellors. Courtesy of Mother Earth.

Figure 5c: Rachael Mellors, *Spring 2016, olive grove and Gargarou beach, daytime. April fire* (2016) from the series *2016 dawn Gargarou beach and dusk olive grove. December fire* (2016), cliff clay, 114 x 8 x 6 cm. Photo by Rachael Mellors. Courtesy of Mother Earth.

the annual cycle of the seasons, and the geological processes of erosion and sedimentation that are inherent to the site; not the artist's personal inclinations. Fifth, instead of privileging self-expression, Mellors materializes the wind, temperature, weather, season, and time of day as they impact her body during the sculpting process. All these sensual connections not only arouse the artist; they imbue her modest artworks with the grandeur of the seaside setting.

Mellors commences her sensuous interactions by collecting clay that falls from the eroding cliffs onto the beach. She then lays the clay sculptures when they are still wet on the beach to allow sand and silt to adhere to these surfaces. Shells and pebbles that lodge in the natural clay are welcomed as enhancements that reveal the site, not impurities that interfere with an artist's aesthetic preferences. The sea factors into the physical forming of the sculptures because Mellors shapes the clay by stepping into the shallow waters offshore and submerging the clay as she sculpts it. The shapes that emerge are evidence of her physical immersion with tidal rhythms, floating seaweed, and churning sand. The remarkable similarity between the clay formed in the artist's hands, and the clay formed exclusively by the churning of sea and sand is reinforced when Mellors creates bases for her sculptures by pairing them with the weathered pottery shards that wash up on the shore.

Another affiliation between art and place occurs when the newly formed artworks are laid on the beach and dried by the sun before firing. Pruned branches from the olive, holly oak, fur, and fig trees also contribute to this creative partnership between the artist and the site because they supply the fuel for firing the sculptures once they dry. This reciprocity is reiterated because the ash leftover from the firings is scattered in the groves to fertilize the trees that supplied the fuel to build the fires. Thus, this artwork not only derives its materials from the site, it contributes resources to the site as well. Mellors expresses this rapport by stating:

> *Being in the presence of what is around me in that moment, moving with sounds I hear, the sea and stones, calls of ravens, owls, jays, wind through the olive trees, bees, geckos, cicadas, even the sounds of chainsaws or tractors in the olive groves around, moving with the olive trees, clouds, waves, mountains, sun and moon rising. They all ignite the rhythms in my body and my sculptures come through me as I move.*[35]

Mellors registers the harmonious alliance with clay, trees, sand, sun, and sea in the title of each series: *2016 dawn Gargarou beach and dusk olive grove. December fire; Spring 2016, olive grove and Gargarou beach, daytime. April fire; and Autumn 2015, olive grove, dusk. November fire*. The specificity of time and place testifies to the significance of these extra-human conditions to influence human sensibilities.

Because of the fluid shapes they assumed as they were being formed, many of these artworks are unsuited to conventional modes of experiencing art. They must be nestled in the palm of the hand. This tactile experience with a few ounces of inert clay enables recipients to resonate with the material and energetic conditions that the sculptures embodied as they were being formed. By equating the physicality of the human organism with the physicality of non-human forms of matter, Mellors integrates the locale into her conception of 'self', which she describes in exuberant terms, 'I love the process of surrendering to making with nature, touching the web of life, allowing whatever comes to be, and giving to the fire to transform and surprise me.'[36]

Because contemporary humans largely rely upon an extensive arsenal of engineered devices to interact with the world around them, activating the tools that belong to the human body currently represents a venture into an exotic domain of possibility. Mellors presents a vivid example of the neglected sensual and perceptual attributes as they unfold in time and space. Because bodily-generated exploration is active and personal, information-gathering ceases to be mere data-collection. It is enriched by immersion, connection, and attunement.

To conclude, activating the body's inherent capacity to gather data is a radical departure from prevailing dependencies upon electronics to 'connect'. The body is a biological operating system that possesses the capacity to scan, scroll, surf, search, and share. Mellors continually reboots her biological central processing unit to improve its capacity to fill her cache with content that is personal, sensual, and vitalizing. Her efforts convey the Eco Material belief that such deepened awareness leads to attending to the vitality of the planet and its inhabitants.

Your opinion: What's next?

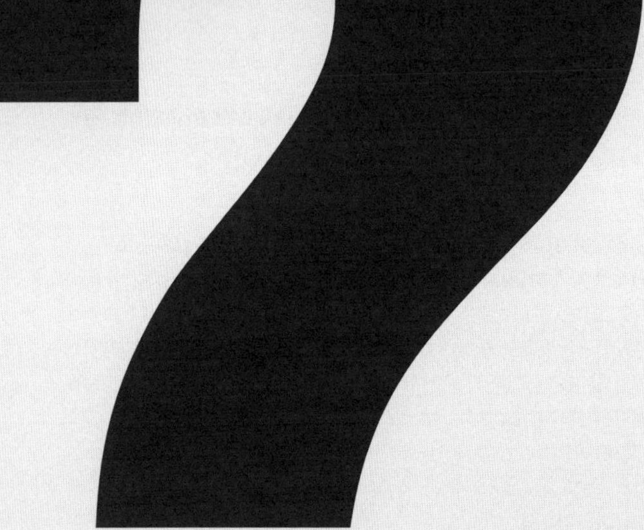

Reader Interaction

Each artwork presented in this chapter relied upon the human body to perform operative tasks. Thwaites's do-it-yourself strategy tested the presumption that solo efforts are not only effective; they offer the possibility of earning esteem associated with virtuosity and genius. Mattingly's do-it-as-community approach explored the operative potential of membership in a community in which the unique skills and knowledge of each individual contributed to a common goal. Alÿs's do-it-as-troop undertaking suppresses personal inclinations to demonstrate the feats that can be accomplished by anonymous contributors to a disciplined pursuit. Mellors's do-it-with-site practice harmoniously integrates human operators with surrounding geological, biological, historical, and meteorological conditions.

The allegiance to the human body manifested by each of these artists may appear more out-of-date than up-to-date. In actuality, it reinforces a current intellectual pursuit that is referred to as 'sensory studies'. This area of study is described as 'research in the social life and history of the senses, perceptual practice, multisensory aesthetics, and the development of technologies for expanding the sensorium in innovative ways.'[37] Such inquiries are being conducted by the Centre for Sensory Studies,[38] the Sensory Studies Network,[39] the *Journal of Sensory Studies*,[40] the Bloomsbury Sensory Studies Series,[41] the Sensory Curricula,[42] and more.

If 'What's next?' takes the form of functional somatic interactions, people will enjoy a reprieve from such monotonously predictable experiences as walking on the pavement, occupying steady-state architectural interiors, and eating packaged foods. Our senses will be reawakened by abandoning dependence upon thermometers to read temperatures, barometers to compute air pressure, and speedometers to measure speed. All four artists in this chapter revamp current modes of production to capitalize on the capacities of the human body to conduct useful tasks. They demonstrate that the senses do not only receive stimuli; they are essential to material operations of all kinds.

In order to encourage readers to activate their sensory apparatus, this chapter ends by diverting attention away from the outputs of human effort. Instead, it directs attention to the sensory inputs that enable such efforts. Because opportunities for sensing are boundless, these inputs have been consolidated into Charts of Material Attributes as reminders that our bodies are mighty tools for exploration and accomplishment. Like the Periodic Table of Elements, these charts provide inventories of physical properties. The traits they identify, like the elements on the Periodic Table, are combined and configured in the physical environment.

These charts differ from the Periodic Table by displaying the 'qualitative' properties of materials as they are experienced, not their 'quantitative' chemical traits that are revealed through objective analysis. Each attribute becomes knowable as it is felt, smelled, tasted, and observed. Each of these sensory attributes carries pragmatic potential. Thus, this chapter ends with a tool for activating the capacity of our human bodies to perform as operant allies of the planet. Create your own Chart of Material Attributes by replacing your own comparisons for the ones provided in the following charts. Thus, your chart will have 'likes' that reflect your sensory interactions.

Charts of Material Attributes

Acknowledge the quantity and diversity of sensations your body is equipped to experience by activating the numerous forms of sensory stimulation that appear on the following charts. Eco Materialism honors this capacity. Eco Materialists utilize them.

<table>
<tr><td colspan="2" align="center">Tangible Attributes</td></tr>
<tr><td>High relief → like a lava field</td><td>like an orange ← Low relief</td></tr>
<tr><td>Hot → like fire</td><td>like ice ← Cold</td></tr>
<tr><td>Wet → like mud</td><td>like fire wood ← Dry</td></tr>
<tr><td>Greasy → like motor oil</td><td>like powder ← Dusty</td></tr>
<tr><td>Beating → like the heart</td><td>like rock ← Still</td></tr>
<tr><td>Silky → like satin</td><td>like burlap ← Rough</td></tr>
<tr><td>Smooth → like an egg shell</td><td>like an unpaved path ← Bumpy</td></tr>
<tr><td>Flat → like paper</td><td>like a ball ← Round</td></tr>
<tr><td>Heavy → like a boulder</td><td>like a feather ← Light</td></tr>
<tr><td>Large → like redwood trees</td><td>like a speck of dust ← Tiny</td></tr>
<tr><td>Slimy → like oil</td><td>like a dog ← Furry</td></tr>
<tr><td>Sticky → like flypaper</td><td>like powder ← Dry</td></tr>
<tr><td>Abrasive → like sandpaper</td><td>like oatmeal ← Mushy</td></tr>
<tr><td>Limp → like silk</td><td>like steel ← Rigid</td></tr>
<tr><td>Jagged → like a saw blade</td><td>like cell phones ← Sleek</td></tr>
<tr><td>Gooey → like taffy</td><td>like chips ← Crunchy</td></tr>
<tr><td>Tingly → like pins and needles</td><td>like wool ← Itchy</td></tr>
<tr><td>Dense → like tar</td><td>like foam ← Frothy</td></tr>
<tr><td>Prickly → like Fiberglas</td><td>like pudding ← Soft</td></tr>
<tr><td>Tight → like gloves</td><td>like a scarf ← Loose</td></tr>
<tr><td>Fluttery → like curtains</td><td>like plywood ← Rigid</td></tr>
<tr><td>Blunt → like a spoon</td><td>like a knife ← Sharp</td></tr>
<tr><td>Chapped → like cold hands</td><td>like nervous hands ← Clammy</td></tr>
</table>

Corrugated → like cardboard	like mustaches ← **Hairy**
Fatty → like butter	like unwashed spinach ← **Gritty**
Prickly → like thorns	like mucus ← **Gooey**
Wrinkled → llike crumpled foil	like rice ← **Granular**
Velvety → like bunny fur	like fat ← **Greasy**
Mushy → like paper pulp	like clay ← **Malleable**
Sharp → like needles	like marine invertebrates ← **Spongy**

Olfactory Attributes

Pungent → like onions	like stone ← **Odorless**
Musty → like an old trunk	like bleach ← **Clean**
Smokey → like smoldering wood	like parsley ← **Green**
Stinky → like a skunk	like roses ← **Perfumed**
Fishy → like sardines	like beeswax ← **Faint**
Acidic → like vinegar	like honeysuckle ← **Flowery**
Foul → like rotten eggs	like newly baked bread ← **Fresh**
Rancid → like mold	like honey ← **Sweet**

Taste Attributes

Salty → like pretzels	like milk ← **Bland**
Sour → like lemon	like honey ← **Sweet**
Bitter → like kale	like chocolate ← **Pleasant**
Spicy → like salami	like banana ← **Soothing**
Smokey → like ham	like sauerkraut ← **Fermented**
Pungent → like onions	like water ← **Tasteless**
Foul → like rancid milk	like an orange ← **Fresh**
Pickled → like pickles	like deer ← **Gamey**
Peppery → like salami	like peaches ← **Juicy**
Savory → like mangoes	like red wine ← **Robust**
Tangy → like ginger	like lemon ← **Tart**

Physical Attributes

High melting temperature → like sand   like snow ← **Low melting temperature**

Malleable → like copper   like flint ← **Brittle**

Expandable → like asphalt   like a bolt ← **Rigid**

Reactive → like lithium   like gold ← **Stable**

Flammable → like paper   like asbestos ← **Inflammable**

Good conductor of electricity → like silver   like cardboard ← **Poor conductor of electricity**

Good conductor of sound → like diamond   like felt ← **Poor conductor of sound**

Good conductor of light → like glass   like lead ← **Poor conductor of light**

Good conductor of heat → like copper   like fiberglass ← **Poor conductor of heat**

Volatile → like alcohol   like corn oil ← **Constant**

Stretchy → like rubber   like a wool sweater ← **Shrinks**

Fast evaporation → like alcohol   like water ← **Slow evaporation**

Volatile → like oxygen   like neon ← **Stable**

Your comments: What's next?

Endnotes

1   'The 2015 Wired 100,' *Wired*, May 23, 2016, accessed August 29, 2017, www.wired.co.uk/article/the-2015-wired-100.

2   'Dronestagram: The Drone's-Eye View,' November 8, 2012, accessed August 29, 2017, booktwo.org/notebook/dronestagram-drones-eye-view/.

3   'Dronestagram.'

4   'Dronestagram.'

5   'Dronestagram.'

6   'Dronestagram.'

7   Dexter Johnson, 'Optical Tweezers Can Now Manipulate Matter on a Nanoscale,' March 6, 2014, accessed August 27, 2017, http://spectrum.ieee.org/nanoclast/semiconductors/nanotechnology/optical-tweezers-can-now-manipulate-matter-on-a-nanoscale.

8   Fox News, '7,000 Feet Beneath Ocean Floor, Deep-Sea Drill Sets Record,' September 7, 2012, accessed August 29, 2017, http://www.foxnews.com/science/2012/09/07/7000-feet-beneath-ocean-floor-deep-sea-drill-sets-record/.

9   Thomas Thwaites, 'How I Built a Toaster – From Scratch,' TED, November 2010, accessed August 29, 2017, https://www.ted.com/talks/thomas_thwaites_how_i_built_a_toaster_from_scratch.

10   Chris Wright, 'Thomas Thwaites's Bespoke Toaster,' *The Boston Globe*, November 13, 2011, accessed August 29, 2017, https://www.bostonglobe.com/ideas/2011/11/13/thomas-thwaites-bespoke-toaster/V63mYKPsXw8vHqEVKjMIeO/story.html.

11   Jennifer Kabat, 'The Rumpus Interview with Thomas Thwaites,' *The Rumpus*, April 12, 2012, accessed August 29, 2017, http://therumpus.net/2012/04/the-rumpus-interview-with-thomas-thwaites/.

12   Kabat, 'The Rumpus Interview with Thomas Thwaites.'

13   Kabat, 'The Rumpus Interview with Thomas Thwaites.'

14   https://www.bostonglobe.com/ideas/2011/11/13/thomas-thwaites-bespoke-toaster/V63mYKPsXw8vHqEVKjMIeO/story.html.

15   Collin Cunningham, *Maker Faire*, July 23, 2009, accessed August 10, 2017, https://makezine.com/2009/07/23/a-homemade-toaster-built-from-scrat/.

16   Cunningham, *Maker Faire*.

17   Cunningham, *Maker Faire*.

18   Mary Mattingly, 'The Waterpod Project, 2009,' accessed August 29, 2017, http://www.marymattingly.com/html/MATTINGLYWaterpod.html.

19   Johanna Björk, 'The Waterpod Project – A Self-Sustaining Ecosystem on the Manhattan Waterfront,' *Goodlifer*, August 31, 2009, accessed August 29, 2017, http://www.goodlifer.com/2009/08/the-waterpod-project/.

20   Robert Voris, 'This Barge Makes Global Warming Look Cool,' *Brooklyn Paper*, accessed August 29, 2017, http://www.brooklynpaper.com/stories/32/32/32_32_rv_waterpod.html.

21   Mary Mattingly, interview with the author, Rhinebeck, New York, May 17, 2017.

22   Interview.

23   Interview.

24   Danahers, 'Artist Interview: Mary Mattingly,' *Rough Copy*, February 17, 2010, accessed August 29, 2017, https://roughcopymag.wordpress.com/2010/02/17/artist-interview-mary-mattingly/.

25   'Artist Interview: Mary Mattingly,' *Rough Copy*, February 17, 2010, accessed August 29, 2017.

26   Waterpod™, *Waterpod Results*, accessed May 25, 2017, http://www.thewaterpod.org/pdf/
      WaterpodResults.pdf.

27   Mattingly, interview.

28   Waterpod™, 'Mary Mattingly,' accessed August 29, 2017, http://www.thewaterpod.org/mm.html.

29   Teammates on board included John McGarvey, Alison Ward, Carissa Carman, Lonny Grafman,
      Dockmaster Frank J. Carnesi, Rik van Hemmen at Martin Ottaway, Evan Korn, and Jessica Rosenfield
      at the NYC Mayor's Office of Citywide Events Coordination and Management, US Coast Guard
      Commander Brian S. Gilda, Sara Reisman at Percent for Art, Ken Hollenbeck, Richard Massey, and
      many others. In the final stages of planning, other invaluable collaborators joined our effort, weaving
      the fabric that made the *Waterpod* work, including co-curator Ian Daniel, Mayra Ciment, Nicole Pilar,
      Kristen Parker, Janet Persia, and scores of talented volunteers.

30   Symbolism.org, 'Natural Places,' accessed August 29, 2017, http://www.symbolism.org/writing/
      books/sp/2/page6.html.

31   Francis Alÿs, 'A Thousand Words: Francis Alÿs Talks about When Faith Moves Mountains,' *Artforum
      International*, 40.10 (2002), accessed August 29, 2017, https://www.questia.com/magazine/1G1-
      87453039/a-thousand-words-francis-Alÿs-talks-about-when-faith.

32   Alÿs, 'A Thousand Words.'

33   Alÿs, 'A Thousand Words.'

34   Mariabruna Fabrizi, 'A Line in the Landscape: Francis Alÿs "When Faith Moves Mountains",' *Socks*,
      March 15, 2014, accessed August 29, 2017, http://socks-studio.com/2014/03/15/a-line-in-the-
      landscape-francis-Alÿs-when-faith-moves-mountains/.

35   Rachael Mellors, email correspondence with the author, August 12, 2014.

36   Rachael Mellors, email correspondence with the author, August 24, 2014.

37   Centre for Sensory Studies, accessed August 29, 2017, http://www.centreforsensorystudies.org/.

38   Centre for Sensory Studies.

39   Nottingham Sensory Studies Network, accessed August 27, 2017, http://www.nottingham.ac.uk/
      research/groups/nottingham-sensory-studies-network/index.aspx

40   Wiley Online Library, *Journal of Sensory Studies*, accessed August 29, 2017, http://onlinelibrary.wiley.
      com/journal/10.1111/(ISSN)1745-459X.

41   Bloomsbury Publishing, 'Sensory Studies Series,' http://www.bloomsbury.com/uk/series/sensory-
      studies-series/.

42   Sensory Studies, 'Sensory Curricula, Syllabi, Etc.,' accessed August 29, 2017, http://www.
      sensorystudies.org/timeline/.

ECO MATERIAL TOOLS

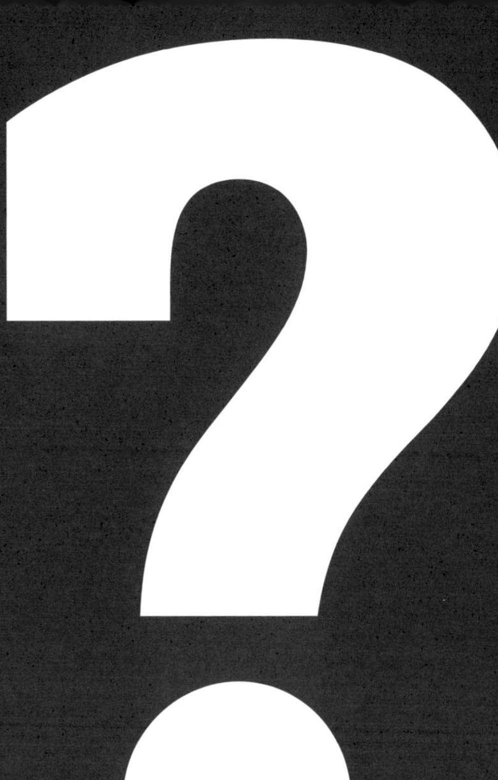

Fire

ECO MATERIAL TOOLS

Fire

Tools

Tools are poorly represented in historic accountings of the great sweep of civilization. Typically, these chronicles feature the end products of human interactions with materials, not the implements and processes that facilitated these effects. Eco Materialists are reversing this pattern by emphasizing the pivotal role the contents of tool chests has played in humanity's mixed record of accomplishment and folly. Breakthroughs in firepower, like the invention of gunpowder and the development of the internal combustion engine, are significant as molders of social structures, economic status, political influence, and ecological impact.

Hammers, derricks, microscopes, lasers, and calculators are all tools because they enable humans to transcend their inherent strength, endurance, precision, and cognitive abilities. To a great extent, civilization is a grand experiment in upgrades and additions to humanity's toolbox. Nonetheless, dependence upon tools is a relatively recent phenomenon when it is inserted into the timeline that begins when hominids first stood upright and began striding across the Earth. From this perspective, humanity's toolbox has only existed for the last four percent of time. This revelation is calculated according to the date hominids first arose (approximately seven million years ago) and the date stone tools first emerged (approximately three hundred thousand years ago).[1] It is hard to imagine that for the preceding ninety-six percent of human history, survival was entrusted to two front appendages for gripping, lifting, throwing, and carrying; two rear appendages for locomotion; a set of teeth for biting and chewing; and eyes, tongue, ears, nose, and skin for gathering information about the surroundings. As such, every material encounter of our earliest ancestors depended upon unaided physical interactions with the fluctuating storehouse of resources they shared with their non-human neighbors.

The material impact of tools escalated when humans transcended dependence upon hands, stone flakes, and sticks, and discovered, one by one, levers, screws, gears, wheels, and pulleys. Greater tooling potential was unleashed when humans learned to harness non-human sources of energy to power these tools. This fuel progression begins with fire and proceeds to cattle, water wheels, windmills, steam engines, internal combustion and jet engines, electricity, rocket technologies, and nuclear power. The history of precedent-breaking innovations is encapsulated in the progression of tools for carving. They commence with fingernails, and proceed to stone flakes, metal blades, power saws, dynamite, sand blasting, pneumatic chisels, hydraulic excavators, laser beams, 3-D printers, and milling robots.

Ever since early humans devised tools to build and shape their lives, tools have been used to build, plow, pave, dam, mine, dig, pump, smooth, separate, lift, sort, rotate, slice, adhere, mix, reach, crumble, measure, align, stack, stir, probe, calculate, scrutinize, and analyze. In the twelve thousand years since 'tool' referred to a stick used for prying roots, humanity's brawn and brain powers have extended the reach of its operations far into the lithosphere, atmosphere, biosphere, and hydrosphere. Thus, when Forbes Magazine invited experts, editors, and readers to review the entire history of tooling and compile a list of 'The 20 Most Important Tools Ever',[2] condoms, syringes, remote controls, and floppy disks were contenders, but not winners. The twenty that were honored as 'most important' seem simultaneously to be 'most ordinary'. They include a compass, pencil, harness, rifle, eyeglasses, candle, scale, and pot. The featured role of tools in Eco Material considerations is revealed by this list because each of these innocuous objects propelled epic transformations in human capacities to manage the material environment.

Today, few Earth contours, textures, colors, sounds, smells, or tempos escape the virtuoso skills and extraordinary acumen of the human species to design tools. Tools deserve substantial credit for humanity's success in negotiating a monopoly position among living creatures. On the one hand, soaring human populations provide evidence of our ecological 'success' as a species. These capacities are also materialized in revered artworks, architecture, technologies, and scientific breakthroughs. But the story of civilization may not continue on this triumphant course. Although the ingenious devices devised for performing feats of manipulation have earned esteem for their inventors, fortunes for their producers, and gratitude from their beneficiaries, all three groups may ultimately be found guilty of reckless endangerment. This is because many of these achievements are registered in oil spills, decimated forests, smog, and toxic waste dumps. Debris from our lavish lifestyles does not merely clutter landfills, gutters, and empty lots. Migrating on air and ocean currents, it settles far from its place of origin. Even remote Northwestern Hawaiian Islands are not immune to such intrusions. Divers from the US National Oceanic and Atmospheric Administration recently collected 114,000 pounds of trash on these islands in just one month.[3] This is a staggering figure because these islands are uninhabited!

Eco Materialists explore the unsettled tension between the material rewards and ecological penalties of the empowering and emboldening capacities of tools. Because contemporary norms of productivity typically ignore the impact of the tool and the tool user upon the surroundings, tools are essential to this book's exploration of Eco Materialism. They link the 'eco' of the planet to the 'materialism' of human culture. The conjoining of these terms raises the question: Will future generations judge humanity's current occupation of the planet as a triumphant or hostile takeover?

Fire

Acknowledging the myriad innovations that expedite functional tasks would burst the bindings of this book. Even representing the various categories of tooling would exceed the capacity of a chapter in a book. These constraints required that a single example of human tooling represent the vast scope of human tooling. By defining tool as anything that expands the human capacity to manage the material environment, this choice proved to be self-evident. Fire was selected because it launched humanity's grand experiment in manipulating materials between 400,000 to 1,400,000 years ago,[4] because it was pivotal in the progression of eras that mark civilization, and because it continues to enable countless tasks.

As a process, not an object, fire combusts fuels and releases gases, producing sparks, flames, ashes, cinders, charcoal, and smoke. None of these material by-products of combustion are waste. Their innumerable tooling capacities are listed at the end of this chapter. While it is generating these by-products, fire provides countless services that account for humanity's comfort and security. Until the modern era, cultures across the globe believed fire embodied the supernatural realms of divinity and magic. Ancient Persians, Egyptians, Mexicans, Jews, Etruscans, Greeks, Romans, and Chinese worshipped fire as the mighty force that embodied the generative power of the Sun and activated inexplicable changes in substances on Earth. Fire's many otherworldly attributes easily justify this belief. Consider, for example, that fire has the power to overwhelm the dark; it can spontaneously erupt; it transforms itself from a dying ember into a ferocious force; it is used to provide protection and to destroy; and it has a voracious appetite even though it has no substance. Ultimately, fire's otherworldly status may be due to its ever-shifting in shape that is not only mesmerizing; it defies gravity in the manner of a spirit, god, or ghost.

Because of these remarkable attributes, many cultures assign fire a mythic origin. Prometheus and Epimetheus, Greek brothers who were Titans, provide a well-known example. In one version of the myth, the brothers were assigned the task of allocating a special attribute to each of the animals as they were first created to populate the Earth. The brothers distributed the gifts of flight, strength, speed, beauty, agility, etc. But when it was humans' turn, there were no attributes left. Humans would have been defenseless prey if it were not for Prometheus who took pity on them by offering them the most empowering capacity of all: fire. Because fire was then the exclusive possession of the gods, Prometheus stole a spark from the thunderbolt of the greatest god of all, Zeus.[5] The myth then describes Zeus's rage because humans now shared his almighty power, asserting that the domestication of fire is a powerful tool that is exclusive to, and universal among, humans.

Early humans were highly skilled fire technicians, capable of controlling its timing, frequency, size, and intensity. This versatile tool rendered Paleolithic lives more secure and comfortable because it aided hunting; hardened wood; melted metals; controlled insects; purged diseases; destroyed enemies; illuminated the night; split stones; kept animals at bay; and warmed cold nights. Over time, fire also transformed wood into tar, resin into pitch and turpentine, grain and grape into alcohol, ash into soap, calcitic rock into lime,[6] and seawater into salt.

The first of three momentous shifts in peoples' interactions with fire occurred at the beginning of the nineteenth century. Until then, the fuels people burned were limited to the availability of local supplies of wood, reeds, vegetable oils, and animal fats. However, once coal, oil, and gas became sources of fuels, people's energy budget was no longer limited by deposits in their local energy bank. Fossil fuels made the entire planet's storehouse of underground energy supplies available for withdrawals. Burning soared.

The second shift marked the disappearance of flame, smoke, spark, and ember from everyday lives. This occurred when candles, oil lamps, fireplaces, hearths, stoves, and torches were replaced by gadgets and appliances with dials, switches, and buttons. Stephen J. Pyne, a scholar of the history and management of fire, observes that prior to the Industrial Revolution, humanity's interactions with fire resembled the care giving of animals.[7] People provided shelter for their fires; they fed and protected them; ensured their supply of oxygen; put them to bed at night; awakened them in the morning; and trained them for hunting, cooking, heating, farming, fighting, herding, clearing land, etc. Like a pet, fires had to be tended or they would either die or grow fierce. With the advent of the Industrial Revolution, fuels for heat and light were delivered from distant sources, ready for burning without effort or engagement.

The third historic shift in the human relationship with fire occurred during the Second Industrial Revolution at the end of the nineteenth century when open flame was replaced by the internal combustion engine, then electricity, batteries, fuel cells, gas turbines, transistors, and so forth. In all these ways, humans severed their connection with a primal component of their planet. Fire, as a central feature of preindustrial lives, has been reduced to decorating birthday cakes.

The artists presented in this chapter offer four examples of fire's tooling capacities: burning, melting, sintering, and cooking. In each case they each apply the power invested in humans by possessing fire: gathering fuel, controlling metals, harnessing solar power, and preparing food. They demonstrate the range of uses of fire by producing heat, implements, weapons, and food. Finally, each demonstrates the environmental impact of fire's varied uses: regenerative, depleting, destructive, and nourishing. Together these artists provide an overview of tooling from an Eco Material perspective.

Fire Burns

Bruce Davies

Bruce Davies (b. 1966, Bollington, United Kingdom) studied art at Northwich Art College, Bath College of Art, and Falmouth College of Art. He has exhibited in Frenchman's Creek – 'This Is Landscape!' Kestle Barton Rural Center for Contemporary Art, Helston Corwall, UK (2011) *and* Happidrome 4 – Space Hopper Travel Scheme (2010) Lizard National Nature Reserve Goonhilly Downs, Helston Cornwall, UK. *Since 2006, he has presented in the Peripherique lectures at the Henry Moore Institute, Leeds, UK. Davies has exhibited his work or performed in Leeds, Liverpool, Finland, Poland, and Norway. He is the curator and chair of the Basement Arts Project, Leeds, UK.*

Contemporary artists who engage with fire have the entire history of civilization at their disposal, since fire has played a role in civilization since its inception. Bruce Davies chose to revisit the extended period when fire was a product of local production and personal tending. *Cut/Stack/Burn* (2007) revises the vernacular conventions of fire-creation that prevailed in the Cornwall region of England prior to both Industrial Revolutions. Then, strategies for survival centered on the annual gathering of furze for fuel. Furze is a perennial evergreen shrub with small leaves and long thorns that thrives in rough pastures, heaths, and rocky places. Historically, Cornwall residents depended upon furze for fuel because few trees grow in this region. Furthermore, the plant catches fire easily and burns well because it has high concentrations of oil in its leaves and branches. Frugality was essential to maintain the energy equation between production and consumption because it takes several years for the shrub to mature. Ironically, this prudent, self-sustaining energy system was replaced by imprudent consumption of fossil fuels that take millions of years to develop. Davies explains, 'This all seemed to further underline how distant [...] this society is from mutually beneficial inter-relationships with the natural resources at our disposal.'[8]

In recent times, this essential resource has been demoted to a 'problem plant' that receives attention when it is being exterminated.[9] Thus, Davies orchestrated a project that reinvested furze with its traditional value by inviting people in the region to re-enact the intensive labors, conducted by hand or with simple tools that were once a routine component of their culture. Over four hundred local farmers, conservation employees, schoolchildren, artists, and others contributed to this communal artwork, titled *Cut/Stack/Burn*.[10] They began by harvesting the furze branches and binding them with natural twine to facilitate stacking. The resulting bundles are called 'fagots'.[11] After drying for several months, the participants transported the fagots to Tremenheere Sculpture Park, Penzance. Davies confesses that he compromised the historic authenticity of this arduous task by using tractors, instead of horses, to pull the silage trailers. Traditionally, the furze would then have been unloaded and stacked into neat piles that supplied each day's fuel for burning. Davies, however, used this opportunity to create a handsome circular sculpture that was fifty feet in diameter, and had six-and-a-half-feet thick walls. The structure ranged in height from eight feet to one foot, in order to accommodate the slope of the hillside. A section of green furze was inserted at the front. Davies explains:

> *Green furze burns exceptionally well and the central point at the front of the work was engineered and intended to be the point of ignition. It felt right to me that all project areas used for fuel gathering were physically represented in the actual work.*[12]

The culmination of these outdated activities occurred at precisely 6 p.m. on the eve of summer, March 25, 2007. That is when Davies and a few assistants donned flame-retardant boiler-suits and set the massive structure aflame by walking the perimeter and ceremonially igniting the furze bit by bit. Orange flames soared sixty feet into the air. At one point they formed a swirling tornado of flame. Then it was over. The massive energy store of furze that Davies and the public had toiled for months to construct was reduced to ash in a mere fifteen minutes. Davies explains the work's impact on the participants:

> *This contrasting final stage of the project was influenced by the flicking of switches, the pressing of buttons – power on demand – and our collective estrangement from the need and means to acquire our own energy at first hand and of how difficult, mundane and time consuming yet satisfying this activity can be.*[13]

Cut/Stack/Burn was not merely a demonstration of independent energy generation. It also offered an example of traditional heath-land management. Although the work appeared to violate good management by increasing CO2 emissions instead of reducing them, Davies deflected this criticism by explaining that the CO2 produced by this fire did not load the atmosphere with excessive carbon. Unlike the burning of fossil fuels, furze is native to the site. As a result, the fire merely returned the carbon that the furze had absorbed in its own growth cycle. There was no net gain.[14] Sharing this insight was an essential aspect of this project, as indicated by the work's subtitle: *Sustainable Sculpture in an Age of Climate Change.* Davies explains:

> *Cut/Stack/Burn was a visual platform for developing debate about the absence of sustainable approaches to the management of land and our natural resources and our ideas about current relationships with non sustainable [sic] energy sources in an age of climate change.*[15]

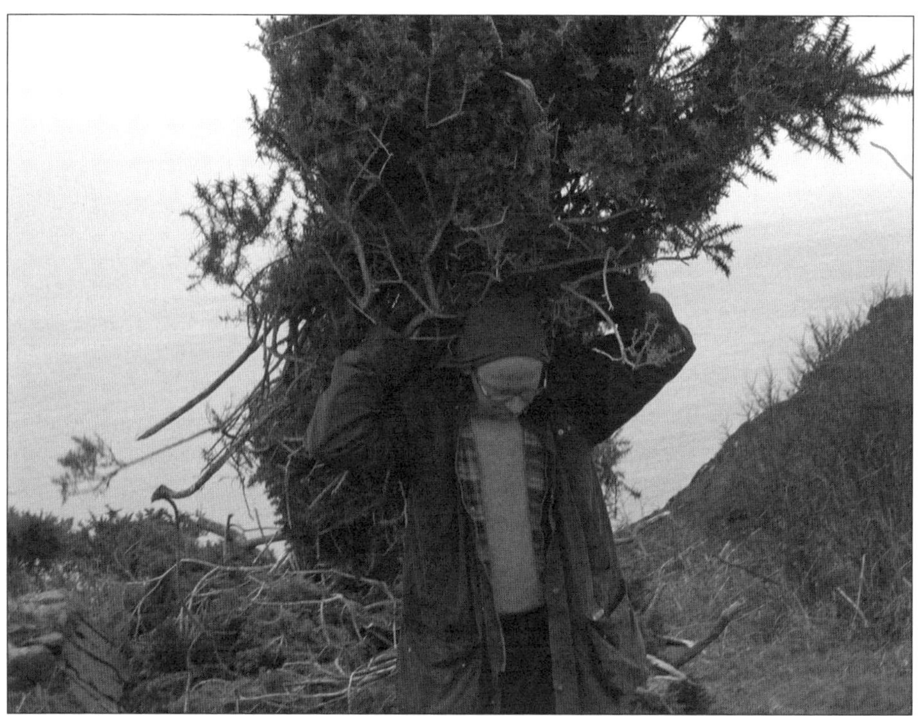

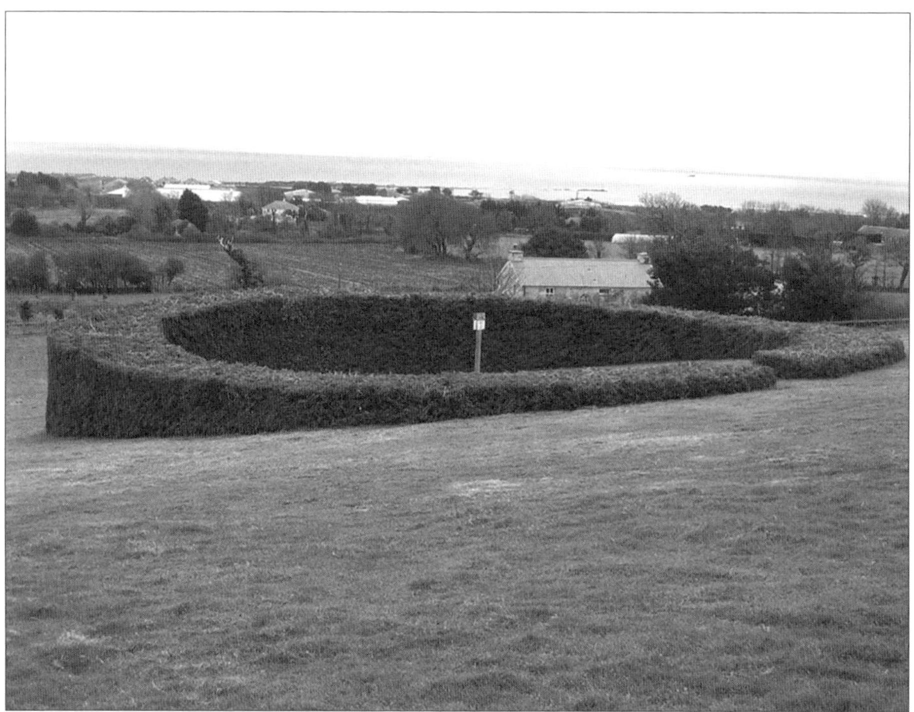

Figure 1a: Bruce Davies, *Cut/Stack/Burn* (2007). Bruce Davies carrying furze. Photo by Bruce Davies. Courtesy of the artist.

Figure 1b: Bruce Davies, *Cut/Stack/Burn* (2007). Prior to the burn. 2 m wide, 2.5 m high at its tallest point, and 15 m in diameter. Photo by Faye Stevens. Courtesy of the artist.

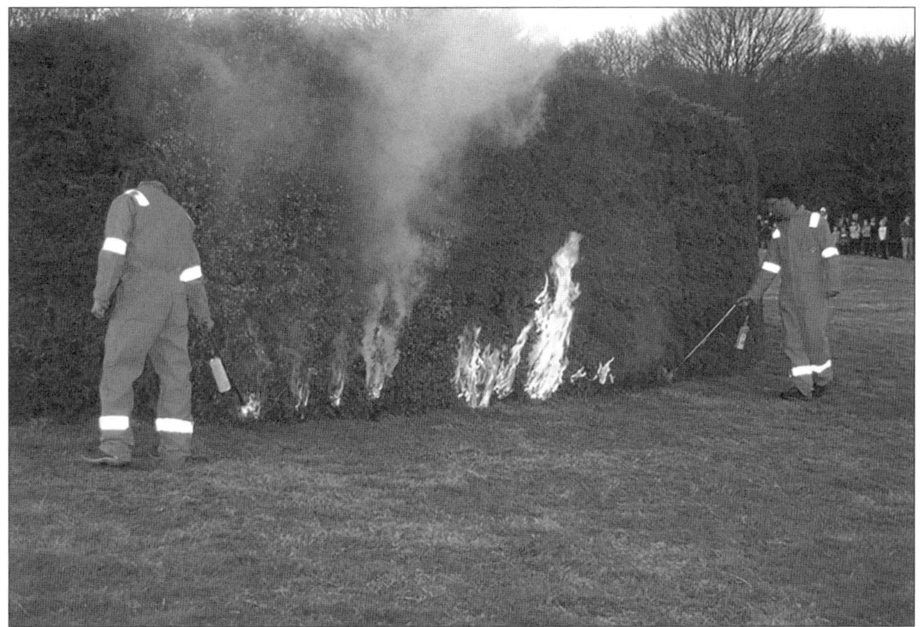

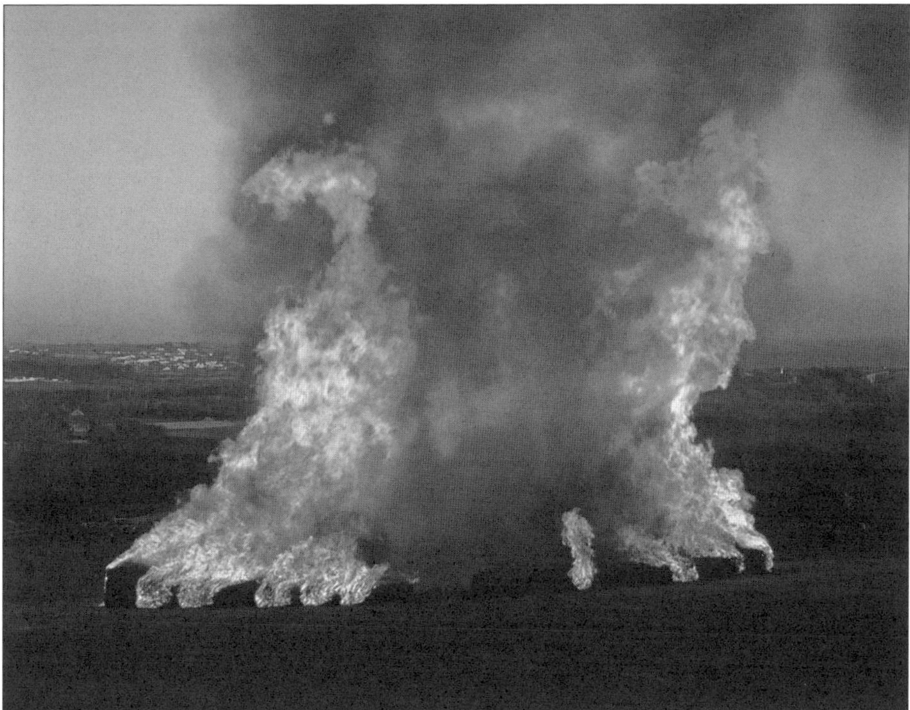

Figure 1c: Bruce Davies, *Cut/Stack/Burn* (2007). Ignition. 2 m wide, 2.5 m high at its tallest point, and 15 m in diameter. Photo by Bruce Davies. Courtesy of the artist.

Figure 1d: Bruce Davies, *Cut/Stack/Burn* (2007). Furze burning. 2 m wide, 2.5 m high at its tallest point, and 15 m in diameter. Photo by Fay Stevens. Courtesy of the artist.

Actual environmental benefits to the local eco system became apparent in the weeks following the burn as the scorched earth became refreshed earth that was reclaimed by populations of rabbits, ants, and beetles.

In addition to exposing local populations to pre-industrial methods of managing their health and gleaning its resources, participants discovered the enrichment that accompanies interactions with a native biofuel. Unlike fossil fuels that are accessed by switch-button technologies and delivered from distant, unknown locations, pre-industrial fires incorporated full-body engagements with gathering, stacking, and tending the fuel they consumed. In addition, it provided the communal bonding that develops from group efforts, an inevitable component of pre-industrial labors. The culminating reward for contributing time and effort was a product of heightened sensitivity to the variable heat of flames, the crackles and sizzles of burning branches, the smell of vaporous smoke, and the visual delights of flickering flames, billowing smoke, glowing embers, and crimson sparks. In all these ways, *Cut/Stack/Burn* added art to fire's traditional role as a tool for sustaining mutually beneficial relationships between necessity, conservation, and community.[16]

Futurefarmers is a rotating group of artists, activists, farmers, and architects, founded by Amy Franceschini, who use various media to destabilize logics of 'certainty'. This design studio was founded in 1995 as a platform to support art projects, a residence program, and to conduct research. Its projects often deconstruct existing food policies, public transportation systems, and rural farming rituals. Through this disassembly, new narratives emerge that reconfigure the principles that once dominated these systems. Futurefarmers are the lead artists of the Flatbread Society, a permanent public artwork in Oslo, Norway. Collectively, Futurefarmers's work has been exhibited at the Guggenheim Museum in New York, the Whitney Biennial in New York, MoMA (Museum of Modern Art) in New York, and the San Francisco Museum of Modern Art. Futurefarmers founder, Amy Franceschini, is the recipient of a 2010 Guggenheim Fellowship and the 2017 Herb Alpert Award for Visual Arts. Franceschini's collaborator, Michael Swaine, is an assistant professor in the 3d4d department at the University of Washington, Seattle.

Fire technologies have propelled human tooling power throughout history. Amy Franceschini and Michael Swaine, members of Futurefarmers, explored its development when they were invited to participate in the SITE: Santa Fe Biennial of the Americas in 2014. Because the exhibition was located near Los Alamos, New Mexico, the remote desert site where a plutonium bomb was first tested on July 16, 1945, the artists undertook a research project to learn more about that fateful event when humanity's tooling power rose to match the destructive force of meteorological storms like cyclones and tornadoes, and geological events like earthquakes and volcanoes. This human-induced blast was equivalent to eighteen thousand tons of TNT. The artists traced this unprecedented intensification of tooling technology back to its humble beginnings when early humans observed copper melting from the rocks that lined their fire pits, noting that metal would hold its new shape after it cooled. The artists' contribution to the Biennial embodied fire's protracted role in formulating humanity's most promising potentials and its most harrowing outcomes. The artwork took the form of three, apparently innocuous handmade nails! *For Want of a Nail* (2014) offers a synopsis of the compelling narrative of fire as a tool of material and social transformation.

The Futurefarmers mission statement expresses Franceschini and Swaine's intention to 'use various media to create work that has the potential to destabilize logics of certainty.'[17] In this instance, the artists destabilized logic by juxtaposing the monumental proportions of the force of an atom bomb with an inconsequential request made by J. Robert Oppenheimer (1904–1967), the renowned nuclear physicist who earned the dubious distinction of being 'the father of the atom bomb'. His achievement resulted in the atomic bomb blasts over Hiroshima and Nagasaki in 1945 that also exploded existing notions of human possibility and morality.

The artists discovered an inconsequential triviality while they were conducting research about the atomic blast – two inter-office memos written by Oppenheimer requesting a nail to hang his hat on! In the midst of concocting the most destructive weapon of all time, Oppenheimer's unfulfilled desire was for a single nail! Franceschini and Swaine capitalized on the disparity between this trivial request for personal convenience and the monumental effort to produce the first atomic weapon for mass destruction. They conveyed this narrative by using a different metal technology to produce each of the three nails. Two nails represent milestones in the historic evolution of humanity's use of fire. The third evokes the era when the atomic bomb was being developed. By presenting this trio of nails side-by-side, Franceschini and Swaine invite the public to join them in speculating which nail Oppenheimer would have chosen for hanging his hat – the one that employs technologies scaled to cooking knives and agricultural shovels, or the one that contributed unfathomable might to civilization's military arsenal.

The first nail evokes a prehistoric era before smelting was developed to separate useful metals from their combined states in ore. Minerals rarely exist in a pure state here on Earth. They are found in mixtures known as 'ores'. Nonetheless, metals were used by humans even before knowledge of extracting minerals from the stockpile of Earth's mineralogical treasures was developed. These early humans acquired pure iron from meteorites that arrived on Earth from outer space. Franceschini and Swaine re-enacted this early strategy by foraging for a fragment of the Canyon Diablo meteorite that slammed into Arizona fifty thousand years ago. This massive meteorite created a crater nearly a mile wide that was so hot, it vaporized upon impact. The artists collected some pieces that remain scattered in the crater. The enormous impact of the meteorite crash offers an extraterrestrial comparison regarding atomic bomb explosions. Indeed, it is estimated to have been 150 times more powerful than the atomic bomb dropped on Hiroshima, Japan.[18]

The artists also searched through history to identify an early technique to transform the foraged meteorite into a functional nail. The strategy they adopted was developed by the ancient Egyptians who developed the technique to create tools and weapons from the 'metal of heaven.'[19] Like these early metal smiths, the artists created the nail by heating the meteorite to soften the iron it contained, which enabled them to hammer the crude lumps of stone into a nail. Hot-working iron is known as 'forging'. The special hearth where this process is conducted is known as a 'forge'. Early blacksmiths were both honored and feared for possessing the seemingly supernatural powers to transform metal through forging. Although this esteem gradually diminished, this five-thousand-year-old process continued to be used for fashioning nails until the beginning of the nineteenth century.

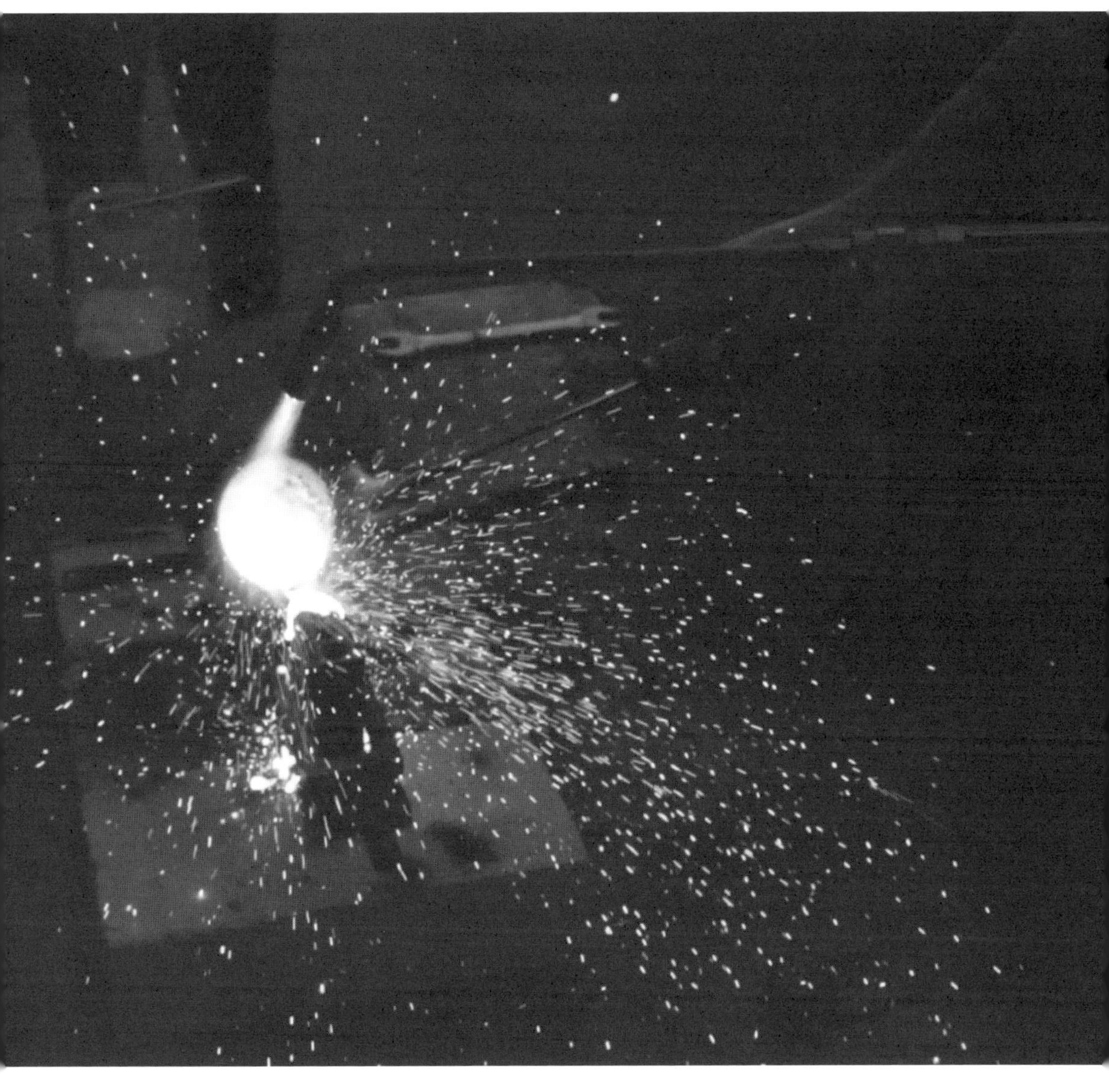

Figure 2a: Futurefarmers: Amy Franceschini with Michael Swaine, *For Want of a Nail* (Forging a nail from a meteorite) (2014), video. Video still by Jin Zhu. Courtesy of Futurefarmers.

An equally transformative technology is relayed in the third nail. Like the first nail, it references an enormous explosion, but this one was not caused by a meteorite; it was a human-induced atomic blast. Even Prometheus might have shuddered if he witnessed the incendiary horror of nuclear detonations erupting as 'firestorms' and 'fireballs' that engulf the surrounding air within seven-tenths of one millisecond from the detonation. The following is a vivid description of an atomic blast:

> *Violent, erratic wind drafts suck everything in their path into the fire. As in all intense conflagrations, radiated heat from the fire is hot enough to melt glass and many metals, and turn street tarmac into rivers of inflamed liquid.*[20]

This horrific accounting is integrated into the third nail because it is formed out of the residue of the infamous plutonium bomb test on the Trinity Site in New Mexico. The fireball from this blast was so hot it fused the desert sand into a jade green substance that was named 'trinitite'. The artists collected some pieces of trinitite and heated it in an electric ceramic kiln. Then they cast the liquid as a nail to support Oppenheimer's hat.

The meteorite nail and the trinitite nail mark an inception and a culmination of firepower. Between these landmarks in the tooling capacity of fire and metal, around 3200 BC, Mesopotamians developed 'casting'.[21] Casting is conducted by pouring metal that has become molten and liquefied into molds where it cools and solidifies. The technique greatly accelerated the pace and efficiency of human productivity because it enabled the production of multiple identical axes, bowls, tools, weapons, figurines, vessels, and many other objects. It is significant that Franceschini and Swain processed this second nail by casting it, even though they only needed one for Oppenheimer's hat. By incorporating this process into their artwork, they marked a tooling landmark that resonates today in the efficiencies of mass production.

The material utilized to create this nail references the cultural conditions in the United States just before the first atomic bombs were dropped. It was cast out of pennies minted in 1943, in the midst of World War II. They are made of zinc-coated steel because copper was being diverted for ammunition. Thus, the absence of copper, as much as the presence of zinc, invokes the magnitude and gravity of war-time mobilization.

The pennies also reinforce the war-time drama alluded to by this nail because the historic record of the first use of atomic firepower as a weapon includes the infamous coin toss between two US pilots. 'Heads you win' determined who would climb into the bomb bay on the US fighter plane and drop the atomic bomb that annihilated Hiroshima.

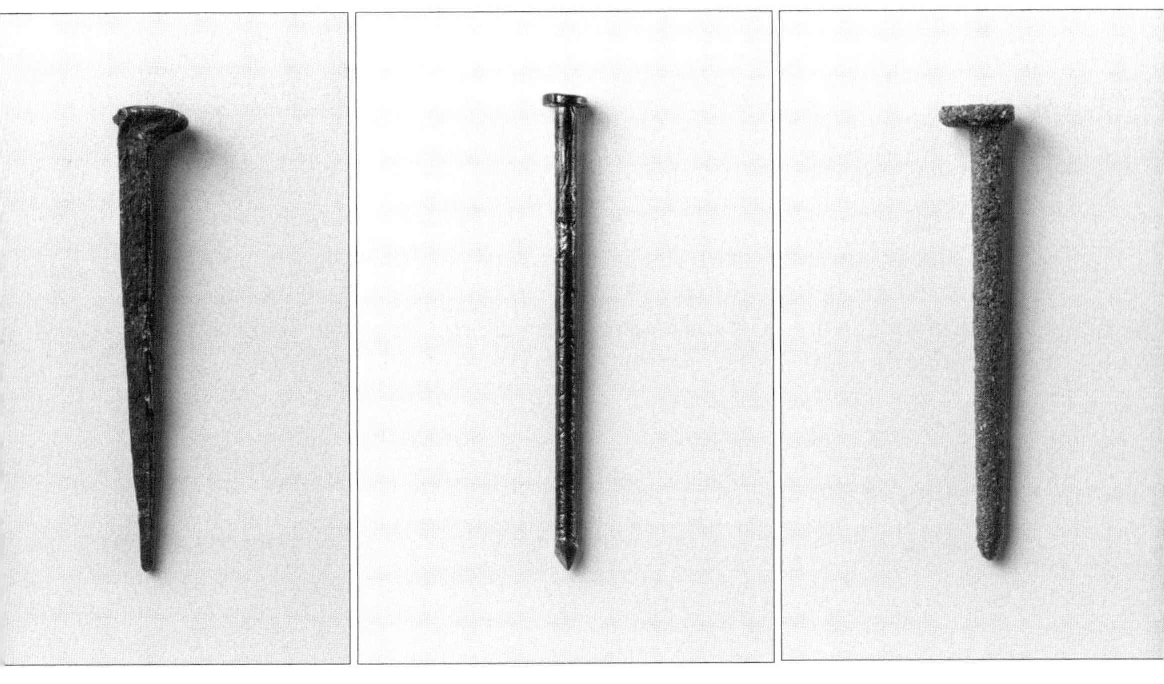

Figure 2b: Futurefarmers: Amy Franceschini with Michael Swaine, *For Want of a Nail* (2014), meteorite, 7 in. x 0.25 in. x 0.25 in. Courtesy of Futurefarmers.

Figure 2c: Futurefarmers: Amy Franceschini with Michael Swaine,, *For Want of a Nail* (2014), 1943 pennies, 7 in. x 0.25 in. x 0.25 in. Courtesy of Futurefarmers.

Figure 2d: Futurefarmers: Amy Franceschini with Michael Swaine, *For Want of a Nail* (2014), trinitite, 7 in. x 0.25 in. x 0.25 in. Courtesy of Futurefarmers.

All three nails that comprise this artwork are products of fire's ability to soften, liquefy, and fuse metals, and the human ability to exploit these transformations. As a trio, these nails dramatize contrasting outcomes. Franceschini expresses the uncertainties that attend to the three evoked by *For Want of a Nail*:

> *An initial impetus was to use this moment we found in the memorandum to draw attention to the many decisions that lead to the making of tools and the consequences of these tools we make and the many steps in the process. If we were to side step, second guess, hesitate, maybe we would have [....] cast different people or forged in a different direction. This small moment shined a light on something overlooked.*[22]

Fire Sinters

Markus Kayser

Markus Kayser (b. 1983, Hannover, Germany) graduated from London Metropolitan University with a BA degree in 3D design in 2008, and with an MA from the Royal College of Art, London in design products in 2011. He is a doctoral candidate at MIT Media Lab in Cambridge, Massachusetts. His current research combines technological and biological processes. His work has been exhibited at the Power Station of Art in Shanghai; the Centre Pompidou in Paris; the Design Museum in London; and the Museum of Modern Art in New York. In 2015, he received the Mediated Matter Emerging Voices Award by the Architectural League of New York.

Imagine hot springs and icebergs, protozoa and antelope, jungles and tundra. The raging fires of the Sun are the source of all these earthly wonders. More specifically, it is the Earth's proximity to these raging fires spewing from the center of the solar system that allows Earth to receive precisely the right amount of solar radiation to nourish and power our entire planet. Other locations in the cosmos gain either too much or too little of the sun's energy. The ideal amount received by Earth amounts to one-half of one-billionth of the Sun's energy output, just enough to account for the miracle of life within the barren expanse of the cosmos. Thus, no discussion of fire and tools is complete without including the Sun, the fire that ignites life.

Fuels from this distant inferno powered early humans' primary source of power – their muscles. Like us, they ate plants that trapped solar energy, and they ate animals that ate the plants. When fire was domesticated, the sun's combustion potential expanded to include flammable materials like brush, dung, and wood that is also fashioned by the photosynthesis of plants. Sun-dependence was not reduced when humanity added the power of wind and water to their catalog of energy sources. This is because wind power is a product of sun-driven atmospheric changes, and water power is dependent upon sun-driven hydrologic cycles. Not even the shift to fossil fuels reduced humanity's reliance upon the Sun. Oil, gas, and coal are deposits in the Earth's solar power bank that were made by plants that lived millions of years ago. That is why gasoline-burning cars, coal-burning power plants, and homes heated by natural gas are indirectly solar powered. These familiar fuel sources are currently being expanded to include wood sludge, railroad ties, pitch, municipal solid waste, agricultural waste, straw, tires, landfill gas, and fish oil. The eccentric items on this list share one important fact: they are all products of the Sun.

The Sun's reputation is extreme in two directions. Self-concerned citizens treat the Sun as a hazard by associating it with skin cancer, cataracts, ozone depletions,

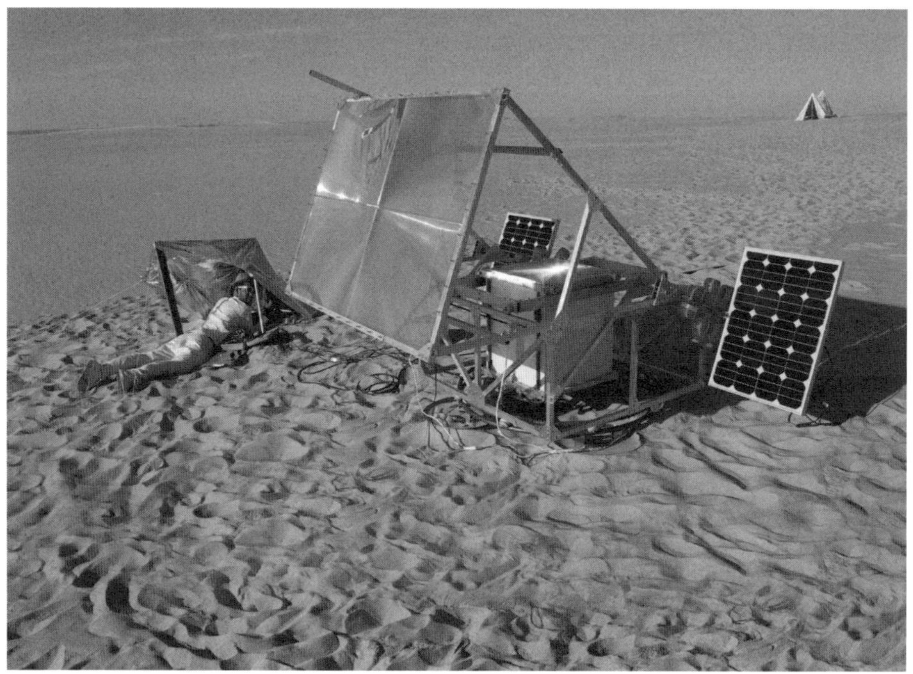

Figure 3a: Markus Kayser, *Solar Sinter* in the Sahara Desert near Siwa, Egypt (2011), photovoltaic panel, sun tracker, Fresnel lens, battery, controlling electronics, and silver tent 'office'. The solar sinter machine is driven by ReplicatorG open-source 3-D printing program. Photo by Amos Field Reid. Courtesy of the artist.

greenhouse gases, and global warming. Eco-concerned citizens regard the Sun's role in the biological and geological processes that sustain life, and they elevate the importance of the Sun as a means to reduce fossil-fuel dependence. The potential of solar energy is disclosed by Solar Impulse 2 – a prototype airplane that made history in July 2016 by completing a round-the-world trip powered only by the Sun![23]

Currently, solar power is blazing a new frontier of tooling right here on Earth. It takes the form of an ingenious application of 3-D printing technology that Markus Kayser developed in 2011 when he was still a student at the Royal College of Arts in the United Kingdom. The device is named *Solar Sinter.*[24] Sintering is the process of forming a solid mass of material by heat or pressure without melting it into a liquid state. Metals, ceramics, and plastics are among the materials that are subjected to sintering in manufacturing processes. Kayser's sintering device resembles a 3-D SLS (three-dimensional selective laser sintering) printer, but it avoids the double environmental jeopardy that conventional 3-D printers pose: *Solar Sinter* does not rely upon powdered resins that contain potentially harmful ultrafine particles

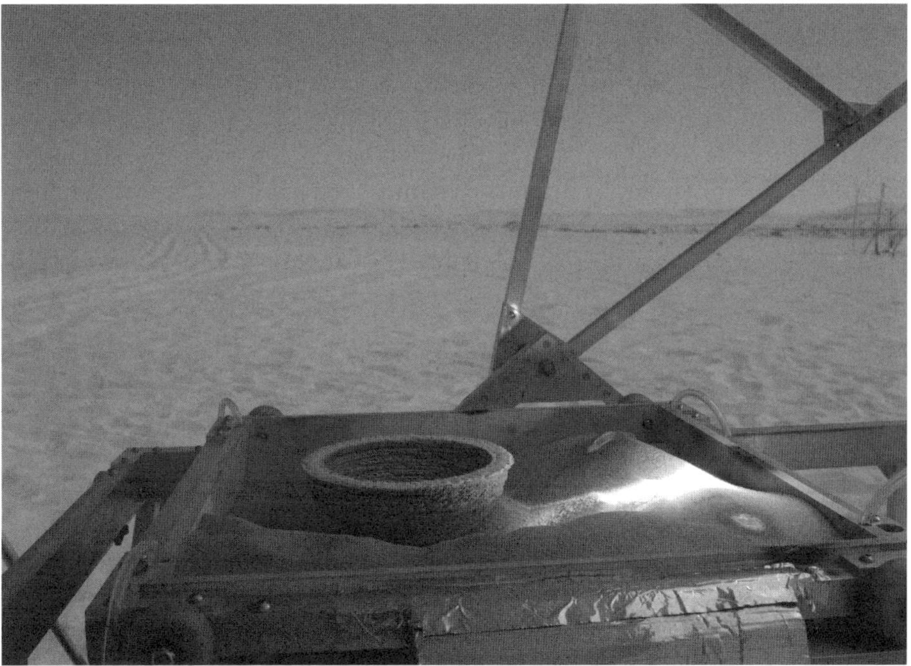

Figure 3b: Markus Kayser, first object printed from a 3-D file with the *Solar Sinter* in the Sahara Desert near Siwa, Egypt (2011), photovoltaic panel, sun tracker, Fresnel lens, battery, controlling electronics, and silver tent 'office'. The solar sinter machine is driven by ReplicatorG open-source 3-D printing program. Photo by Amos Field Reid. Courtesy of the artist.

and produce toxic fumes; moreover, it does not consume significant quantities of electricity to melt the materials these printers heat, extrude, and deposit. Instead, *Solar Sinter* coaxes the Sun to perform a sequence of actions off-grid:

- Raw sunlight is concentrated as it passes through a Fresnel lens and heats sand to a melting point.

- Digital designs are downloaded from a computer onto a simple memory card and transferred to the *Solar Sinter*. The sinter follows these instructions by moving the sand box at a speed calibrated to the temperatures needed to melt the sand.

- An electronic sun-tracking device moves the lens vertically and horizontally while rotating the entire machine about its base. It tracks the Sun as it moves across the sky to maximize light capture and optimize heat generation throughout the day. These rays are concentrated by a Fresnel lens.

- Two photovoltaic panels provide electricity to charge a battery, which in turn drives the motors and electronics of the machine. The motors move the box along the computer-controlled path.

- The 3-D printer passes the lens over the bed of sand. The top layer heats and melts into glass. Once it cools in the desired shape, a new layer of sand is sprinkled on top and the process repeats. An object is gradually built up, layer by layer, within the sand box.

The Sun also powered the video that recorded the dramatic transformation of the inert sand as the intense heat caused it to soften, bubble, then cool and harden into a desired shape. Kayser describes the reciprocity of the process by stating:

> *The primary energy to produce the heat to melt the sand is purely concentrated sunlight and the secondary energy is converted by photovoltaic cells producing electricity needed for electronics and motors to drive the machine. The concept is not to harness sunlight, convert it to electricity, store it, and then turn it into heat again. Instead, Solar Sinter's concept is to directly use the sunlight where it occurs [...] in the given environment and the cycle of day and night. This allows for the fabrication of products using advanced technology that lives symbiotically with nature.*[25]

This symbiosis is evident in the solid objects that are produced by the *Solar Sinter* that lack the slick perfection of typical laser-fashioned objects. Their crude appearance results from variations in solar intensity that varies with the time of day and weather. Kayser recognizes that his machine is more a breakthrough than an accomplishment. He states, '....the creation of artifacts made by sunlight and sand is an act of pure experimentation and expression of "possibility", but what of the future? I hope that the machine and the objects it created, stimulate debate about the vast potential of solar energy and naturally abundant materials like silica sand. These first experiments are simply an early manifestation of that potential.'[26]

Because the ceramic objects this device produces are extremely stable, Kayser refers to them as 'an act of pure experimentation and expression of "possibility".'[27] He describes the possibility he envisions by stating, 'Printing directly onto the desert floor with multiple lenses melting the sand into walls, eventually building is ecture in desert environments, could also be a real prospect. Experiments in 3D printing technologies are already reaching towards an architectural scale and it is not hard to imagine that, if partnered with the solar-sintering process demonstrated by the Solar-Sinter machine, this could indeed lead to a new desert-based architecture.'[28]

In 2014, a group of postgraduate students, led by their tutor and supervisor, Qiu Song, accepted this challenge. They explored the possibility of using *Solar Sinter* technology to manufacture building components. Their efforts were channeled along previously unexplored program paths suggested by this technology, constructing metropolises in uninhabitable desert lands where sun and sand were plentiful. Two looming crises related to global warming could be averted. One involved relocating populations displaced by rising waters and political turmoil. A related problem exacerbated this problem – habitable land was shrinking. Where will the homeless go? This team of Chinese designers imagined the potential of sintering with sand and sun to satisfy these looming requirements. They are as follows:

> **Qiu Song** *(b. 1964, E Zhou, Hu Bei Province, China): design director and chief designer.*

> **Kang Pengfei** *(b. 1988, Beijing, China): designer in charge of the parametric design and modeling.*

> **Bai Ying** *(b. 1987, Beijing, China): designer in charge of graphic design.*

> **Guo Shen** *(b. 1989, Wuhan, China): designer in charge of form.*

> **Ren Nuoya** *(b. 1990, Nanjing, China): designer in charge of research, data collection, and text translation.*

The design team envisioned transforming uninhabitable land (deserts) into cities utilizing Earth's most abundant resource (the streaming energy of the sun) to transform an abundant, neglected material (sand) into a stable construction material. They named the futurist towers that embody this ingenious solution *Sand Babel Towers* (2014)[29], a reference to the Biblical Tower of Babel that would be so tallit could reach all the way to Heaven. Artists throughout the ages have imagined the Tower of Babel as a soaring spiral. *Sand Babel* adds to this Biblical narrative the promise of a miraculous transformation of grains of sand into a sustainable metropolis in the desert.

By locating the towers in the deserts of the Middle East and North Africa, entire city complexes could be manufactured on-site with an abundant, untapped local resource – sand. Because this strategy eliminates construction waste and the costs of transportation and energy, it introduces an affordable, 'green' construction option into vast unoccupied territories.

The team supported this visionary concept by developing a functional program that is comparably futuristic. The solution they envisioned takes the form of an intelligent network of skyscrapers distributed across a desert landscape. The swirling shapes of the towers are not aesthetic affectations. Their spiral skeleton structure is modeled after the natural forms of desert rocks. By directing air

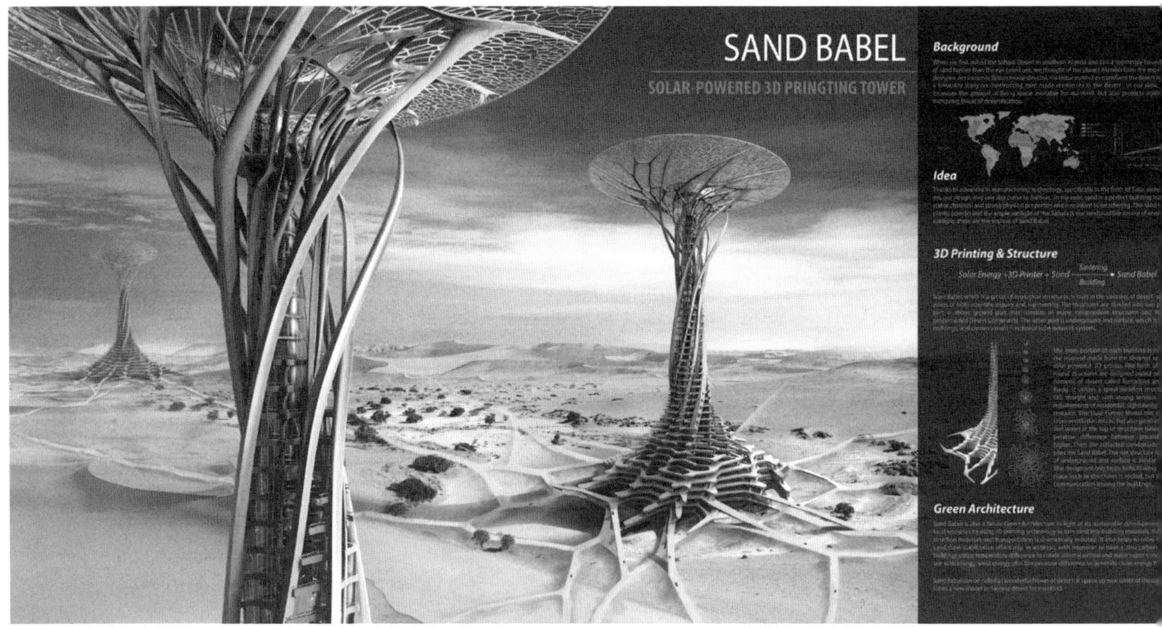

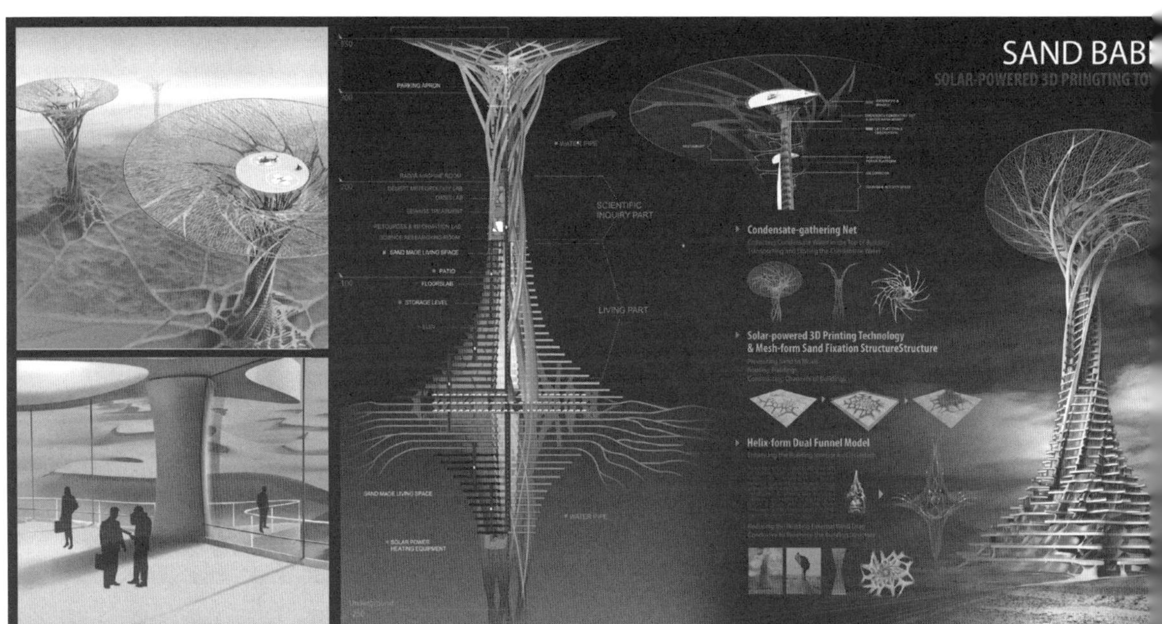

Figure 4a: Qiu Song, Kang Pengfei, Bai Ying, Guo Shen, Ren Nuoya, *Sand Babel Towers* (2014). Sintered sand. Courtesy of the artists.

Figure 4b: Qiu Song, Kang Pengfei, Bai Ying, Guo Shen, Ren Nuoya, *Sand Babel Towers* (2014), diagram. Courtesy of the artists.

flow, a sufficient temperature differential is created to condense air vapor, and thereby produce enough water for use by the building's occupants. Furthermore, these structures are topped with sun-tracking mirrors that focus and concentrate sunlight to generate clean energy. In this manner, the towers' self-powered machines for living off the grid function as self-contained desert ecosystems.

Deserts currently comprise one-third of the land's surface area.[30] Desertification is increasing due to the degradation of ecosystems by unsustainable farming, mining, overgrazing, and the clear-cutting of land.[31] The sintering technologies that Kayser introduced, and that the Chinese designers expanded, do not merely offer a promising solution to a growing crisis; they expand the primeval dependence of humans upon the fire of the Sun as a tool to defend and protect.

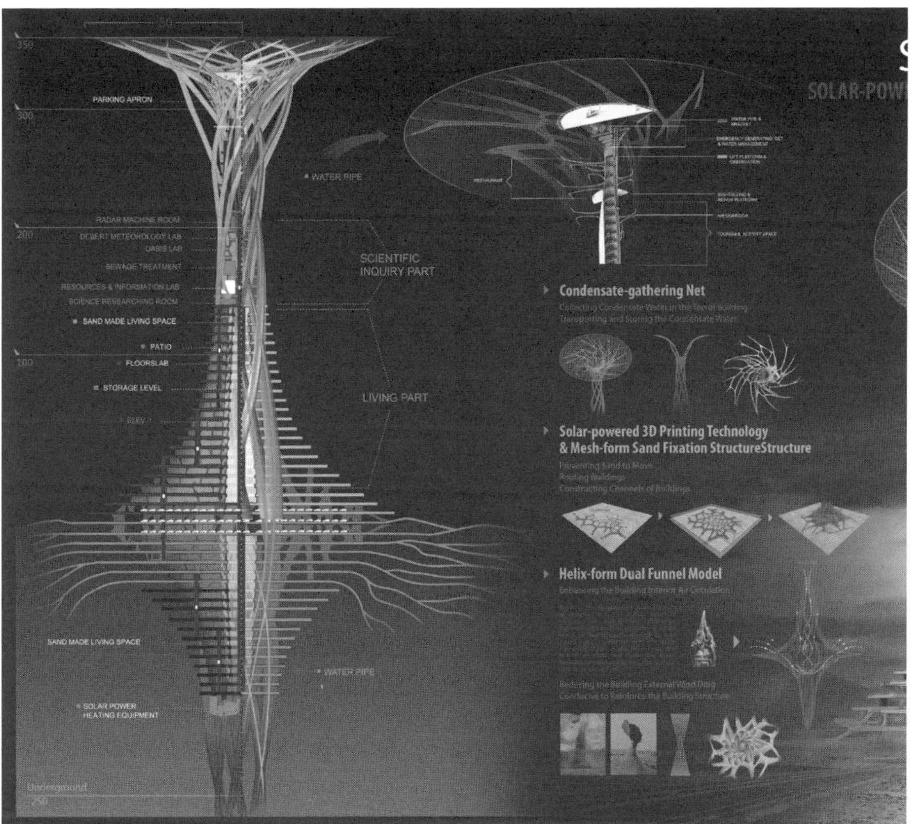

Figure 4c: Qiu Song, Kang Pengfei, Bai Ying, Guo Shen, Ren Nuoya, *Sand Babel Towers* (2014), detail of diagram. Courtesy of the artists.

Fire Cooks

Spurse

Spurse is a design collective founded in 1998 in the United States that focuses on social, ecological, and ethical transformation. The collective's name identifies their practice: 'spur' refers to an indirect path that offers new solutions and possibilities. The members, who prefer to be known only as Spurse, span the fields of science, art, and design. They replace their individual biographies with their shared mission: to empower communities, institutions, infrastructures, and ecologies with tools and adaptive solutions for system-wide change. The collective co-develops pirate projects, such as the Eat Your Sidewalk Cookbook[32]*.*

Cooking, according to the definition devised by Spurse, is any action that utilizes heat to transform raw food. The food preparations conducted by its members contrast the sterile efficiencies of contemporary technologies with the lively interactions of early humans. These ancestors either ate raw foods they foraged as they encountered them, or they ate foods 'cooked' through the natural process of rotting. Spurse explain, 'Rotting is humans' first exposure to cooking. Our pre-human ancestors were gifted with leftovers from kills by predatory animals like lions and wolves. We can imagine our ancestors eating the partially decomposed meat that was left. They were scavengers.'[33]

Eating changed when cooking involved flame, which commenced almost two million years ago. The first form of cooking with fire involved roasting meat above open fires.[34] Gradually cooking strategies expanded to include steaming food over hot embers by wrapping it in wet leaves, toasting wild grains on flat rocks, and using shells, skulls, or hollowed stones to heat liquids. These culinary advances revolutionized community as well as diet. Unlike foraging, which is a solitary endeavor, cooking food with fire requires communal cooperation. People spend time together to gather fuel, prepare the fire, oversee cooking, and share the meal. They linger around the hearth to tell stories and sing songs. In ancient Rome, fire became so intertwined with home that travelers carried glowing embers from their hearth with them to retain the comforts and security it had afforded them at home.[35]

The culinary role of fire changed again when cooking moved indoors where it was confined in stone fireplaces, and later in iron and steel stoves. Gradually, flame was replaced by gas released from metal piping and electric currents running through wires in stoves. The microwave oven represents a radical decommissioning of the socializing opportunities and sensual pleasures that culinary flame provides. Despite their contrasting technologies, in all of these contexts cooking involves heat-induced material changes. Butter melts. Starches gelatinize. Eggs coagulate. Meat shrinks. Bread browns.

Few of these examples of humanity's alliance with food and heat were represented in preparations for the *Eat Your Sidewalk* dinner that Spurse orchestrated in 2017 in upstate New York.[36] This interactive art event ventured into the unfamiliar zone of dining in which combustion occurs, heat is produced, and food is cooked – but all these changes occur in the absence of flame. Flameless foodways typically emerge in regions that lack fuel for burning, such as frozen tundra. Serving eggs roasted in the heat generated by decaying organic matter exemplifies one of the eating adventures that the artists offered to twenty guests. They served tiny portions of twenty dishes to a sustainability expert, a builder, an economist, a farmer, a musician, a magician, a poet, and several artists. Spurse announced at the outset of the meal, 'We don't prepare food. We initiate heat exchanges.'[37]

The elemental heat-generating processes they employed predate the use of fire. They even predate the evolution of humans. They originated when life on Earth first appeared over three billion years ago. That is when 'combustion' began to occur in living cells. This process, known as metabolism, involves the cells' ability to combust fuel by burning calories. Over eons of time, two forms of metabolism evolved enabling organisms to derive the energy they require to maintain life functions. One is respiration. The other is fermentation. Spurse exploited the culinary potential of both.

Respiration is the form of metabolism that occurs in the presence of oxygen. It functions like fire by breaking the chemical bonds of fuel, thereby producing heat and water vapor. Plants generate heat as a secondary process of their cells' respiration. Animals generate heat by breaking the chemical bonds of the foods they consume. In this manner, all forms of life burn fuel in the form of carbohydrates. Like fire, respiration depends upon the presence of oxygen. Fermentation, the other metabolic process, occurs when oxygen is not present. Sugars serve as the fuel for this type of metabolic combustion. Fermentation is a function of bacteria and fungi that transform sugars into acids, along with gases or alcohol.

By harnessing the heat generated by these metabolic processes, the *Eat Your Sidewalk* dinner was not cooked with usual sources of heat like gas, electricity, and charcoal. Instead, it was cooked by living microbes. In this manner, the eating adventure prepared by Spurse exploited the tooling potential of flameless fires.

All skilled cooking involves balancing temperature and duration. In this instance, the ideal conditions were determined by the populations of microorganisms that Spurse enlisted to process the foods on the menu. Thus, Spurse's kitchen skills not only addressed the amount of warmth delivered to a food and the length of exposure; it included feeding microbes and managing their wastes. Spurse delicately orchestrated these variables to coax out the essence of each ingredient. The artists explain that their culinary interactions 'released the "agency" of each

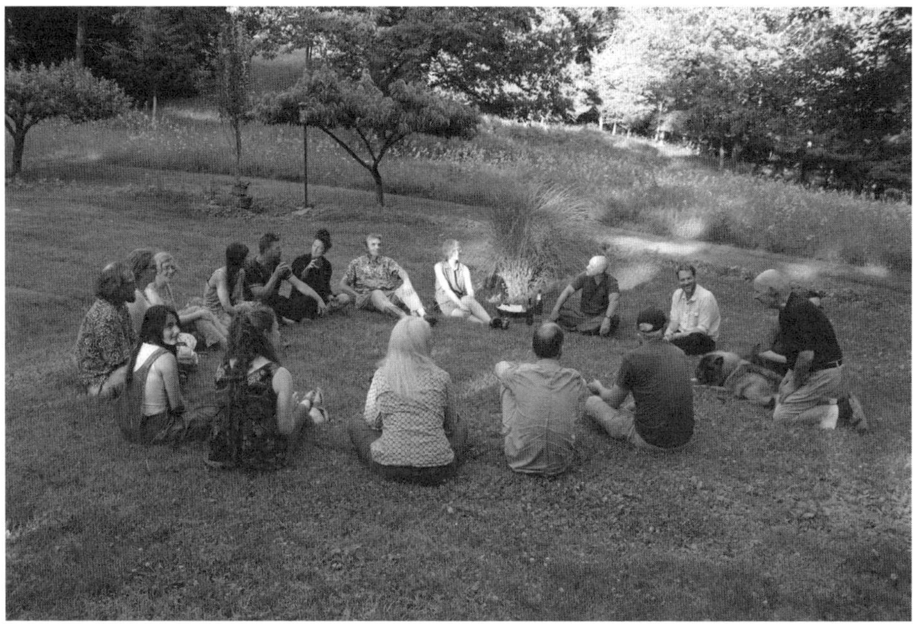

Figure 5a: Spurse in collaboration with Linda Weintraub, Ruth Borgenicht and SUNY New Paltz MFA students: Amanda Heidel and Emma Chandler, *Eat Your Sidewalk* dinner (2017), guests seated in a meadow. Photo by Lindsey Guile. Courtesy of the artists.

ingredient.'[38] Such attention paid to invisible processes that occur on microscopic scales reveals the artists' conviction that people do not have a monopoly on 'agency'; even invisible microbes exert influence on their surroundings. The artists explain, 'Our starting point is to ask about anything, "What is it doing?" instead of "What is it?" "Doing" is different from "being", because through doing you become embedded in a system.'[39]

The cultivation of 'embeddedness' explains why gathering ingredients for this banquet not only excluded purchased foods and fuels (because both originate in distant places); it also excluded the vegetables growing in the garden on the property (because they were introduced and cultivated). Not only were the ingredients on this adventuresome menu wild and local, the microbes that processed these foods were also foraged on-site. In this way, the dinner introduced guests to eating 'of' place, as well as 'in' place. In all these ways they answered

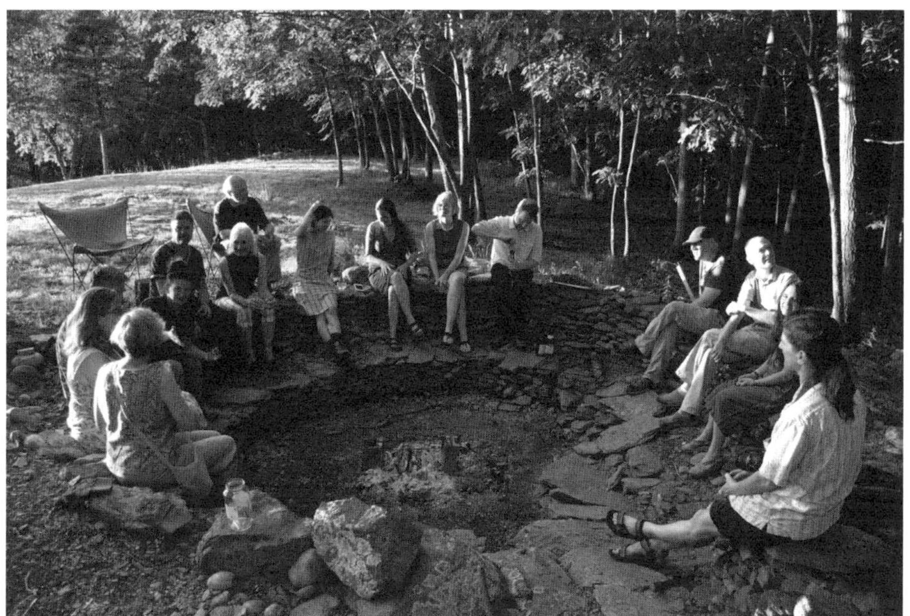

Figure 5b: Spurse in collaboration with Linda Weintraub, Ruth Borgenicht and SUNY New Paltz MFA students: Amanda Heidel and Emma Chandler, *Eat Your Sidewalk* dinner (2017), guests seated around a fire pit. Photo by Lindsey Guile. Courtesy of the artists.

the question posed and answered by Spurse: 'How can we actively be of that system? Our interest is in place-making.'[40]

Spurse staged the meal to optimize these connections with place by dispensing with dining tables and chairs. The first course was eaten seated on the ground in the meadow. Then guests were relocated indoors where they sat on the floor. Dessert was served around an outdoor fire pit. Flat stones and irregularly shaped ceramic forms replaced dinnerware, and silverware was rejected in favor of eating with the hands. The artists explain:

> *We didn't sit in chairs because chairs separate us from the environment. It was important that we used our hands because they are covered with bacteria. We actively added our own biome to our food. These bacteria joined the plants we ate. We undermined the procedures that produce the upright scanning brain with procedures of entanglement. The basic*

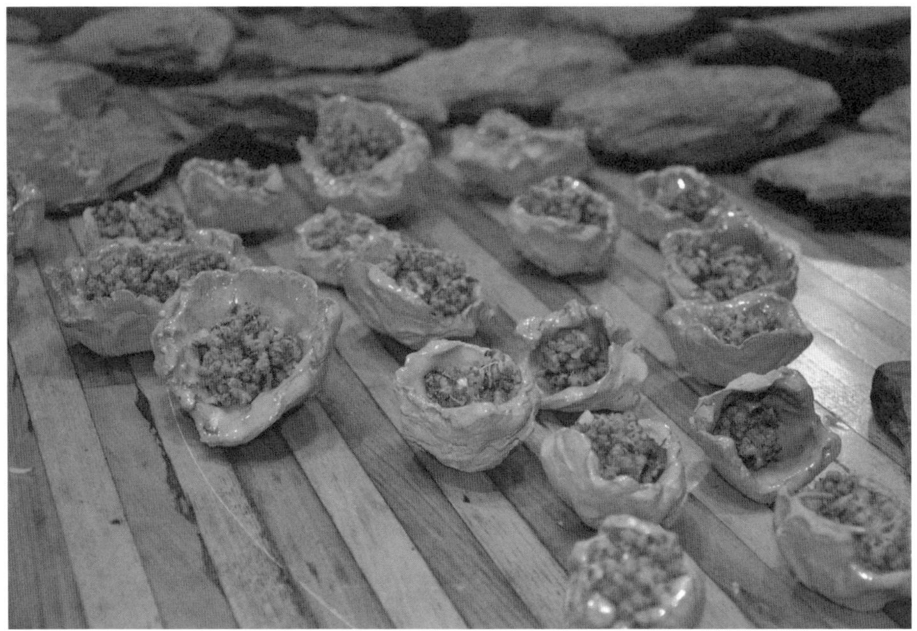

Figure 5c: Spurse in collaboration with Linda Weintraub, Ruth Borgenicht and SUNY New Paltz MFA students: Amanda Heidel and Emma Chandler, *Eat Your Sidewalk* dinner (2017), wild fermented apple cider vinegar pickled milkweed blossoms, wood sorrel pods, and purslane served in algorithmically generated ceramic ware. Photo by Lindsey Guile. Courtesy of the artists.

> *way things are in the world isn't conceptual or intellectual. It is through affect, feeling, sensing. Materialism pushes us into the sensual world of stuff.*[41]

By constructing an eating experience that necessitated sensual exchanges, the artists intentionally undermined key western assumptions about functional expediency and cleanliness and objective knowledge. Instead of seeking to discover what the world is, Spurse explores how humans behave in the material world. If a fundamental aim of science is prediction, the ultimate aim of Spurse's methods is to experience, in as intimate a manner as possible, how things change as we participate in ecosystems.

What follows is a list of heat-generating strategies that were utilized to cook the foods served in this atypical meal. Significantly, the artists avoided the use of heat they refer to as 'Promethean' that is manifested in conventional cooking technologies that depend upon harnessing fire's power. This assertive use of fire

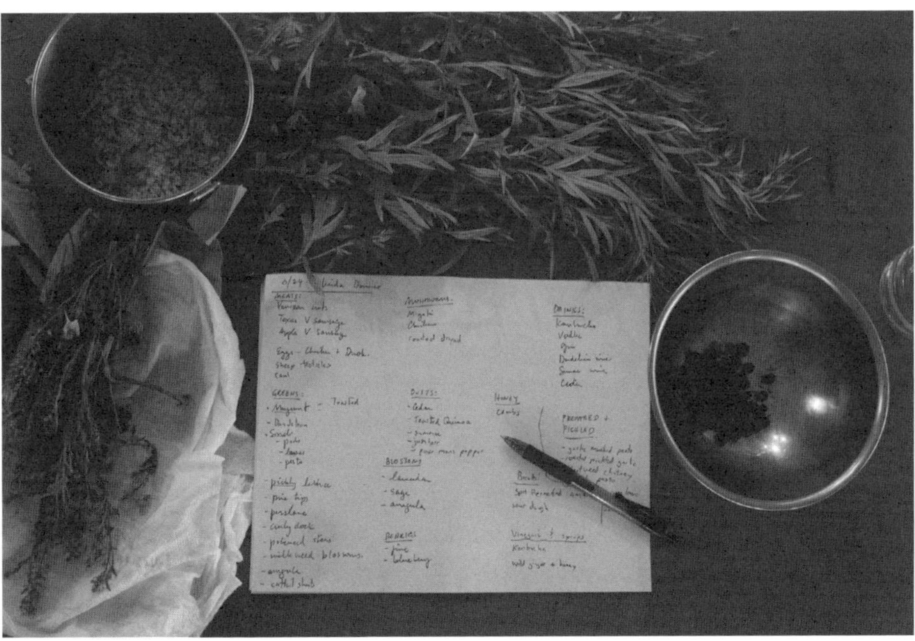

Figure 5d: Spurse in collaboration with Linda Weintraub, Ruth Borgenicht and SUNY New Paltz MFA students: Amanda Heidel and Emma Chandler, *Eat Your Sidewalk* dinner (2017), dinner preparation: June berries, wild arugula, poor man's pepper, pickled wild garlic aerial bulbils. Photo by Lindsey Guile. Courtesy of the artists.

separates and distinguishes humans from other forms of life. Instead, the artists note, 'Our question for the dinner: how could we use heat to move out of our historical paradigm and towards an alternative model (of relational distributed agency). We term this alternative "the multi-species commons".'[42]

Photosynthesis: The giant nuclear fusion reactor at the center of the solar system was honored at the outset of the dinner as the source of all metabolic functions conducted by living entities. Plants are the originators of this Earth-wide system of energy exchanges. They accomplish this feat by manufacturing their own food, glucose, through the process of photosynthesis. Thus, the meal commenced by honoring plants and their ability to derive heat energy from the sun. Guests were assembled in the meadow and instructed to pick and taste a variety of plants growing around their feet that would become ingredients for the salads on the menu: plantain, lambsquarters, dandelion, etc. These plants, which are commonly demonized as weeds and invasives, were welcomed by the artists whose connections with place are inclusive, not discriminatory. They comment,

'We don't check the passports of animals and plants to see if they belong.'[43] Spurse displayed its appreciation of weeds by preparing knotweed chutney and garlic mustard pesto – two plants that appear prominently on lists of unwelcome invasives.

Boiling: Blanching pokeweed demonstrates that Spurse also rejects western fears that apply to plants considered to be poisonous. Pokeweed, for example, is typically avoided because ingesting it can cause convulsions, nausea, heart block, and low blood pressure. Instead of omitting it from the menu, Spurse served small portions after dipping the leaves in boiling water and rinsing them several times to reduce their toxins. The artists explain their willingness to serve a known toxin, 'A certain amount of anything can cross the threshold of safety and will kill you. That means we must negotiate quantity.'[44] Boiled pokeweed was served with smoked cat tail and fermented aerial garlic, and dressed in vinegar. Everyone delighted in the taste and texture. No one became ill.

Fermenting: Fermentation is a metabolic process occurring mostly in yeasts that convert sugars into acids and gases. While these single-celled, microorganisms regularly perform these tasks in the guts of living organisms, Spurse invited them into the kitchen. The artists captured wild yeasts by placing a mixture of water and sugar in a warm place, which attracts the populations of invisible spores that already occupy that environment. The heat that the yeasts and bacteria generate to conduct their own life processes was utilized to cook food for humans. Spurse explain, 'This process of livingness equals cooking. It shows that cooking is not always about killing anything. Heat exchange can allow life's continuation.'[45] The fermented foods included on this menu were sumac wine, dandelion wine, apple cider, vinegar, and several forms of mushrooms. The highlight of the fermented foods on the menu was sourdough bread activated by the microorganisms in the artists' spit. Spurse explains the advantages of spit over commercial yeasts: 'Commercial fermentation only has yeast, and only one variety of yeast. It is like a little mono culture that ensures a predictable industrial product. Spit provides a whole variety of yeasts along with bacteria. Each one does a job.'[46] By introducing a diversity of living microbes to conduct food preparation, Spurse defies the assumption that microorganisms are dangerous 'germs' and that precautions must be taken to banish them. Instead, they are welcomed as creative partners.

Aging: Otherwise known as controlled rotting, aging is the process of maturing a food to enhance its digestibility and flavor. This transformation is performed by bacteria and fungi. Although rot is associated with repulsive sights and foul odors, even grocery-store beef is aged to break down the tough muscle tissues and make it tender enough for human consumption. Spurse refers to this natural process as 'excessively collaborating with the ecosystem.'[47] For this dinner, the artists aged the venison by coating it with salt and natural preservatives that included foraged cedar, juniper, and sumac berries. The artists explain, 'Aging is a prime example of livingness of cooking because several forms of life join in the process.'[48] The menu included aged raw venison served with arugula, wood sorrel pods, pumpkin seed oil, and homemade vinegar.

Composting: Several days before the dinner, the meadow was mowed and the grass was heaped into a large pile to allow time for microorganisms to activate the decomposing process that would transform the grass into compost. Eight hours before mealtime, raw eggs in their shells were inserted into this heap of grass that had reached a temperature of 160 degrees, ideal for roasting eggs. At mealtime, the eggs were retrieved, shelled, and served without embellishments so the guests could savor their creamy texture and succulent taste. The artists delighted in this process because

> *it resembled the incubation of eggs by the warmth of the hen's body, a quintessential example of cooking with livingness. In fact, the process of cooking eggs at a low temperature for a long time also resembles the digestion process that occurs in our stomachs. As in the pile of grass, our intestines warm things up to be broken down by bacteria.[49]*

Pickling: Pickling is another form of fermenting. Fast pickling is the process of preserving food with vinegar. Traditional pickling uses salt to remove the liquid in the food being pickled. The transformation of food converts carbohydrates without the need for oxygen. This process depends upon lactic acid bacteria to produce acid, which inhibits the growth of undesirable organisms. The pickled foods on the menu included kombucha made with wild honey, and pickled dandelion leaf served with fermented wild garlic that was sprinkled onto the tartar venison.

Salivating: The guests' mouths served as the heat activators when the artists offered each a morsel of frozen, raw venison topped with little balls of frozen herbed butter. Because the venison defrosted on the tongue more slowly than the herbed butter, the meat was gradually bathed in a savory, buttery sauce. Chewing and swallowing was delayed until both foods attained body temperature. Spurse described this manner of eating to their guests as 'your aliveness, your energy activated the transformation. You feel that on your tongue.'[50]

Because cooking helps shape culture, making tools, food, and fire integral to the context where they are consumed provides a tangible example of 'going local' to audiences mostly familiar with 'going global'. By activating living sources of heat, all seven cooking techniques used to prepare the *Eat Your Sidewalk* dinner also expand conceptual considerations of food and functional interactions with fire. Spurse explain:

Many active environmentalists ignore connection. They are interested in saving the whales and the Amazon. This is important, but it is about saving another place. They are not becoming actively entangled with their place. We wish to provoke people into speculating with us, not intellectually, but through doing things, to consider inventing new cosmologies, through new practices, that don't separate nature and culture. Nature is stuff. Culture is what stuff means to us. Those values shape us as we shape them.[51]

Spurse developed five material principles to guide their rapport-building repertoire of exchanges with food:

- Everything is 'embodied' – it has a concrete form.

- Everything is 'enacted' – it has the power to perform.

- Everything is 'embedded' – it is part of a system.

- Everything is 'extended' – it influences other beings and other things.

- Everything is 'cosmological' – it defines a cultural context.

Your opinion: What's next?

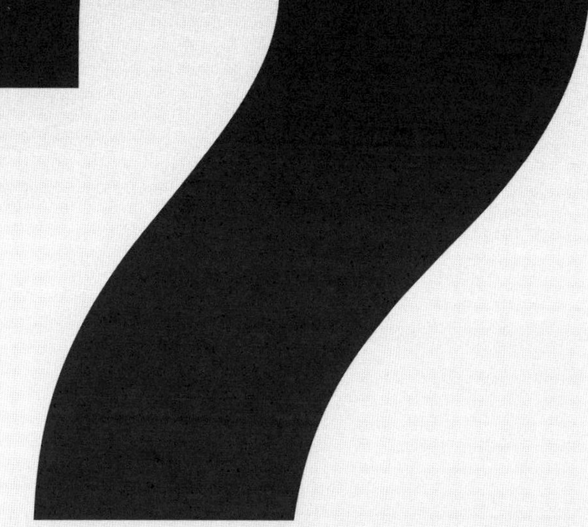

Reader Interaction

The Tooling Versatility of Fire

The utilization of fire as a tool to alter the form, chemistry, and qualities of materials to satisfy a need is commonly recognized as a defining attribute of humans. Enumerating these tooling strategies reveals the versatility of fire. At the same time, it testifies to the ingenuity of humans who discovered these uses. Currently, fire-related ingenuity seems more likely to address fire prevention, suppression, and extinction than its functionality. In order to revive our ancestors' appreciation of fire, a list of primeval tooling actions with fire was prepared. While many are not applicable to current circumstances, there are some that can be revived. Please activate one of the following functional interactions with fire that was practiced by your distant ancestors, or add one that is missing from this list.

Primeval Uses of Fire as a Tool: Heat/Flame

- Straighten spindles and shafts.

- Cauterize wounds.

- Remove seed chaff.

- Modify wood, stone and bone for use as tools/handles.

- Soften materials to bend them and to make them easier to modify with tools.

- Make scarification tattoos.

- Melt hardened substances for use.

- Create ash (alters soil composition, polishes gemstones or wood).

- Transform wood into charcoal (freshens air, mulch, prevents tools from rusting, pigment, medicine, drawing device).

- Produce smoke (keeps leather flexible, preserves meat, controls pests, transmits signals, waterproofs leather, flavors foods).

- Produce smudge from a smoky fire (repels mosquitoes).

- Produce soot from smoke (pigment for inks).

- Create sparks (initiates combustion).

- Create light (lengthens the workday).

- Produce embers (restarts fires).

- Create water vapor (assists breathing, blocks the sun, creates fog, and transports water).

- Heat tar and pitch (used as adhesives).

- Heat bone (for shaping into tools).

- Shatter stone (for quarrying and carving).

- Create flame (clear brush, melt wax, shape bowls and canoes, repel predators, heat shelters, remove unwanted vegetation to create trails and control weeds, regenerate habitat, enhance fertility, generate light to extend the workday).

- Evaporate liquids (concentrates solids).

- Shrink materials (tightens rawhide for drums).

- Harden wood (for tools and arrows).

- Flush animals out of the bush (for hunting).

- Frighten animals (defends against invaders).

- Kill enemies (weapons).

- Kill germs and bacteria (sanitize).

- Produce heat (enable comfort and survival).

- Roast, barbecue, boil food (provides nutrition and conserves the body's energy).

- Forge metals (create implements).

Your comments: What's next?

Endnotes

1	Charles Q. Choi, 'Human Evolution: The Origin of Tool Use,' *LiveScience*, November 11, 2009, accessed September 11, 2017, https://www.livescience.com/7968-human-evolution-origin-tool.html.

2	David M. Ewalt, 'The 20 Most Important Tools Ever,' *Forbes Magazine*, March 15, 2006, accessed September 11, 2017, www.forbes.com/2006/03/14/technology-tools-history_cx_de_06toolsland.html.

3	Gina-Marie Cheeseman, '57 Tons of Garbage Removed from NW Hawaiian Islands,' *Triple Pundit: People, Planet, Profit*, February 7, 2015, accessed September 11, 2017, http://www.triplepundit.com/2014/11/57-tons-garbage-removed-northwestern-hawaiian-islands/.

4	Johan Goudsblom, 'The Civilizing Process and the Domestication of Fire,' *Journal of World History*, 3.1 (1992): 2–11.

5	Aeschylus and Percy Bysshe Shelley, *Prometheus Bound* (New York: Heritage Press, 1966), 500.

6	Stephen J. Pyne, *Fire: A Brief History* (Seattle: University of Washington Press, 2001).

7	Pyne, *Fire*.

8	Bruce Davies, 'The *Cut/Stack/Burn* Project: A Land Based Art Collaboration Involving Agriculture, Conservation, and Local Community,' October 2007, accessed September 4, 2016, http://www.brucedaviesartist.com/info/journal%20article/page10.html.

9	IVM Technical Bulletin: Gorse, accessed March 10, 2018, http://members.efn.org/~ipmpa/Noxgorse.html.

10	A film of the *Cut/Stack/Burn* project by Bruce Davies was shown alongside material collected from the aftermath of the burn.

11	Fagots are bundles of sticks and twigs. The word is unrelated to a similar word that insults homosexuals.

12	Bruce Davies, email correspondence with the artist, December 16, 2017.

13	Bruce Davies, *Cut/Stack/Burn*, accessed December 12, 2017, http://www.brucedaviesartist.com/info/page7.html.

14	Davies, *Cut/Stack/Burn*.

15	Davies, *Cut/Stack/Burn*.

16	Davies, *Cut/Stack/Burn*.

17	'Projects,' *Futurefarmers*, accessed September 11, 2017, http://www.futurefarmers.com/.

18	'Famous Meteorites,' *Fall of a Thousand Suns*, accessed September 11, 2017, http://www.fallofathousandSuns.com/famous-meteorites.html.

19	Colin Schultz, 'The Ancient Egyptians Had Iron Because They Harvested Fallen Meteors,' *Smithsonian.com*, May 30, 2013, accessed September 11, 2017, http://www.smithsonianmag.com/smart-news/the-ancient-egyptians-had-iron-because-they-harvested-fallen-meteors-86153874/.

20	Wikipedia, 'Firestorm,' accessed September 11, 2017, https://en.wikipedia.org/wiki/Firestorm.

21	*History of Metal Casting: A Brief Timeline*, http://www.metal-technologies.com/docs/default-source/education/historyofmetalcasting.pdf?sfvrsn=8.

22	Amy Franceschini, email correspondence with the author, March 18, 2017.

23	Damian Carrington, 'Solar Plane Makes History after Completing Round-the-World Trip,' *The Guardian*, July 25, 2016, accessed September 11, 2017, https://www.theguardian.com/environment/2016/jul/26/solar-impulse-plane-makes-history-completing-round-the-world-trip.

24   Sintering is the process of converting a powdery substance into a solid form via a heating process. In recent years sintering has played a central role in design prototyping known as 3-D printing or SLS (selective laser sintering).

25   Markus Kayser Studio, *Out of Hand: Materializing the Postdigital*, ed. Ronald T. Labaco (London: Black Dog Publishing, 2013, p 42).

26   https://www.dezeen.com/2011/06/28/the-solar-sinter-by-markus-kayser/.

27   https://www.dezeen.com/2011/06/28/the-solar-sinter-by-markus-kayser/.

28   https://www.dezeen.com/2011/06/28/the-solar-sinter-by-markus-kayser/.

29   'Sand Babel: Solar-Powered 3D Printed Tower', published by eVolo Magazine, March 20, 2014, http://www.evolo.us/competition/.

30   Fraser Cain, 'What Percentage of the Earth's Land Surface is Desert?,' *Universe Today*, December 24, 2015, accessed September 11, 2017, https://www.universetoday.com/65639/what-percentage-of-the-earths-land-surface-is-desert/.

31   United Nations, 'Desert, Desertification, Soil, Drylands, Ecosystem, Environment, Land, Climate Change,' accessed September 11, 2017, https://aabgu.org/drylands-deserts-and-desertification/.

32   Spurse, *Eat Your Sidewalk Cookbook* (CreateSpace Independent Publishing, 2017).

33   Spurse, interview with the author in Rhinebeck, NY, August 30, 2017.

34   Rebecca Rupp, 'A Brief History of Cooking with Fire,' *The Plate*, September 2, 2015, accessed September 11, 2017. http://theplate.nationalgeographic.com/2015/09/02/a-brief-history-of-cooking-with-fire/.

35   Joshua J. Mark, 'Vesta,' *Ancient History Encyclopedia*, September 2, 2009, accessed September 11, 2017, http://www.ancient.eu/Vesta/.

36   The dinner occurred on the Weintraub homestead in Rhinebeck, New York on June 24, 2017. Twenty guests were in attendance.

37   Spurse, interview with the author in Rhinebeck, NY, July 13, 2017.

38   Spurse, interview, July 13, 2017.

39   Spurse, interview, July 13, 2017.

40   Spurse, interview, July 13, 2017.

41   Spurse, interview, July 13, 2017.

42   Spurse, email correspondence with the author, July 31, 2017.

43   Spurse, interview, July 13, 2017.

44   Spurse, interview, July 13, 2017.

45   Spurse, interview, July 13, 2017.

46   Spurse, interview, July 13, 2017.

47   Spurse, interview, July 13, 2017.

48   Spurse, interview, July 13, 2017.

49   Spurse, interview, August 30, 2017.

50   Spurse, interview, July 13, 2017.

51   Spurse, interview, August 30, 2017.

ECO MATERIAL CREATIVITY

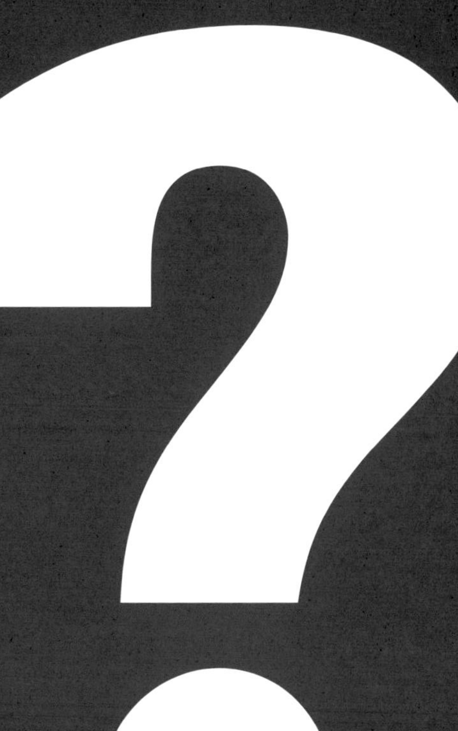

Water

ECO MATERIAL CREATIVITY

What is creativity? Is it inventiveness? Insight? Originality? Intelligence? Psychologists tend to think of it as subjectivity. Philosophers might call it 'spirit'. Theologians refer to it as 'soul'. Neuroscientists call it 'consciousness'. Artists refer to it as 'inspiration'. A creative act involves billions of neurons, millions of muscle fibers, a huge sensory capacity, and a great quantity of memory. Through their coordinated efforts an intention is formed, insight is generated, action is planned, resources are gathered, skills are acquired, and something new is created. Neuroscientists report that these elusive operations lurk in the crevices and connectors throughout our neural networks. When we create, convoluted cerebral events churning in our skulls cohere into a capacity to innovate. Nonetheless, creativity is not easily accounted for by mechanical and chemical processes. Instead of having a specific brain address, creativity seems to travel along and among the brain's intricate networks. Along the way it loosens the moorings of habits and sets the mind free to release memories, gain knowledge, and form intentions.[1] Human creativity may be best known through its remarkable accomplishments that mark the course of civilization. These milestones explain why the functions of the human mental apparatus are considered the crowning achievement of 3.5 billion years of evolution.

Throughout history, the human potential to create has been most conspicuously expressed by members of society who forge humanity's future – designers, activists, engineers, scientists, inventors, and artists. Despite the diversity of their accomplishments, few challenged the anthropocentric assumption that it was humanity's right to tame the unruly forces and extract the substances of the planet. Within this worldview, expansions of the existing techniques for extraction and domestication are honored as evidence of progress. Human creativity has generated so many instances of anthropocentric progress that there is hardly a territory on Earth that is not marked by its interventions. Solids, liquids, and gases the world over are imprinted with evidence of human creativity.

Anthropocentric creativity is evident in the manifold means humans have devised to manage outcomes that privilege the human species over other species and features of the planet. The products of human creativity are evident in highways, dams, cell towers, industrial sites, parks, farms, docks, cornfields, planters, storage tanks, towers, orchards, tunnels, runways, gutters, sidewalks, playgrounds, toothbrushes, and so forth. Although each represents a triumph of human cognition and imagination within the context of anthropocentrism; within the context of Eco Materialism, many of these markers of human creativity are implicated in a distressing litany of ecosystem disruptions. They indicate that humanity's creative acumen can be bold where caution is warranted; ambitious where restraint is prudent; willful where care is needed; and headstrong where it should be responsive. Oil spills, decimated forests, and toxic waste exemplify instances where creativity went awry. These endangering pitfalls tend to result from two intentions that epitomize anthropocentric creativity. One involves privileging humans over the well-being of the countless organisms that co-habit Earth. The other entails pursuing immediate benefits while ignoring eventual detriments.

Current conventions of art seem rooted in anthropocentrism. They are evident in the belief that gifted individuals perform autonomous acts that generate discrete artworks intended exclusively for human viewers. Eco Materialism reverses these norms by championing interdependence that may involve engaging non-humans as creative partners. It cultivates responsiveness to fluctuating material conditions that may entail originating, activating, or participating in dynamic occurrences, not enduring objects. Such cooperative and interactive actions may welcome interdependent feedback mechanisms that cannot be controlled or anticipated. In all these ways, Eco Material artists do not claim independence, nor favor self-expression, nor assert command over the outcomes of their creative efforts. Instead, they funnel humanity's impulses, desires, habits, insecurities, ambitions, imaginations, fears, and passions into behaviors that benefit all species, now and in the future.

Because Eco Material creative processes and creative outcomes reverse the conventions of art that have prevailed for centuries, they are not intended as provocations. Nonetheless, they invite such questions as: Is it art if it doesn't sit on a pedestal or hang on a wall? Is it art if it isn't the product of inspiration? Is it art if it doesn't express an emotion, or reveal individuality, or involve manual skill? Is it art if it isn't visually pleasing, or even visually interesting? Indeed, these departures from custom are not only new within the history of recent art; they indicate values that are new within the history of civilization. Since existing means of artistic expression reflect anthropocentric values, a new language with a new syntax and grammar is required to communicate the new worldview conveyed by Eco Material values.

Water

Water is the substance and the theme that unites the artists exploring Eco Material creativity in this chapter. Of all the substances on Earth, water was selected to signify creativity because it is constantly recreating itself, relinquishing its liquid state to become a solid or a gas. It epitomizes creativity as an agent of change by assuming a multiplicity of forms: flakes, frost, pellets, hail, ice, cloud, fog, mist, vapor, humidity, steam, etc. As a liquid, water can be ultimately passive, yielding to gravity and to the forces of the moon; it can be meddlesome, actively seeping into crevices in the Earth and permeating the expanses of the atmosphere; it can be gregarious, constantly dissolving and transporting any substance it receives; and it can be ferocious, wielding the power to kill and destroy. These contrasting qualities explain why water is essential to life and health, but also a conveyor of death and disease. As a quintessential tool of transformation, water serves as a poignant symbol of creativity.

In addition to signifying 'creativity', water also signifies 'creation'. The ultimate beginnings of life have been imagined within waters of potentiality since the earliest myths and cosmologies arose. Today's biologists support this contention by theorizing that life first emerged from the dark and teeming waters that enveloped the Earth some 3,500 million years ago. They explain that over the succeeding eons, droplets of this primordial ocean became enclosed within wombs, eggs, and seeds. These capsules accomplished a marvel of creation by engendering the planet's multiplicity of life forms. Thus, as a symbol and a substance, water represents the life-giving amniotic fluid of the planet. This connection has special significance for Eco Materialists because it evokes the harmonious accord between humans, energetic forces, and Earth's myriad forms of life.

As a component of anthropocentric creativity, water is often subjected to techniques that control its inherent ebbs and flows. The catalog of engineered strategies that command water and extract its uses currently includes dams, levies, bulwarks, cisterns, drains, sewers, mills, gutters, and reservoirs. Recent additions to this catalog include forcing water into injection pools to crack deep-rock formations that release natural gas or petroleum; and converting water into high-pressure steam that drives turbines in thermoelectric generating facilities. Eco Materialists note that these achievements of human creativity not only conscript waters for human advantage, they afflict them with mining refuse, sluice from mills, waste from fish-traps, seepage from septic tanks, oil from spills, toxins from industrial production, runoff from agriculture, and overflow from storm water. They also note the irony that comparable feats of human creativity are inspired by the need to correct these environmental offenses. Inventive solutions take the form of treating tainted water with algaecides, antifoams, biocides, coagulants, corrosion inhibitors, disinfectants, flocculants, and oxidants.

The following text tracks five examples of Eco Material creativity activated by artists who engage water as a theme and as a medium. Simparch creates a solution to the problem of contaminated water by making use of its ability to condense. Dove Bradshaw creates an opportunity to display the power of minute droplets of water to dissolve mineral matter. Pierre Huyghe creates a freshwater biome that supports diverse forms of aquatic life. Jason deCaires Taylor creates a marine habitat that is vitalized by his sculptures. Red Earth creates a ceremony that revives the age-old veneration that water once inspired.

Simparch is an artist collective created in 1996 by Matt Lynch (b. 1969, Indianapolis, Indiana, United States) and Steve Badgett (b. 1962, Champaign, Illinois, United States). Badgett received a BFA at the University of Illinois. He lives and works in Chicago. Lynch earned a BFA at Ball State University and an MFA at Syracuse University. He lives and works in Cincinnati. Simparch's practice involves large-scale, usually interactive installations and works that, as the groups name suggests, examine simple architecture, building practices, site specificity, and make use of materials that may be salvaged, or recycled, and DIY practices. In 2004, Simparch's work was included in the Whitney Biennial at the Whitney Museum of Art, and Rodadora Frontera (Border Tumbleweed), commissioned by the US General Services Administration in 2014.

In the late eighteenth century, many British farmers abandoned rural life to find employment in factories. Artists and intellectuals within and beyond England's borders recoiled from the sudden incursion of machine smoke, noise, and pollution. They reacted by extolling the wonders of uncorrupted nature as a gateway to transcendence and spiritual truth. Furthermore, they resisted the stultifying routines of factory labor by producing personal, spontaneous, imaginative, and emotional embodiments of creativity. Their opposition to the blight imposed by the sudden industrial takeover inspired many of that era's treasured masterworks.

Environmental offenses and contaminations are, once again, stimulating artistic creativity. Simparch provides an example of Eco Material creative responses that occupy the opposite end of the creativity spectrum from Romanticism. Instead of clinging to a vanishing ideal of a pristine landscape, Lynch and Badgett seek practical solutions to existing environmental problems. Commitment to creative problem-solving is embedded in the collective's name. 'Simparch' is a contraction of 'simple' and 'architecture'. 'Simple' suggests that the artists seek solutions that are accessible, straightforward, and unadorned. 'Architecture' indicates that their mission involves planning, designing, and constructing physical structures. Lynch and Badgett directed this mission to water laced with toxic substances and nasty microbes.

'Drinking Water Fresh from the Sewer'[2] is the headline of a story reporting the *Dirty Water Initiative (DWI)* undertaken by Simparch after they travelled from the Midwest to Tijuana, Mexico in 2003. They were invited to contribute to *InSite05*, an exhibition of artworks installed at the US/Mexico border. With the support of the US-based community development organization, Fundación Esperanza, Badgett and Lynch visited Tijuana to study the city's municipal sewage management program. They discovered that the percentage of collected waste

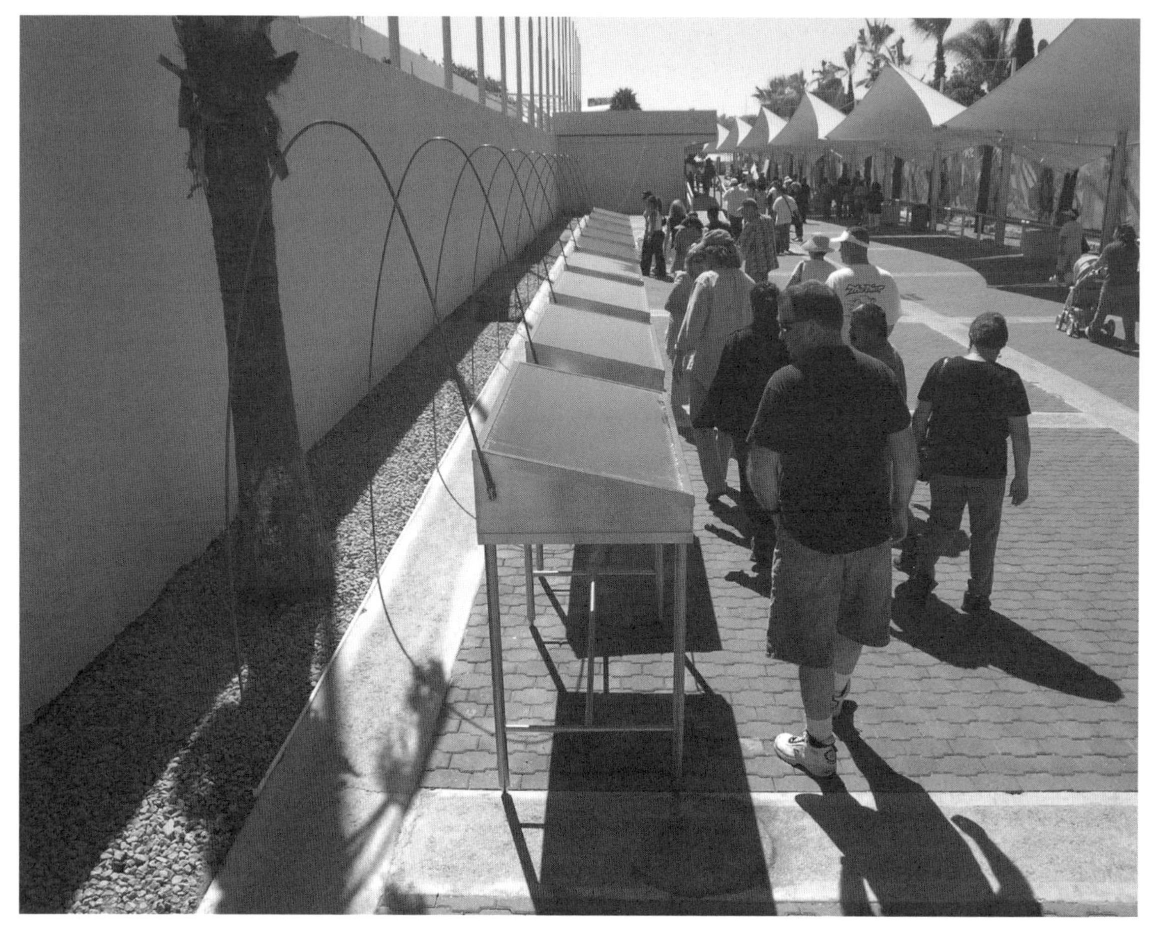

Figure 1a: Simparch, *Dirty Water Initiative* (2005), stainless steel, glass, silicone, plastic bottles, hardware, 84 × 80 × 6 ft. Photograph by Simparch. Courtesy of the artists.

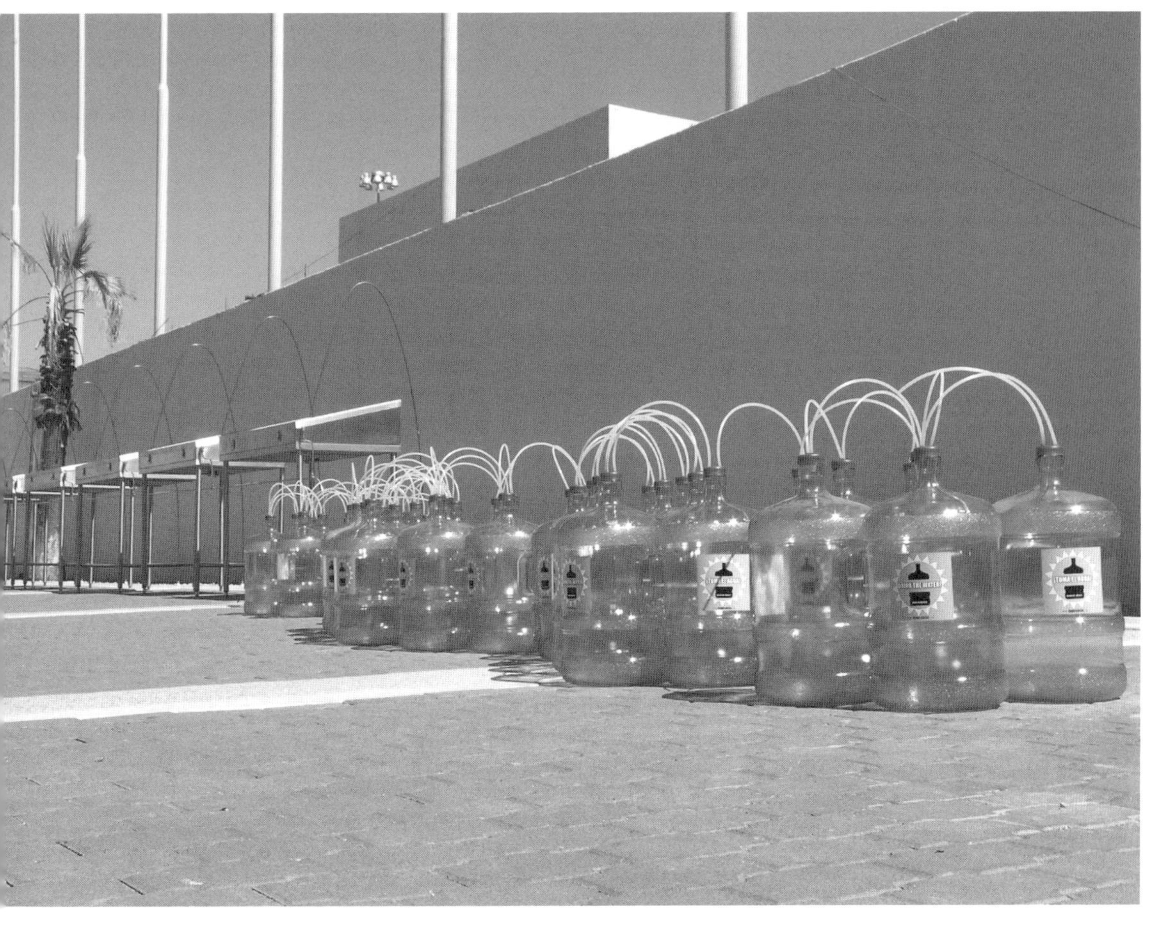

Figure 1b: Simparch, *Dirty Water Initiative* (2005), stainless steel, glass, silicone, plastic bottles, hardware, 84 × 80 × 6 ft. Detail of the ready-to-drink water after the distillation process. The standard five-gallon containers are daisy-chained together. Photo by Simparch. Courtesy of the artists.

water that is treated has risen in recent years; nonetheless, the area remains vulnerable to water-borne illnesses and groundwater contamination. A 2000 Mexican census estimated that between seventeen and fifty percent of Tijuana's more than 1.4 million residents live without indoor plumbing. Many homes dump wastewater directly into the streets. The resulting pollution contains heavy metals, viruses, litter, human feces, and sediment, subjecting the population to water-borne illnesses and groundwater contamination.[3] The artists describe the project they undertook for the exhibition as a 'good-will gesture towards a correction of the problem, however minor.'[4]

Simparch's goal was to eliminate the need for the signs posted to tourists and residents warning 'Don't drink the water!', which are ubiquitous just south of the border. The artists not only offered the people pure water to drink, they demonstrated how, with simple DIY technology, the region's abundant sewage can actually be used to reduce the country's scarcity of drinking water. The solution Simparch crafted highlighted the striking disparity between the state of water on either side of the US–Mexico border.

Any artistic motive that is pragmatic instead of romantic requires a creative process that is methodical, not emotional. Thus, the artists commenced this two-year-long preparation by making extended visits to Mexico to study how squatters' communities manage when municipal sewage treatment systems are inadequate or non-existent. They discovered that these haphazard settlements 'somehow exist beyond catastrophe with an order beyond comprehension for most people in the United States.'[5] Although these populations endure, the artists committed to demonstrating a way to remediate tainted waters. Strategies for handling hazardous water that the artists researched and rejected included reduction, isolation, and removal. The strategy they chose was decontamination using solar distillation technology. Their plan involved enlisting the sun to conduct a simple solar distillation to purify untreated wastewater, thereby transforming it from a health hazard into a beneficial resource. Once this basic research was complete, the artists undertook the process of building and testing their own prototype stills for decontaminating water that would comprise their installation in the art exhibition.

In 2005, Badgett and Lynch unveiled the *Dirty Water Initiative*, a solar distillation device that resembled an ingenious public drinking fountain. It functioned by channeling contaminated tap water into nine, glass-topped, stainless steel basins. The basins were lined with black silicon to increase the sun's heat and augment evaporation. The evaporated vapor rose to the top of the still where it condensed and trickled down an angled piece of glass. From there, the water flowed through tubes that led to large, plastic water containers, the kind that typically provide drinking water in office settings. Because each still was outfitted with a float, only the cleansed water arrived at this destination. Each still was capable of producing up to two gallons of clean water a day, depending on the sun's intensity. In this manner, Simparch created a solution to Mexico's long-standing shortage of pure water. Because their solution utilized inexpensive technology that could be easily installed, the system could be replicated by citizens without the intervention of industry or government, or the need for infrastructure and ordinances.

By attaching the adjective 'functional' to the noun 'art', Simparch's intervention may seem like a heresy to advocates of 'art for art's sake' who believe art's intrinsic value is corrupted by utility. True art is insular, and the true artist is autonomous. Eco Material artists object to 'art for art's sake' because it promotes anthropocentric estrangement from the forces and substances of the planet. Simparch exemplifies this Eco Material alternative by applying its artistry to real tainted water treated with actual remedial regimens, which quenches the thirst of real people without endangering their health. When such functions assert themselves as components of art, the adjectives that are allocated to art merge with those that occupy the pragmatic sphere. The empirical/technical domains that satisfy human wants and needs are enfolded into the arts/humanities that engage metaphysical realms of perception, imagination, and emotion. A divorce between these divergent zones of human pursuits was initiated 2,500 years ago when Aristotle canonized the idea of dividing knowledge into disciplines. It persists to this day within the construction of academic departments and research protocols. Simparch manifests this Eco Material merger by rejecting the aesthetics of 'landscapes', and adopting functional engagements with 'ecosystems'.

Simparch installed the *Dirty Water Initiative* along a highly trafficked pedestrian walkway on the Tijuana side of the border known as the San Ysidro Port of Entry. The artwork announced the pragmatic solution to this endemic problem. The artists hoped passers-by could also enjoy the refreshing drink of water. However, the nervous organizers of the exhibition prohibited the artists from greeting visitors crossing the border into Tijuana with potable water offerings instead of dire water warnings. They were thwarted by the double jeopardy of Mexico's impure water, and North Americans' fixation with ultra-pure water, despite the fact that the contaminated water remediated by the *Dirty Water* stills 'will be much purer than the US's carefully treated recycled toilet flush.'[6] Although their artistic intention was not fulfilled, people observed these mini water purification plants arrayed along the main pedestrian walkway. They saw contaminated water entering the stills and clean water exiting. Lynch explains that their initiative fulfilled the hope 'to inspire others to see how well (the stills) work and how easy it is to build their own.'[7]

The disposition of the stills after the exhibition closed was calculated into Simparch's Eco Material considerations. The artists explain, 'The stills were relocated after the exhibit as a small gesture to assist several families in water-challenged communities that survive on the "technophagic fringe" of development, where people fulfill basic needs without services and infrastructure that Americans expect.'[8] 'Technophagic' is typically associated with the out-of-body experience had at 'techno-raves'. It refers to using technological means to unleash celebratory joy. Applied to drinking water, the word suggests the extreme joy of satisfying basic, neglected bodily needs. Badgett explains the Eco Materiality of the joy they offer, 'We'd like to see people doing more for themselves [...] in terms of limited resources, being more resourceful, to be conscious of use [...] Our works involve keeping in mind a second use, doing a permanent thing rather than just the metaphorical concept.'[9] In this manner, Badgett and Lynch demonstrate an Eco Material form of art creativity that diverges from self-expression and abstraction in favor of pragmatic solutions: 'material and process are tangled up in the conceptual development.'[10] The artists demonstrate Eco Material creativity by explaining:

> *It's a thrill to see people involved with a piece in a physical way. That was always our intention to get people on the piece – not just in the room with the piece. We want to please, to be generous.*[11]

Lynch and Badgett summarize their pragmatic approach to artistic creativity by stating, 'To be useful and practical is good, but we also want (our work) to remain art [...] The art world can exist as a model where ideas can be tested.'[12]

An ironic addendum to this narrative involves the fact that the artists' well-meaning solution, once it was installed in needy communities, actually aggravated their water hazard because residents became more careless with their sewage once adequate supplies of pure water were assured by the solar distillation technology.

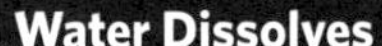

Dove Bradshaw (b. 1949, New York, United States) is an interdisciplinary artist based in New York. In 1969, Bradshaw integrated the concept of Indeterminacy by enlisting the unpredictable effects of time, weather, and erosion on natural, chemical, and manufactured materials. From this starting point, Bradshaw creates chemical paintings that change with the atmosphere. She's been awarded grants from the National Endowment for the Arts (1975); the Pollock-Krasner Foundation (1985); Furthermore (2003); and the National Science Foundation (2006). Her work has appeared in the 6th Gwangju Biennale, South Korea. She is represented in the permanent collections of the Metropolitan Museum of Art; MoMA (Museum of Modern Art), New York; the National Gallery of Washington; the Art Institute of Chicago; the British Museum, London; Centre Pompidou, Paris; and the Russian State Museum, St. Petersburg. Bradshaw received her MFA from the School of the Museum of Fine Arts, Boston.

As an artist, Dove Bradshaw creates opportunities to discern the transformative effects of water that are rarely detected by most peoples' radars of consciousness. Few contemporary humans are skilled in discerning water's formative role in interchanges between organic and inorganic matter. This means that few recognize water's contribution to transmissions between human and non-human forces, and few appreciate its crucial contributions to the planet's ongoing, all-inclusive, creative becoming.

Metaphor is not required to link water and creativity. Long before humans created tools and techniques to dominate the planet's waters, water manifested its creative force by carving the topographies of the planet as it rushed down from the mountain peaks, splashed into the valleys, seeped into crevices, hurtled against obstacles, bathed the plains, sliced channels through the forests, and carved the land into a myriad curves and angles. These are remarkable acts of formation for a substance that is colorless and tasteless, has no shape, acquiesces to the forces of gravity, and is ruled by temperature. While the carving effects of water are dramatized by flood surges, monsoon rains, tidal waves, and tsunamis, Bradshaw ignores the manic side of water's temperament in order to focus on miniscule drops and gradual seepages. Individually, these minute changes are imperceptible, but their cumulative effects over the course of millennia are stunning.

Presenting the incremental shape-shifting force of water that transpires over vast spans of time poses a challenge to artists. Writers might leave gaps in their temporal descriptions to incorporate decades or centuries. Filmmakers might focus their camera on the neglected phenomenon for a long time and then speed the film when it is projected. Bradshaw's Eco Material approach avoids such common techniques. She allows water to conduct its own actions, according to its natural tempo. Her sculptures are formed by water – dripping, condensing, seeping,

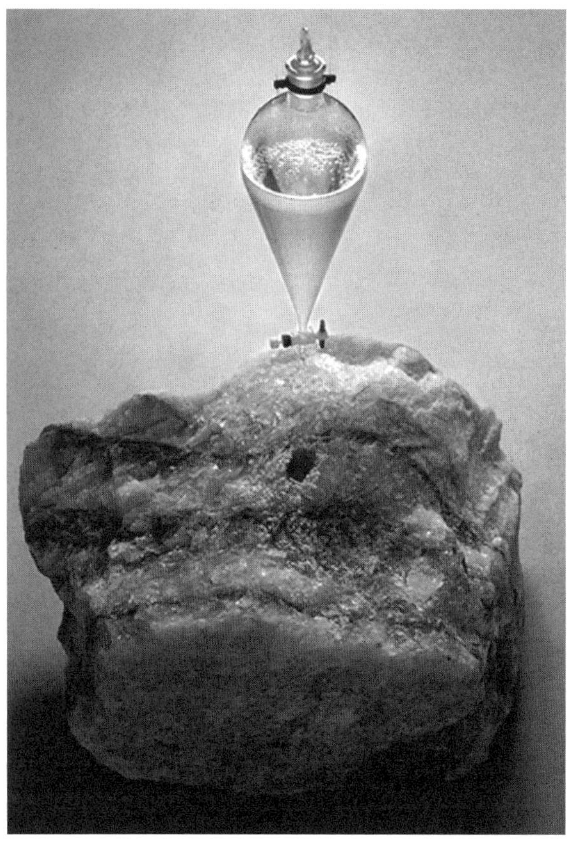

Figure 2a: Dove Bradshaw, *Negative Ions I, first boulder* (1996), rock salt, 1,000 ml separating funnel, water, 16 × 24 × 20 in. Photo by John Bessler. Courtesy of the artist and the Sandra Gering Gallery, New York, NY.

evaporating, and suspending – not by the artist. These incremental and ongoing changes are evident in *Negative Ion*, an ongoing sculptural series that Bradshaw initiated in 1996. In this series, creativity takes the form of meticulously creating circumstances for water to display its multiple, minute effects upon matter. First, Bradshaw selected a substance that would register the effects of the drips of water that fell upon it. To create *Negative Ions*, Bradshaw selected salt because of its extreme solubility. Next, she calculated the speed and volume of the flow of water upon it so that effects occurred just barely within the threshold of human perception. Then she determined the distance between the water source and its destination, which would also influence the effect of water as it accumulated on the salt. Lastly, she considered light and temperature in the environment to account for water's sensitivity to these fluctuations. All these decisions optimized water's influences without interfering with them. As a result, the creative act that

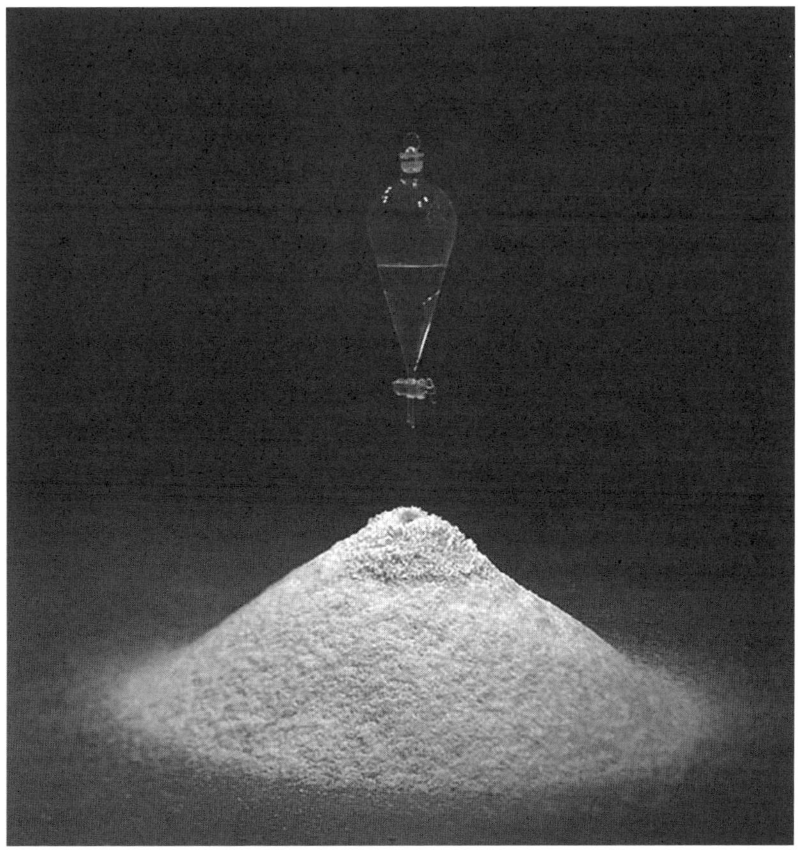

Figure 2b: Dove Bradshaw, *Negative Ions II* (1996–ongoing), 200 lbs salt, 1,000 ml separating funnel, water. Photo by Bent Ryberg. Courtesy of the artist and the Stalke Gallery, Copenhagen.

forms these sculptures is more closely aligned with geological processes, than human will. They enlist the geological events that form deltas, beaches, cliffs, valleys, drainage basins, floodplains, lagoons, marshes, and gullies that occur in geological time scales. Bradshaw explains, 'The slowness of action constituting a return to real time and real materials – a response to the art world's overwhelming infatuation with shock value entertainment and virtual reality.'[13]

While some versions of *Negative Ions* incorporate rock salt, those that utilize salt granules are formed by pouring the particles of salt around a single point forming a pile approximately three-feet high. The resulting pristine cone shape is due, entirely, to gravity. In addition to the salt, the installation includes a glass laboratory funnel suspended from the ceiling directly above each salt pile. The funnel is calibrated for the slowest possible flow. Five drops per minute alters the salt at the rate of about two inches a month.

The sculptural changes that can be observed reflect the simultaneous actions of dissolution (water dissolves salt), evaporation (dissolved salt becomes re-crystallized), and erosion (material is transported and deposited elsewhere). While these categories of action are predictable, it is not possible to anticipate the particular successions of events that occur in these sculptures. This is because each individual crystal of salt responds to moisture according to its individual impurities and its unique placement in the heap. Some crystals re-crystallize into miniature fractal forests that develop at the mouth of the hole, while others form at the base. These sculptures, thereby, yield to the ever-changing unpredictability of geological occurrences.

In addition to providing an opportunity for humans to observe the subtle dictates of water upon physical topographies, the title of this sculptural series, *Negative Ions*, invokes the influence of water upon psychological states. Negative air ions are oxygen atoms or molecules that have gained an electron.[14] These ions enter our bodies as we inhale them from the surrounding atmosphere. Then they are transported via our bloodstreams to our brains. Although negative ions are odorless, tasteless, and invisible, they are known to alleviate depression, relieve stress, boost energy, increase alertness, stimulate our senses, improve appetite, and increase sexual drive.[15] Water droplets that create negative ions are responsible for the calm that is experienced by the seashore or beside waterfalls. Venerable mystics of the East recognized the remarkable effects of moving water when they advised students to practice breathing exercises to perfect their bodies and minds near a waterfall.[16] This connection is confirmed in Hinduism and Buddhism where the energies of water, manifested as the sacral chakra,[17] are associated with sensuality and creativity. Neuroscientists are experimenting with electroencephalography (EEGs) and magnetic resonance imaging (MRIs) to disclose the biological basis of these mood enhancers.[18]

By conjoining the effects of moving water and human creativity, Bradshaw goes beyond demonstrating how such connections affect the physical environment. Her sculptures also manifest the way these connections shape creativity in art. By extension, her creative process reveals how minds and emotions are also shaped by our interactions with water. In all these ways, Bradshaw resurrects connections between humans and the planet that are often severed in urbanized, industrialized, digitized societies. This book provides forty examples of artists who cultivate these connections as prerequisites for alleviating humanity's impact on the planet's fragile systems. Bradshaw sums up these Eco Material principles by commenting, 'Poetry is everywhere evident, all one has to do is present materials.'[19]

**Water Forms
eshwater Biomes**

Pierre Huyghe

Pierre Huyghe (b. 1962, Paris, France) is a New York-based artist who works across a variety of media including film, sculpture, public interventions, and living systems. In the 1990s, Huyghe became known for his 'post-production' technique, or the reuse of film and mass-media images. More recently, he has mounted work outside of museums and traditional art venues where he has engaged public art and living systems. Huyghe is also known for multimedia collaborations with other artists, including Philippe Parreno and Douglas Coupland. In 2002, Huyghe won the Hugo Boss Prize. His recent solo exhibitions have been held at the Los Angeles County Museum of Art (2014) and the Centre Pompidou (2013). He studied at the École Nationale Supérieure des Arts Décoratifs, Paris.

Creative acts that share the features and functions of ecological systems differ from human-centric cultural contexts. The latter assumes that artists perform autonomous acts upon malleable mediums that become resistant to change at the completion of the creative process. The French artist, Pierre Huyghe, dispensed with these conventions by adopting a dynamic ecosystem as his model of creativity. Instead of a static object, Huyghe welcomed diverse forms of life into his repertoire of creative practices, allowing them to conduct their life functions according to their own needs and dispositions.

To create *Nymphéas Transplant* (2014), for example, Huyghe reformulated artistic production by abandoning the intention of recording a visual observation. Instead, he delved beneath surfaces to explore the 'how' and 'why' that accounts for outward appearances. Huyghe shares this impulse with historic artists who studied the skeletal and muscular structure of the human body to better understand its outward appearance. But Huyghe is not a traditionalist. The surface he penetrated in this artwork was not flesh; it was a pond. The pond he chose is renowned throughout the world. It is the pond, located in Giverny, France, that inspired the revered paintings by Claude Monet (1840–1926), the esteemed master of Impressionism. Monet not only captured in paint the colorful impressions aroused by the pond's beauty; he designed and constructed it. As a result, Monet's artistry is revealed in two types of surfaces: one exists where water lilies bud and bloom on the rippling water; the other surface belongs to the canvas on which he painted his impressions of these sumptuous and fleeting visual phenomena.

Nymphéas Transplant shifts attention away from appearance and representation to engage the underwater functions that account for the beauty of the pond. The installation consists of three aquariums that contain water from Monet's actual pond. The aquariums are installed high enough to raise the pond water to eye level, directing the viewer's attention away from the Impressionist's interest in facades, to reveal the ecologist's attention to life dramas occurring beneath the surface. Viewers can observe the teeming underworld where insects, crustaceans, and invertebrates enact actual responses to the actual conditions that occur in this bit of Monet's pond. Ides fish and axolotls newts swim among the water-lily roots. Water beetles, scorpions, and salamanders struggle up to the surface. A variety of crustaceans, insects, and fresh water amphipods swim among the aquatic plants.[20] The Eco Material focus on function over appearance is suggested by the equipment that Huyghe installed to replicate pond conditions. Aquarium lights, pumps, heaters, and filters establish the ecosystem conditions that enable these aquatic inhabitants of the pond to survive within an art gallery.

Huyghe reinforced the alliance between his pond aquariums and the pond paintings by reinterpreting Monet's earnest attempts to capture the fleeting perceptions of light. While the Impressionist translated each photon into a dab of paint, Huyghe encased the aquariums in switchable glass that made it possible to both conceal and reveal their contents. The lighting sequence was also calculated to provide viewing opportunities and to maintain the living biotope – eight hours for viewing, eight hours to optimize ecosystem functions, and eight hours of dark to conform to the animals' circadian rhythms. When the light appeared, it was programmed to replicate the actual weather conditions that existed between 1914 and 1918, enabling visitors to actually 'see' the pond as Monet saw it when he painted his water-lily masterworks. For example, the sequence for one aquarium replicates and then condenses the mixture of sunshine and fog that occurred on the shortest day of the year in 1914. Despite this authentic recreation, Huyghe articulates the difference between Eco Materialism and Impressionism by commenting:

> *It's true, in the work and in the practice, I'm more interested in looking for something transitory than in producing a conclusion or turning a resolution into an object. It's the set of stations as a whole that I'm looking at, the dynamic chain of events.*[21]

Indeed, Huyghe did not merely 'look at' dynamic chains of events; he incorporated them into the living systems he created.

Color provides another opportunity for comparison. Monet lavished his canvases with chromatic arrays of paint in order to capture impressions of the shifting light upon the water's rippling surface. Huyghe's somber palette diverges from the visual pleasure offered by his esteemed Impressionist predecessor. The muted tones of underwater browns and greens that characterize his aquariums reveal that Huyghe renounced aesthetic visualizations in favor of incorporating actual ecosystem functions. Because the experience he provides is unmediated by

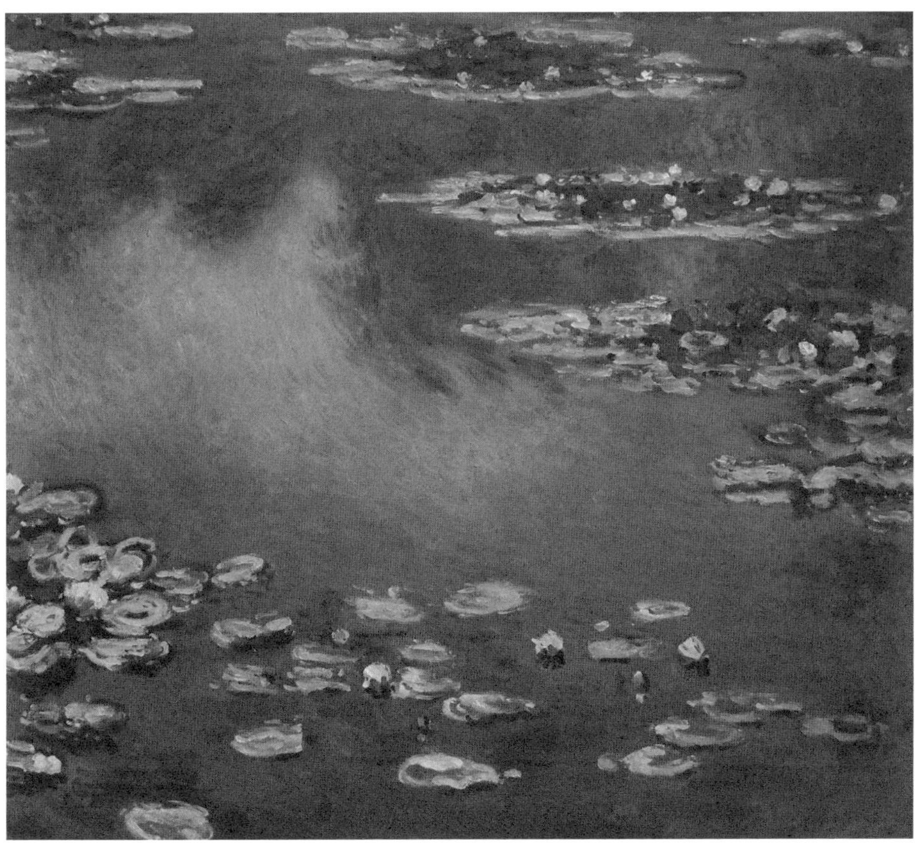

Figure 3a: Claude Monet, *Waterlilies* (1906), oil on canvas, 89.9 x 94.1 cm (35 3/8 x 37 1/16 in.). Mr. and Mrs. Martin A. Ryerson Collection, Art Institute of Chicago.

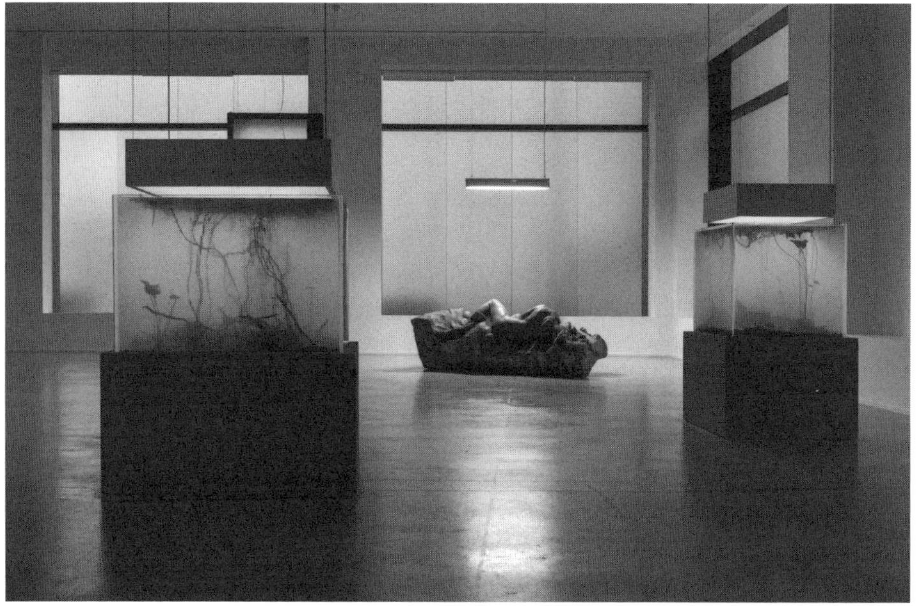

Figure 3b: Pierre Huyghe, *Nymphéas Transplant* (2014), live pond ecosystem, light box, switchable glass, concrete. Overall: 193 × 143 × 128 cm, Tank: 163 × 143 × 128 cm. Exhibition view, *In. Border. Deep*, Hauser and Wirth, London, September–October 2014. Photo by Alex Delfanne. Courtesy of the artist, and Hauser and Wirth.

representation, the artwork establishes an alliance between the viewer and actual underwater events.

Likewise, Huyghe relinquished Monet's status as a master artist by deferring to the aquatic and botanical organisms that create the visual components of his artwork. Sharing creativity with this multispecies community actualizes the integrated, self-organized relationships that constitute ecosystem dynamics. *Nymphéas Transplant* demonstrates that when feedback mechanisms are integrated into an artist's creative process, the resulting artwork is responsive, not autonomous. Engaging non-humans as creative partners confirms three essential Eco Material principles: humans are not independent agents of change; humans are not the

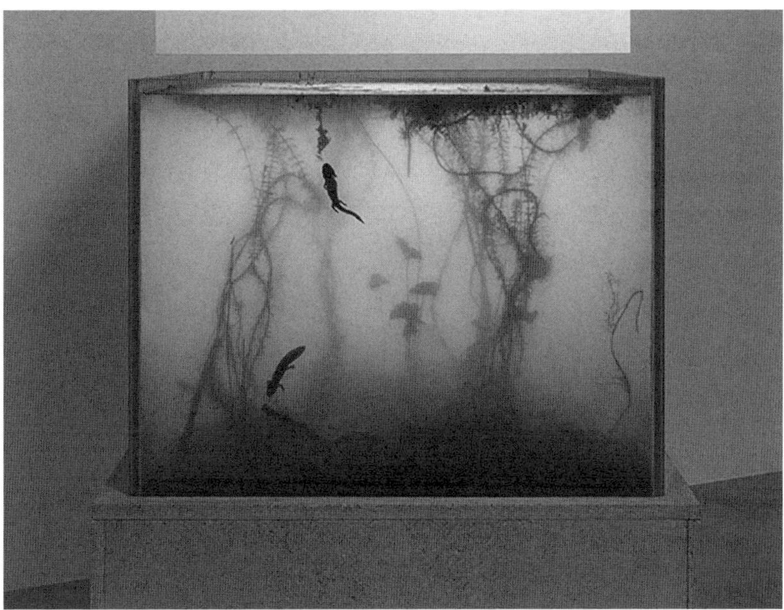

Figure 3c: Pierre Huyghe, *Nymphéas Transplant (14-18)*, live pond ecosystem, light box, switchable glass, concrete, 74 3/8 × 56 1/2 × 50 5/8 in. (188.9 × 143.5 × 128.6 cm). Courtesy of the artist, Hauser and Wirth, and the Marian Goodman Gallery.

exclusive initiators of change; humans are not essential for change. Huyghe summarizes this form of Eco Material creativity by explaining:

> *I'm interested in the indifference of address toward a viewer who enters. Not in a negative way, but in that elements exist and evolve whether someone is there to see it or not, with or without a discourse. It doesn't matter; it lives on.*[22]

The essence of this work is the creativity of the system, not the artist.

Water Provides Marine Habitats

Jason deCaires Taylor

Jason deCaires Taylor (b. 1974, Dover, United Kingdom) is a British sculptor, environmentalist, and professional underwater photographer. In 2006, he created the first underground sculpture park, the Molinere Underwater Sculpture Park. *Other major projects include* Museo Atlantico *(2016);* The Rising Tide *(2016); and* Ocean Atlas *(2014). He has received numerous sculpture and photography awards. In 2014, he was awarded that year's Global Thinker by* Foreign Policy *magazine.*

Coral reefs are one of the major wonders of the world. They are wonderfully beautiful because of their exotic shapes and spectacular colors (which only matter to humans). And they are wonderfully functional as initiators of the aquatic food chain (which tends to be unappreciated by humans). Both categories of 'wonder' are due to the creative output of millions of coral polyps. These tiny animals resemble upside-down jellyfish. They are equipped with a remarkable ability to contribute to the construction of reefs by transforming the excess carbon dioxide in the water into limestone. Within the nooks and crannies of the reef structures that they generate, twenty-five percent of all ocean species acquire a place for spawning, nurseries for raising their young, refuges for protection, and food for sustenance. In all these ways, reefs perform the vital functions of habitats.

Two additional reef 'wonders' exist. Environmentalists 'wonder' about the rarity of reefs; only two percent of the ocean bottom consists of reefs.[23] Second, they 'wonder' about their survivability because reefs are not only rare, they are extremely fragile. The hypersensitivity of this habitat makes them vulnerable to warming ocean temperatures, to the intrusions of snorkelers, to toxic wash-off from land, and to fishermen who slam the coral reefs with crowbars to shake out stunned fish. Many reefs that once teemed with life are turning into underwater wastelands. Scientists predict that between fifty and seventy percent of the world's coral reefs are now endangered and could succumb if predictions of global warming materialize.[24]

Jason deCaires Taylor explains how he decided to devote his art career to help relieve the plight of reefs: 'I felt disillusioned that my works were just about creating art – I wanted to do something that maybe went beyond that and was actively beneficial.'[25] Initially, this 'something' involved constructing small artificial reefs. Then, he says, he 'found out about how a lot of conservation was about controlling people's movements.'[26] This insight led Taylor to create monumental artworks that function as reef rescue stations. Unlike Pierre Huyghe – who introduced art-audiences to an underwater ecosystem by relocating a bit of a pond to an

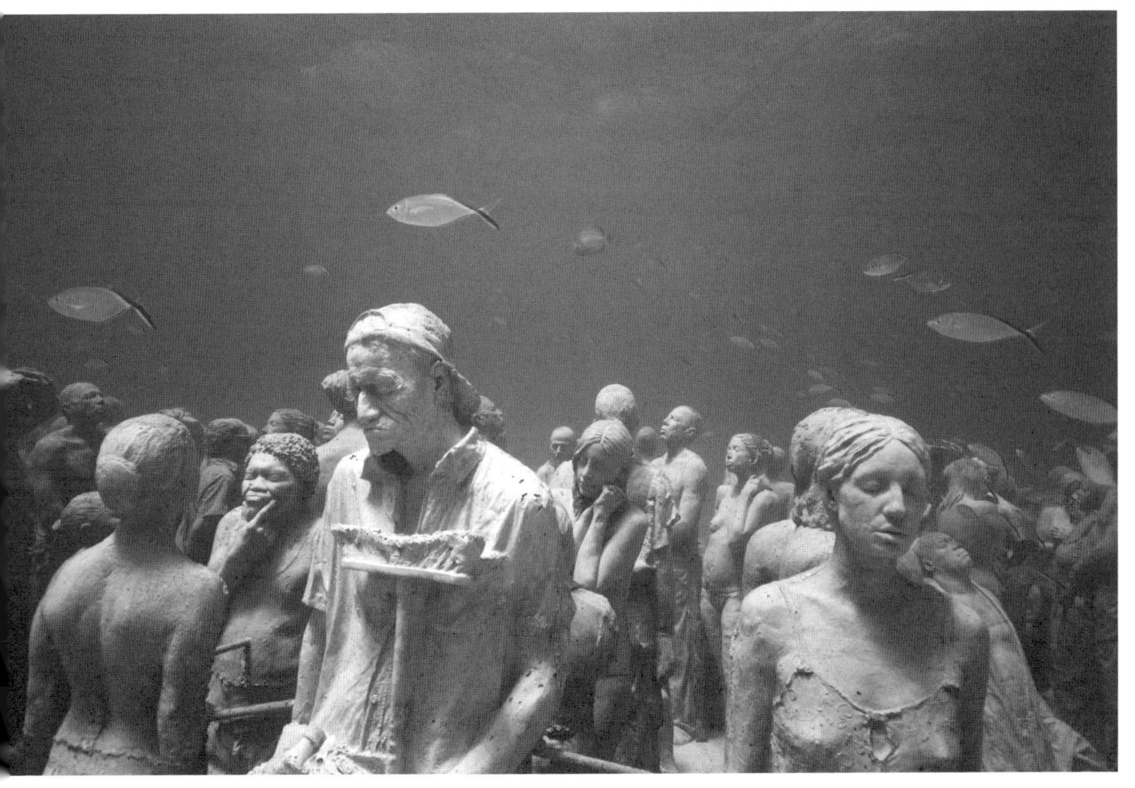

Figure 4a: Jason deCaires Taylor, *The Silent Evolution* (2010), 450 life-sized figures made of pH neutral marine cement. Underwater sculpture, Museo Subacuático de Arte, Cancun, 2011. Photo by Jason deCaires Taylor. All rights reserved, DACS/Artimage 2018.

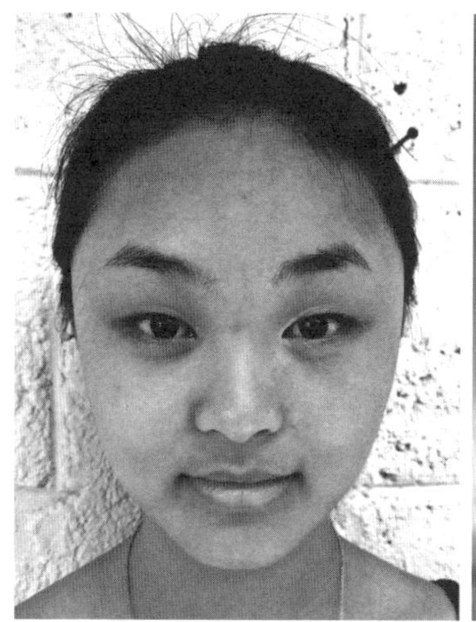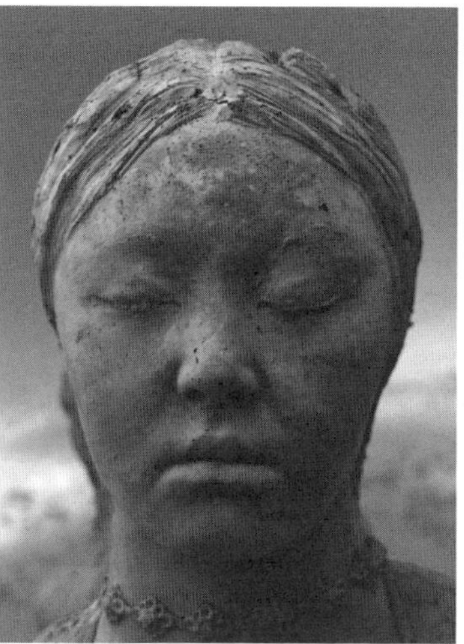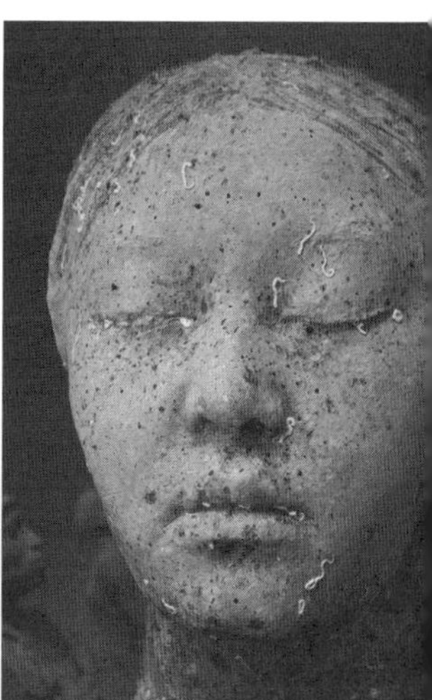

art gallery – Taylor exhibits his sculptures 'on-site' upon the ocean floor. The dimensions of his sculptures reveal this unconventional art venue. Besides height and width, they include depth. Some of his sculptures exist twenty-six-feet deep, immersed in the fluid subterranean environments beneath the surface of the ocean. Art-viewing is, therefore, reserved for divers and snorkelers.

Each life-sized sculpture in Taylor's ambitious underwater installations is formed by making a meticulous cast of a person from the local community. A broad range of ages, body-types, and demeanors are included. Some figures are assembled in groups. Others appear among sculpted books, bottles, a couch, a television, a desk, a sleeping dog, and other artifacts of their Earth-bound existence. An amusing scene depicts suited businessmen who kneel with their heads buried in the sand; small fish and crustaceans have taken up residence in their raised buttocks!

When these vivid depictions of oxygen-dependent lives are first installed within an underwater environment, they appear to be oddly misplaced. But their incongruous terrestrial status gradually diminishes as calcium-producing organisms transform them into safe-haven reefs for countless aquatic life forms. The medium Taylor selected to create these lifelike representations of humans provides the hard surfaces required to attract algae, barnacles, corals, and oysters. As these aquatic organisms attach themselves to the sculptures' surfaces, the artworks become overlaid with intricate structures. As new reefs are formed in this manner, innumerable species of ocean fish are granted protective habitats for occupation. Taylor's artistic intention is fulfilled when the sculptures he painstakingly creates

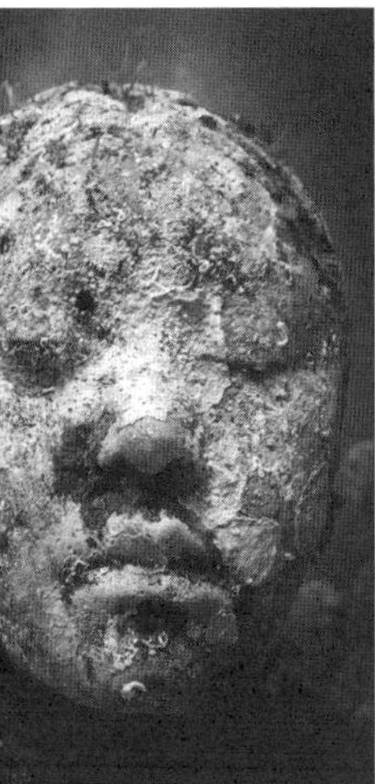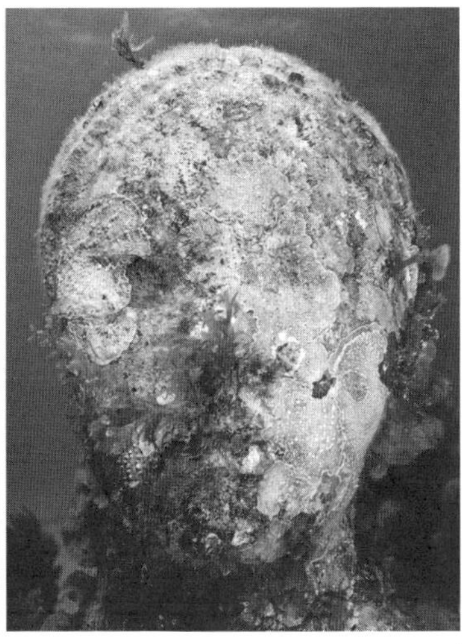

Figure 4b: Jason deCaires Taylor, *The Silent Evolution* (2010), pH neutral marine cement. Underwater sculpture showing the model and the sculpture's gradual occupation by aquatic organisms. Photo by Jason deCaires Taylor © by Jason deCaires Taylor. All rights reserved, DACS/ Artimage 2017.

are overgrown and disappear beneath the biomass. At this juncture, the sculptural representations of Earth-bound humans relinquish their status as art objects intended for human audiences. They acquire a new identity as functioning reefs for marine wildlife. Such interspecies rapport provides a vivid example of Eco Material creativity.

While Taylor has created monumental submerged sculptures worldwide, the underwater museum MUSA (Museo Subacuático de Arte), launched in 2009,[27] has special significance within the context of Eco Material creativity. MUSA exists in the waters surrounding Cancun, Isla Mujeres, and Punta Nizuc off the coast of Mexico. The region is renowned for its spectacular reefs. By attracting 750,000 tourists every year, their popularity benefits the tourist industry, but threatens the reefs' existence. Taylor committed to alleviating the damage inflicted by thousands of snorkeling vacationers upon the region's natural coral reefs.

MUSA represents the diversionary tactic Taylor developed to lure visitors away from the region's precarious natural reefs, and allow these vulnerable aquatic habitats to regenerate. To accomplish this mission, he resolved to produce an art installation that was so spectacular it would entice visitors away from their usual reef destinations. MUSA is the result of his ambitious undertaking. This sprawling underwater museum has become a major tourist attraction. It currently exhibits five hundred life-sized sculptures, but the museum's collection keeps expanding.[28] That is because new figures need to be added to compensate for those that disappear as they are subsumed by reef-constructing organisms. Taylor has begun inviting other artists to contribute to the museum's art collection.

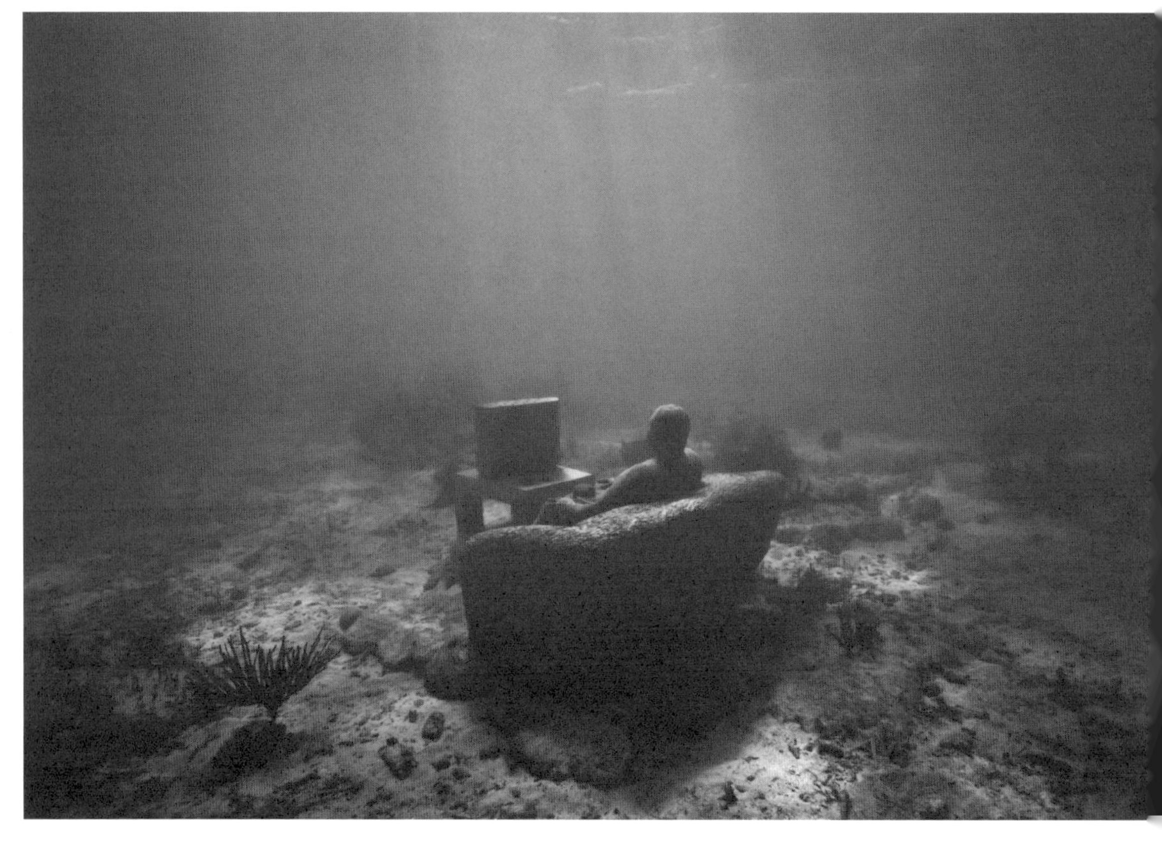

Figure 4c: Jason deCaires Taylor, *Inertia* (2011), concrete, 320 × 210 × 200 cm. Photo by Jason deCaires Taylor © by Jason deCaires Taylor. All rights reserved, DACS/Artimage 2017.

The algae Taylor welcomes to colonize his sculptures serve two audiences. Human visitors delight in the visual display these tiny organisms produce as they deposit patinas in yellow, green, brown, purple, and brown-red upon the sculptures. At the same time, the entire ocean food chain benefits from the capacity of algae to carry out photosynthesis. They provide food for the larvae of fish and marine invertebrates that feed forage fish, which feed larger predatory fish, which then feed humans.

As habitats for underwater life forms, these sculptures are anathema to museum settings where steady-state climate controls and artificial illumination ensure the art's protection. In contrast, MUSA displays its collection within the perpetually shifting conditions of the sea bottom where aquatic creatures reproduce, acquire food, and seek refuge. The light that illuminates this collection filters through waters that are sometimes murky and sometimes clear. The space in which the sculptures are installed teems with bubbles ascending, seaweed floating, and particulates drifting by. By cultivating responsiveness to changing conditions, the underwater habitats that MUSA establishes diverge sharply from the isolating, steady-state environments of conventional museums that provide habitats for fine art.

The concept of a 'permanent' collection marks another difference between MUSA and above-ground museums. By initiating the creative interactions of aquatic organisms, Taylor's monumental effort is designed to be taken over and occupied, not stored and preserved. As a result, 'permanent' is irrelevant to this submerged museum. Taylor comments, 'I would like to point out that this installation is by no means over [...] the second phase is dependent on nature's artists of the sea, to nurture, evolve, and apply the patina of life.'[29] Taylor's installations become 'permanent' by being occupied, replenished, and revitalized.

MUSA also serves as a catalyst for human transformation. Taylor explains that it offers an opportunity for instilling 'a closer understanding of our [human] relationship with the natural marine environment and [an appreciation of] the need to value and protect this fragile ecosystem in order to save ourselves.'[30] This statement suggests that although the ocean occupies seventy-one percent of the world's surface, it is as unfamiliar to most people as a distant planet. Those who claim to be sea-lovers are often content to frolic in shoreline waves and observe the sunset beyond the ocean horizon. While snorkeling to reefs is popular, few have visited underwater kelp forests and sea grass meadows, or explored the mountains, ridges, volcanoes, trenches, canyons, plateaus, and plains that are submerged beneath the surface of the ocean. Taylor comments:

If we walked past a forest that was disintegrating every day and with animals dead by the side of the road, we would be much more aware of our actions. But underwater, it is out of sight and is a problem so easily ignored. So, a big part of my work is to bring people's focus and their awareness to this destruction of our seas and of the natural world.[31]

A barren patch of seabed unoccupied by marine species provided MUSA's original location. Now millions of microscopic organisms that drift down toward the sea bed have attached themselves to the lifelike sculptures. They are colonizing its hard surfaces as if they were rocky outcrops. These tiny plants and animals are attracting fish, mollusks, crustaceans, echinoderms, sponges, tunicates, sea urchins, star fish, sea worms, mollusks, turtles, and sharks that breed and take refuge in its enclosed spaces. To date, the area has attracted over fifty aquatic species, which may explain why the National Marine Park and the Cancun Nautical Association have agreed to co-sponsor Taylor's artwork. The merging of art and conservation missions is demonstrated by the way MUSA utilizes the funds it collects from admission charges and donations. Unlike land-bound museums, an underwater museum is spared the expense of insurance, electricity, guards, and fire protection. Thus, Taylor can invest MUSA's revenues in marine-conservation initiatives and alternative employment for local fishermen whose practices endanger the reefs. In this manner, every visitor to the museum contributes to the safeguarding of reefs.

Water Inspires Reverence

Red Earth

Red Earth: Caitlin Easterby (b. 1963, Bedford, United Kingdom) and Simon Pascoe (b. 1960, Newport, United Kingdom) both studied at Sussex University and are co-directors of Red Earth, based in the South Downs National Park in East Sussex, United Kingdom. They have created work in landscapes across Europe, Mongolia, Japan, and Indonesia.

To early humans, rivulets and rapids must have appeared as mysteries that rebounded to the far side of celestial occurrences. A delicate spring shower and a torrential downpour provided equal opportunities for wonder. Although this was long before the environment was scrutinized with the rigor of the scientific method, early humans recognized water as a quintessential tool of transformation, wielding forces that were equally destructive and creative. For this reason, water is associated with the origin of life in myths and cosmologies from around the world. Chalchiuhtlicue (Aztec), Acionna (Celtic), Gong Gong (Chinese), Tefnut (Egyptian), Vedenemo (Finnish), Nerites (Greek), Apam Napat (Hindu), Pariacaca (Incan), and Suijin (Japanese) are among the many gods that embodied water's powers as the life-giving amniotic fluid of the planet, and a fearsome substance that unleashes tidal waves and floods. Such awe marks the origin of ceremonies venerating water. Purification rites are conducted with water; sanctification practices anoint with water; and transcendence beyond the phenomenal world is manifested by walking on water. These traditions arose before dams, levies, bulwarks, cisterns, drains, sewers, mills, gutters, and reservoirs broke the inherent rhythms of water's ebbs and flows, and robbed it of its autonomy.

In order to resuscitate water's esteem as a powerful shaper of mental and geographical topographies, artists Simon Pascoe and Caitlin Easterby, known collectively as Red Earth, created a series of events in which water's inherent powers, un-augmented by humanity's technological prowess, were summoned and exalted. These attributes are striking in the location the artists selected for these special events; they took place on the stony beach that separates the sea from the gleaming white cliffs along the southern coast of England. This dramatic meeting point between land and sea culminates in the renowned Seven Sisters Cliffs that rise near the eastern edge of this region. The relentless crashing of waves has been carving and eroding the shoreline's dramatic geography since prehistoric times. But, in even more ancient times, water was also responsible for the original creation of the cliffs that the waves are now wearing away. This history began ninety million years ago when layers of shells began to build up and consolidate into rock.[32] Now, accelerated by climate change and human

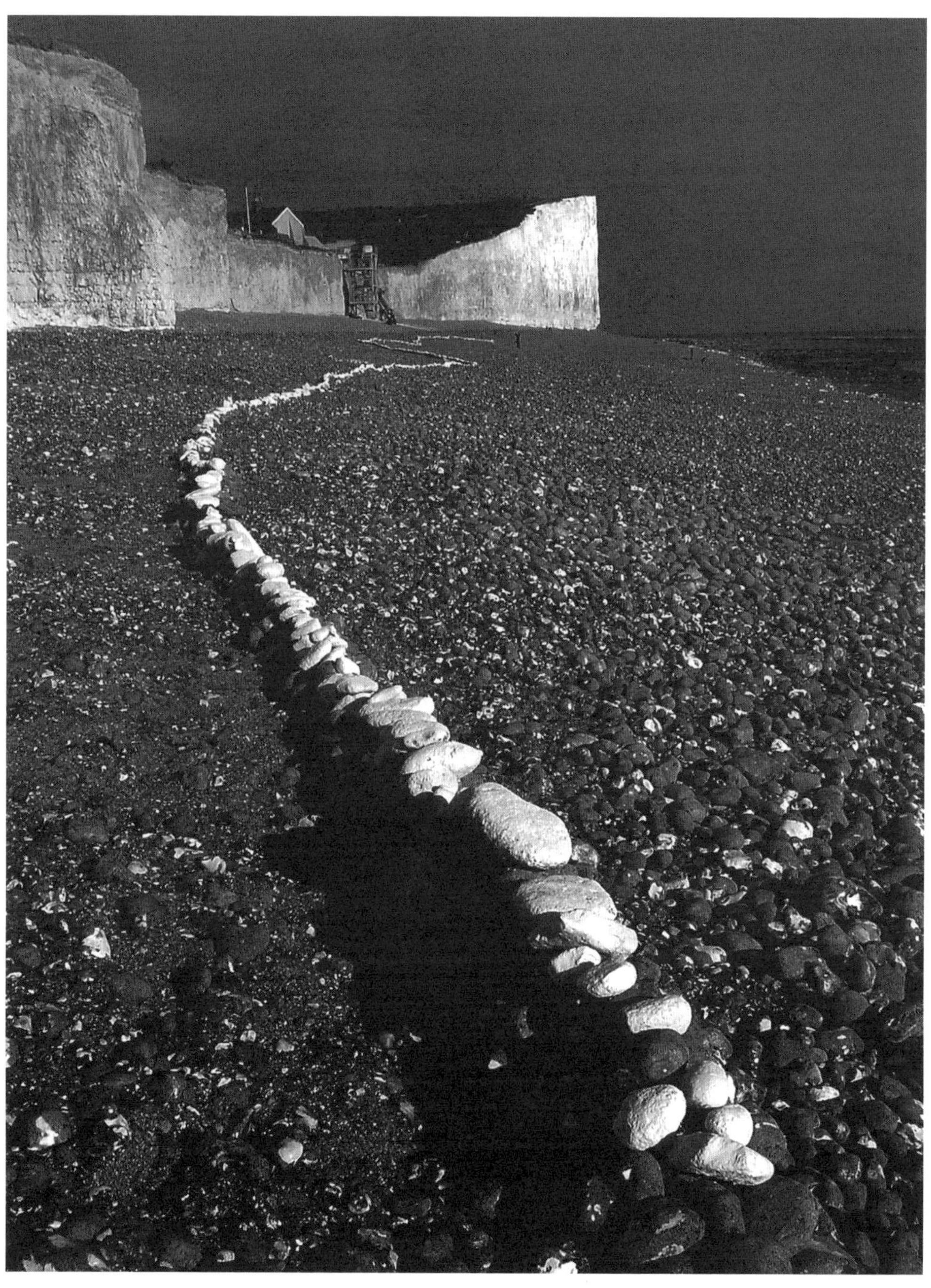

Figure 5a: Red Earth, *Geograph, Trace* (2005), chalk and flint line created by the public marking a future erosion line, approx. 90 m. Performance. Ritual. Alchemy. Photo by Red Earth. Courtesy of the artists.

mismanagement, the cliffs are falling back into the sea at a rate of up to a meter per year.[33] As a force that both removes and deposits matter, water is a quintessential tool of material transformation. Easterby describes this process as 'the geological tension between the forces of the sea and the forces of the land.'[34]

Red Earth paid tribute to the connection between these reciprocal processes by orchestrating three ceremonial events between May and September of 2005. Known collectively as *Geograph: Drawing the Landscape*, these events revealed that the cliffs' present condition represents an interlude along a trajectory of continuous geological change. Thus the artists created a ceremony in three acts to register these geological occurrences. Easterby then explains, 'Erosion is about constant transformation. We're always in a very liquid state, which can never be fixed. Fixing is a completely artificial notion.'[35]

In recent history, however, the ongoing geological process of erosion has been intensifying. The artists were intent on physically articulating how the impact of human interventions and climate change are dramatically accelerating this process. They explain:

> *Climate change was such an important theme [...] On the Sussex coast over the past seven thousand years erosion rates were between two and six centimeters. Now they average twenty-two to thirty-two centimeters per year. This is due to rising sea levels, fiercer waves and storms and human interventions along the coast.*[36]

Red Earth's event incorporated these temporal and material dimensions by bracketing the cliffs' present condition between markers of its past and future formations between events that began in May 2005 and ended four months later. The first event, *Trace,* marked the past. The second, *Vanishing Point* – installation and performance – featured the present. The third, *Geograph,* was a closing ritual that anticipated the future. This ceremonial context commemorated the force of water's actions upon the land.

For *Trace*, over fifty participants ventured onto the beach between high and low tide to collect stones – pieces of flint and chalk – and arrange them in a long curved line that mapped the place where the cliffs rose thirty years earlier. The artists invited the painter Sax Impey to evoke the ghostly presence of the vanished cliffs by creating a series of chalk and water paintings along this line. Simultaneously, four thousand miles away on the tsunami-ravaged Javanese coastline, Red Earth collaborator and movement artist Parmin Ras performed a parallel water ritual. The distance between the two events dramatized the fact that the planet's water is finite. People on opposite sides of the Earth are united by the forces of the sea to persistently wear away Earth matter.

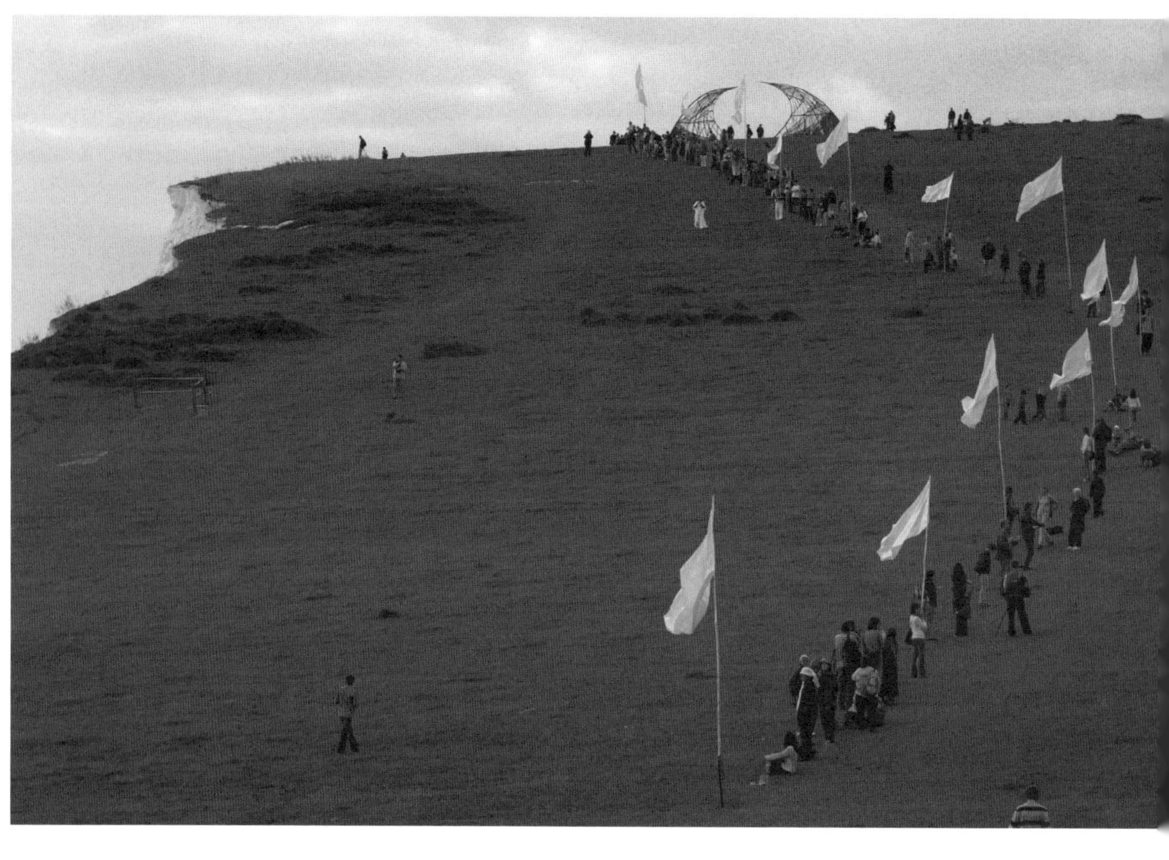

Figure 5b: Red Earth, *Geograph, Vanishing Point* (2005), chalk and flint line created by the public marking a future erosion line. Performance. Ritual. Alchemy. Photo by Charlotte Macpherson. Courtesy of the artists.

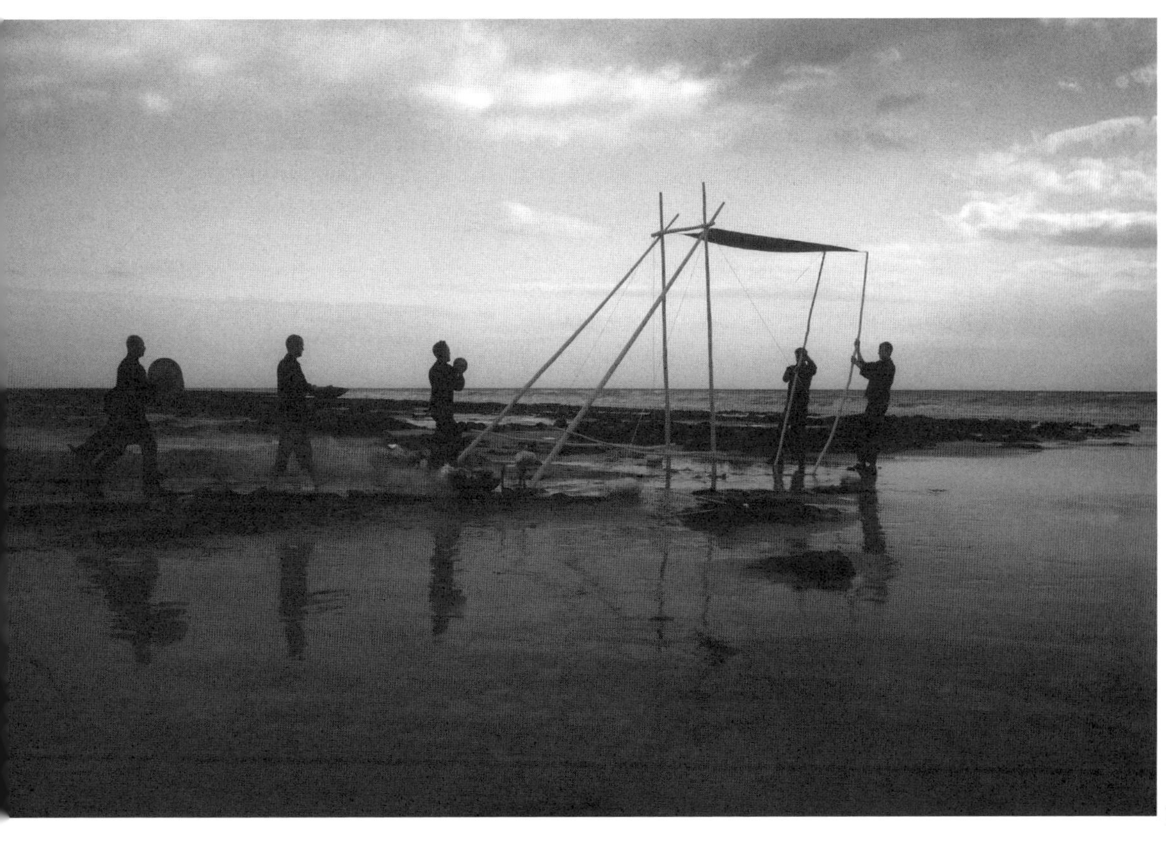

Figure 5c: Red Earth, *Geograph, Trace* (2005). Final action, toward the sea. Performance. Ritual. Alchemy. Photo by Charlotte Macpherson. Courtesy of the artists.

Weeks later, Caitlin and Simon installed *Vanishing Point*, a striking sculpture on the cliffs above the beach, made from greenwood, wattle, and chalk daub. Shaped like two interlocking waves, the form echoed the surging power of the sea. Then, at the end of September, the closing ceremony was conducted. Entitled *Journey*, this event projected the work's temporal dimensions into the future by positioning ten large white flags in the landscape above the cliffs to demarcate the anticipated retreat of the cliffs twenty years into the future. Easterby, Pascoe, Parmin Ras, and musician Ansuman Biswas, guided the public on a procession that led them through the woven sculpture that served as a gateway into the heightened domain of ritual, which the artists describe as 'the liminal point between the mundane and the heightened experience of the "other".'[37] As the sun set, the performance moved along the beach to the water line that was illuminated with bonfires. The closing water ceremony. Everyone gathered on the shoreline to witness the return of a bowl of chalk water to the sea. This ritual of meeting and return occurred amidst the smoke and flicker of the fires and the haunting sounds created by the musicians.

While awe for the power of water and the sea and its effect on the formation of the landscape received ceremonial acknowledgement in the first two components of *Geograph*, the culminating event introduced participants to an Eco Material response to this fluid force. The public participated in the geological processes by symbolically offering chalk back to the sea. This action embodied three aspects of Eco Material creativity: people acquiesced to, honored, and fostered a geological force. Easterby explains:

> *There is an ecological imperative behind this performative work. Deep green activism is connected to indigenous tribes and animistic societies. Those societies developed personal relationships with the landscape and its elements. Theirs was a living relationship, built on necessity, not just respect. That is something as a modern society we have lost. Our arrogance and technical domination of the environment means we have become separated from a spiritual connection to landscape. For indigenous societies performance is a pay-back to nature for what nature has given them. As Gary Snyder so eloquently puts it, 'Performance is currency in the deep world's gift economy.'*[38]

By relinquishing human-centeredness and acquiescing to Earth processes, Red Earth acknowledges erosion as the essential counterpart to accretion. The creative processes that continue to form the living and non-living components of the planet are always complementary acts. Eco Material creativity applies such mantling and dismantling to geological events, biological entities, and human cognition.

Eco Material Creativity

The artists in this chapter respond to the anthropocentric assumptions of autonomy and separateness that are linked to fears of social and ecological cataclysms, by cultivating inter- and intra-connectedness with all forms of matter and all forms of life. Eco Materialistic creativity invents ways to act as affiliates, not commanders; and allies, not competitors. This willingness to cooperate has introduced three principles that reverse the anthropocentric ascendance. One is 'Cross-Species Altruism': this occurs when humans curb their desires for the benefit of other species. Another is 'Substance Altruism': this occurs when humans curb their desires for the sake of a substance, such as water. The third is 'System Altruism': this occurs when humans curb their desires for the sake of a system within which that substance exists. All three forms of altruism are nurtured by the environmental mantle that protects humans as well as non-humans. This is the goal of Eco Material creativity.

In all these ways, Eco Material creativity constitutes a menu of hopeful prospects that is being conducted by ecologically minded artists, designers, scientists, and architects. Room needs to be cleared for them to migrate from this cultural fringe into the chambers of government, science, politics, and culture. When this occurs, the baseline of human creativity may become the front line of humanity's integration into the complex adaptive systems on Earth. Creativity that manifests such an alliance is constructed within networks; it is not self-induced; it is structured in terms of feedback loops; it is not isolated; it acknowledges that every action has rippling effects; it is not self-contained; it concedes to external environmental conditions; it is not self-generated. Such expansions of consciousness might burst the lid on human potential. The new frontier of human creativity emulates and embraces the non-human processes, substances, and forces that shape our planet. Instead of control, suppression, or exploitation, Eco Materialists welcome collaborators, partners, and mentors. Implicit in these Eco Material efforts is the hope that by revising the goal of humanity's creative endeavors, future environmental predicaments can be avoided.

Your opinion: What's next?

Reader Interaction

BIOLOGICAL NEEDS					
	Negligible	Slight	Moderate	Strong	Excessive
Air					
Food					
Water					
Shelter					
Warmth					
Sex					
Sleep					
Other					

My Needs Versus My Wants

Material habits in the art studio typically mirror material habits in all other facets of an artist's life. The current norms in both contexts overwhelmingly reflect easy access to consumer goods that enable people to indulge in wants that exceed needs. As explained in the preceding text, the Eco Material alternative introduces three principles: 'Cross-Species Altruism', 'Substance Altruism', and 'System Altruism'. All three forms of altruism are nurtured by the environmental mantle that protects humans as well as non-humans. This is the goal of Eco Material creativity. The following charts are opportunities for you to examine your personal want/need ratios – emotional and material. This accounting may aid you in taking environmentally responsible actions regarding your studio activities, as well as your clothing, career, diet, mobility, ambition, and leisure activities.

BIOLOGICAL WANTS					
	Negligible	Slight	Moderate	Strong	Excessive
Air					
Food					
Water					
Shelter					
Warmth					
Sex					
Sleep					
Other					

SAFETY AND SECURITY NEEDS					
	Negligible	Slight	Moderate	Strong	Excessive
Protection against sunlight					
Financial security					
Law & order					
Liability insurance					
Health insurance					
Sanitation					
Door locks					
Other					

KINSHIP AND AFFECTION NEEDS					
	Negligible	Slight	Moderate	Strong	Excessive
Workplace colleagues					
Family					
Personal friends					
Social media friends					
The public					
Life partner					
Pet					
Other					

SAFETY AND SECURITY WANTS					
	Negligible	Slight	Moderate	Strong	Excessive
Protection against sunlight					
Financial security					
Law & order					
Liability insurance					
Health insurance					
Sanitation					
Door locks					
Other					

KINSHIP AND AFFECTION WANTS					
	Negligible	Slight	Moderate	Strong	Excessive
Workplace colleagues					
Family					
Personal friends					
Social media friends					
The public					
Life partner					
Pet					
Other					

ESTEEM NEEDS					
	Negligible	Slight	Moderate	Strong	Excessive
Public recognition					
Social status					
Financial status					
Praise from acquaintances					
Rewards/ awards					
Popularity					
Self- actualization					
Other					

COGNITIVE NEEDS					
	Negligible	Slight	Moderate	Strong	Excessive
Learn					
Teach					
Create					
Imitate					
Invent					
Design					
Interpret					
Other					

ESTEEM WANTS					
	Negligible	Slight	Moderate	Strong	Excessive
Public recognition					
Social status					
Financial status					
Praise from acquaintances					
Rewards/ awards					
Popularity					
Self-actualization					
Other					

COGNITIVE WANTS					
	Negligible	Slight	Moderate	Strong	Excessive
Learn					
Teach					
Create					
Imitate					
Invent					
Design					
Interpret					
Other					

SENSORY NEEDS					
	Negligible	Slight	Moderate	Strong	Excessive
Visual					
Tactile					
Aural					
Olfactory					
Taste					
Kinesthetic					
Temperature					
Other					

SPIRITUAL NEEDS					
	Negligible	Slight	Moderate	Strong	Excessive
Ritual					
Obedience					
Music					
Nature					
Physical stress					
Meditation					
Prayer					
Other					

SENSORY WANTS					
	Negligible	Slight	Moderate	Strong	Excessive
Visual					
Tactile					
Aural					
Olfactory					
Taste					
Kinesthetic					
Temperature					
Other					

SPIRITUAL WANTS					
	Negligible	Slight	Moderate	Strong	Excessive
Ritual					
Obedience					
Music					
Nature					
Physical stress					
Meditation					
Prayer					
Other					

Water As Creative Opportunity

The Eco Material frontier of human creativity emulates and embraces the non-human processes, substances, and forces that shape our planet, diverging from anthropocentric efforts to control, suppress, or exploit them. These post-anthropocentric creative endeavors are being explored to resolve current, and avoid future, environmental and social predicaments.

The following list enumerates many attributes of water and suggests ways for artists to engage them in their creative practices. While the kinds human/non-human alliances listed here involve water and artists, they are transferable to many substances and many situations. It is hoped that they inspire future material interactions that are beneficial to the individual, to future generations, and to the planet.

Water as theme: Water related environmental issues include melting glaciers, rising seas, acid rain, eutrophication, etc. Works of art can utilize water as a theme.

Water flows: As a fluid medium, the particles suspended in water can be distributed evenly over broad surfaces. Works of art can be created with flowing water and pigments made from earth, minerals, and biological materials.

Water carves: Flowing water that carries grit can carve very hard substances like rocks and glass. This is called scouring. The result of water flowing over granular material is called erosion. Water can also carve hard substances like stone when it carries a load of chemicals capable of dissolving the stone. This is called etching. Erosion, scouring, and etching are subtractive processes that can create artworks by removing material.

Water is transparent and colorless: It reflects if it is still, sparkles if it ripples, and refracts (magnifies) objects that are submerged. Artists can exploit the visual qualities.

Water can be translucent and opaque: This occurs when loosely bonded, soluble compounds dissolve in it to form solutions. Their opacity depends upon when these solutions are diluted or concentrated. Artists can vary the hue by altering the degrees of saturation of the dissolved matter.

Water emits odors: This occurs when aromatic substances dissolve in it. Water allows artists to add smell to the experience offered by their artworks.

Water conducts heat: By serving as a conduit for heat, water enables artists to incorporate dynamic change into their artworks.

Water creates shapes: The visual patterns generated by water include straight vertical lines when it falls unimpeded; ripples, spirals, concentric rings when it is disturbed; complex interference patterns when multiple rings intersect. Artists can explore water as a generator of form.

Water evaporates: Evaporation leaves behind the residues it dissolved and transported. If water is reintroduced, these substances will dissolve again and can be transported again. Artists can incorporate such ongoing temporal interactions into their creative process.

Water creates sound: Some sources of water-induced sounds include dripping faucets, roaring waterfalls, rippling streams, pulsating waves, gurgling drains, and hissing teapots. Artists might create a symphony of water sounds.

Water flows: Water not only flows downward due to gravity, it creeps upward due to capillary action. Both of these actions are occurring all over earth, all the time. Artists can utilize the contrasting abilities of water to flow up and flow down as they paint.

Water beads: When water is placed on a surface such as wax paper or oil, it beads instead of puddles. Artists can take advantage of the capacity of water to form droplets.

Water alters the shape and dimension of materials: Artists can utilize conditions of dry, damp, wet, and saturated to alter the shape of substances.

Water creates buoyancy: The power to support a body so that it floats allows water to serve as a means of transport. Thus water offers artists an opportunity to introduce movement into their artworks.

Water cleans: Because water is a solvent, artists can create forms, patterns, and textures by exploiting the power of water to dissolve loosely bonded compounds off dirty surface.

Water darkens: The color of absorbent objects darken when they become wet. Artists can create ephemeral patterns by selecting a material that displays differentiating light and dark states when it is dry, damp, wet, and saturated.

Water activates processes: Rusting, dissolving, shrinking, growing, are all processes activated by water. The source of water that conducts these processes might be a faucet, river, rain, or your mouth. Artists can tap many sources of water and apply them to generate material changes.

Water is essential for life: Water carries nutrients that make organic growth possible. All the chemistry of life is performed in water. Artists can manifest water's essential role in maintaining life.

Water kills: While no living organisms can survive without water, it is also true that every living organism is susceptible to death from water. Water kills in the form of tsunamis and tidal waves of sea water, and avalanches of snow. It kills if it carries disease-causing microbes or poisons. It kills if there is too much or too little. Artists can create artworks that convey water as a danger.

**Your comments: What's next?

Endnotes

1   Scott Barry Kaufman, 'Beautiful Minds: The Real Neuroscience of Creativity,' *Scientific American*, August 19, 2013, accessed July 12, 2016, http://blogs.scientificamerican.com/beautiful-minds/the-real-neuroscience-of-creativity/.

2   Claire Caraska, 'Drinking Water Fresh from the Sewer,' *Voice of San Diego*, October 13, 2005, accessed May 2, 2017, https://www.voiceofsandiego.org/topics/science-environment/drinking-water-fresh-from-the-sewer/.

3   Rob Davis, 'Leaky Border: Efforts to Stop the Flow of Pollution from Tijuana Have Bogged Down in a Mess,' *High Country News*, September 12, 2008, accessed March 1, 2018, https://www.hcn.org/issues/40.17/leaky-border.

4   Claire Caraska, 'Drinking Water Fresh from the Sewer.'

5   Matt Lynch, telephone interview with the author, July 8, 2017.

6   Mark Harris, 'Interview with Simparch,' *Miser Now*, 2013, accessed December 10, 2016, http://markharrisstudiocom.ipower.com/markharrisstudio/wp-content/uploads/2013/01/Miser-Now-Simparch-interview.pdf.

7   Matt Lynch, telephone interview with the author.

8   Simparch, 'Dirty Water Initiative.' Technophagy refers to operations that apply technologies to a mix of scientific knowledge, artisinal lore, and political actions.

9   Claire Caraska, 'Drinking Water Fresh from the Sewer.'

10   Harris, 'Interview with Simparch.'

11   Mark Harris, 'Simparch,' accessed September 1, 2017, http://markharrisstudio.com/simparch-interview-miser-now/.

12   Harris, 'Simparch.'

13   Dove Bradshaw, *Salt: Works from 1996–2006*, accessed September 1, 2017, http://www.dovebradshaw.com/Catalogues/6.%20SALT%201996.pdf.

14   'Health Benefits of Negative Ions: The Official Healthy Line,' accessed January 10, 2018, https://healthyline.com/negative-ions-benefits/.

15   Denise Mann, 'Negative Ions Create Positive Vibes,' *WebMD*, accessed September 1, 2017, http://www.webmd.com/balance/features/negative-ions-create-positive-vibes?page=2.

16   Health Benefits of Negative Ions.

17   'Chakras' are meeting points on the body where energies are channeled. Each is associated with an element.

18   Comtech Research, 'Negative Ions: More Scientific Research,' accessed September 1, 2017, https://www.negativeiongenerators.com/negativeionsresearch.html.

19   Dove Bradshaw, *Salt: Works from 1996–2006*, accessed September 1, 2017. http://www.dovebradshaw.com/Catalogues/6.%20SALT%201996.pdf.

20   The species of animals and plants present in the *Nymphéas Transplant 14–18* aquariums include ide fish, axolotls, great diving beetles, water boatmen, water beetles, water scorpions, water skaters, fairy shrimps, copepods, water lilies, and a variety of crustaceans, insects, freshwater amphipods, and other aquatic plants.

21   Doug Aitken, 'Pierre Huyghe by Doug Aitken,' *Bomb*, accessed September 1, 2017, http://bombmagazine.org/article/2669/pierre-huyghe.

22   Ruba Katrib, Jameson Fitzpatrick, Gregory Volk, and Edgar Arceneaux, 'Let the Light In: Pierre Huyghe in Los Angeles,' *Art in America*, accessed September 1, 2017, http://www. artinamericamagazine.com/news-features/interviews/let-the-light-in-pierre-huyghersquos-first-us-retrospective-comes-to-los-angeles/.

23   Emily Frost, 'Corals and Coral Reefs,' *Smithsonian Ocean Portal*, August 30, 2016, accessed September 1, 2017, https://ocean.si.edu/corals-and-coral-reefs.

24   D. Bryant, L. Burke, J. McManus, et al., *Reefs at Risk: A Map-Based Indicator of Threats to the World's Coral Reefs* (Washington DC: World Resources Institute, 1998).

25   Hannah Ellis-Petersen, 'Underwater Sculptures Emerge from Thames in Climate Change Protest,' *The Guardian*, September 2, 2015, accessed September 1, 2017, https://www.theguardian.com/ artanddesign/2015/sep/02/underwater-sculptures-thames-london.

26   Ellis-Petersen, 'Underwater Sculptures.'

27   MUSA was founded by Roberto Díaz Abraham, former president of the Cancun Nautical Association, and Jaime González Cano, director of the National Marine Park. They invited Jason deCaires Taylor to create permanent life-sized and monumental sculptures.

28   The museum has begun to add the work of other Mexican sculptors to its collection. The goal is to expand to the permissible limit: 1,200 structures in ten different underwater areas within the National Marine Park.

29   Outposts, 'Artist Completes Artificial Reef, '"The Silent Evolution" installing 400 Sculptures Underwater', *Los Angeles Times*, November 1, 2010, accessed September 2, 2017, http://latimesblogs. latimes.com/outposts/2010/11/jason-decaires-taylor-artificial-reef-silent-evolution-underwater-sculptures.html.

30   David Sim, 'Museo Atlántico: Europe's First Underwater Art Museum Opens to the Public,' *International Business Times*, January 11, 2017, accessed September 1, 2017, http://www.ibtimes.co.uk/ museo-atlantico-europes-first-underwater-art-museum-opens-public-1600510.

31   Ellis-Petersen, 'Underwater Sculptures.'

32   'The Seven Sisters,' *Voyages of Discovery*, accessed October 5, 2017, http://www.solarnavigator.net/ geography/sussex/seven_sisters_chalk_cliffs_sussex_coast.htm.

33   Gaby Bissett, 'Massive Chunk of 250ft Cliff Plummets into the Sea at Seven Sisters Beauty Spot Visited by Thousands Every Year in Biggest Landslide in a Decade,' *Daily Mail*, May 24, 2016, accessed October 5, 2017, http://www.dailymail.co.uk/news/article-3606507/Massive-chunk-250ft-cliff-plummets-sea-b.

34   Oliver Lowenstein, '"Land, Sea and Sky: Red Earth's Seven Sisters Ritual Performance",' *The Fourth Door Review, Unstructured .4,* accessed October 23, 2017, http://www.fourthdoor.co.uk/unstructured/ unstructured_04/article4_5.php.

35   Lowenstein.

36   Red Earth, email correspondence with the author, October 7, 2017.

37   Red Earth, email correspondence with author. Quote from Gary Snyder, *Back to the Fire* (Berkeley, California, Counterpoint; First Trade Paper Edition, 2007), 32.

38   Red Earth, email correspondence with author.

ECO MATERIAL TECHNOLOGY?

Electronics

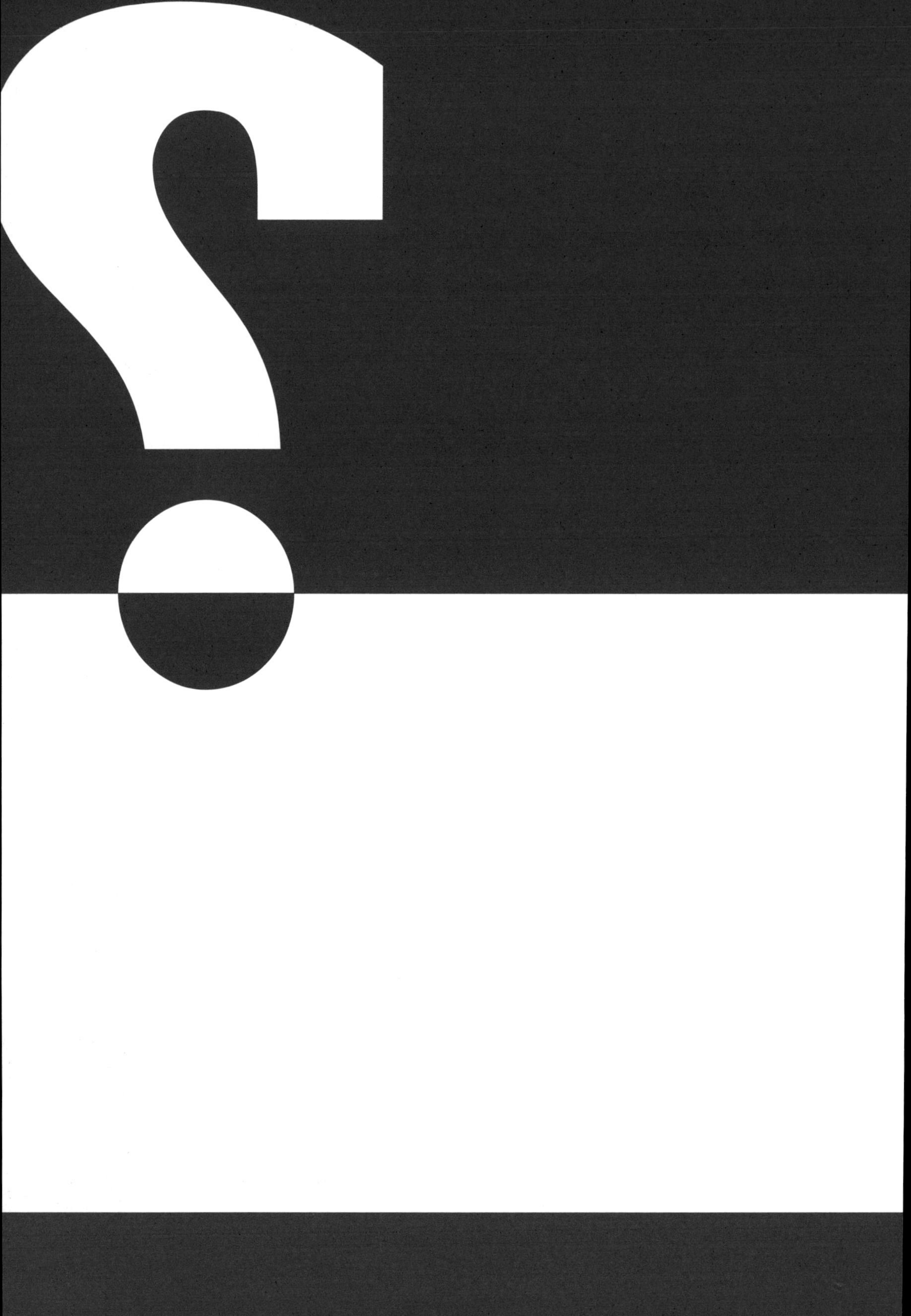

How many digital devices do you interact with during a typical day?

e.g. cell phone, computer, automobile engine and dashboard, thermostat, microwave, radio, LED bulb, ID cards, cameras, games consoles, MP3 players, ATM, stove, traffic lights, security systems, e-book reader, CAT scanner, grocery scanner, MRI, alarm, printers, GPS, calculator, TV monitor, clock, headphones, toys, tools, etc.

What is the average age of your devices?

What do you do with obsolete devices that still function?

If any part of a device you own were to break:

Could you fix it?

Would you fix it?

Would you replace it?

Do you equate digital technology with 'progress'?

Do you believe electronics represent clean technology because they reduce energy, conserve resources, and minimize waste?

By featuring the materiality of electronic mediations, the Eco Material artists in this chapter manifest the shift in definitions for terms such as farms, streams, clouds, and tweets that once evoked images of idyllic pastoral settings. Now 'farms' refer to internet servers; 'streams' provide access to programming; the 'cloud' is where digital data travels; and 'tweets' are communications from people, not birds. These metaphors indicate how deeply electronic communications are embedded in contemporary ideas, and how broadly these transmissions have dematerialized contemporary experience. This Eco Material playbook regarding digital technologies diverges from tempting upgrades, snazzy design, and affordability to attend to the physicality of computer parts and processes.

The Eco Material artists in this chapter challenge the common assumption that electronic technologies are safe for the environment because they are weightless, wireless, and frictionless. Gebhard Sengmüller converts bytes of data into bits of matter to reveal the physical underside of dazzling virtualized accomplishments; he exposes the consumption of invisible energy by digital technologies. Germaine Koh confronts people with the ubiquitous presence of electronic transmissions and connections that they may not initiate. Raphael Perret focuses on the unsavory consequences of e-manufacture and e-use that take the form of e-waste. SEAD, the collective founded by Angelo Vermeulen, redesigns the systemic functions of electronic devices by adding biological systems to the microchips, semiconductors, motherboards, cathode ray tubes, housing, wires, batteries, and other concrete components that transmit dematerialized data. Alice Anderson evokes the drawbacks of allowing digital devices to usurp human mental functions. Kal Spelletich challenges the blind faith that technologically advanced societies lavish upon electronic devices.

In all these ways, these Eco Material artists are conducting a full accounting of the environmental and cultural consequences of global signal networks. By including tangible transmitters and receivers in their calculations of the nebulous cloud, these artists are questioning the esteem bestowed upon the creators of immaterial ideas, codes, languages, mathematics, theories, and images. Bolstered by the realization that there is no escape from the material world, their diverse approaches suggest that technologies that are prized for delivering communications among humans, also deliver endangering jolts to habitats and their diverse populations.

E-Energy
Consumption

Gebhard
Sengmüller

Gebhard Sengmüller (b. 1967, Vienna, Austria) is currently working in the field of media technology. Since 1992, he has developed projects and installations that focus on the history of electronic media; creating alternative ordering systems for media content; and constructing auto-generative networks. His work has been shown in Europe, the United States, and Asia. Notable exhibition venues include Ars Electronica Linz; the Venice Biennale; the Institute of Contemporary Arts, London; Postmasters Gallery, New York; the Museum of Contemporary Photography, Chicago; Microwave Festival, Hong Kong; and the ICC Center, Tokyo. He holds an MA from the University of Applied Arts, Vienna.

When refueling a car, we physically apply pressure to a nozzle, observe the meter measuring the transfer of gas from the pump to the car, and smell the evidence of the fuel being transferred. Energy consumed by smartphones, tablets, and other digital devices, on the other hand, is imperceptible and automatic.[1] Owners of digital devices commonly decry the effect of big oil in producing energy, but ignore the fact that big data guzzles a great deal of this energy, and coal, as the generator of electricity, also runs the digital universe.[2] Consider, for example, the following statistics: the global carbon footprint from spam annually is equivalent to the greenhouse gases pumped out by 3.1 million passenger cars using 7.6 billion litres (two billion gallons) of gasoline in a year.[3]

The artist Gebhard Sengmüller, in collaboration with Franz Büchinger, created a compelling portrayal of the correspondence between digital transmissions and the energy needed to activate them. *A Parallel Image* (2009–10) is an absurd apparatus in which a camera captures the images that are then projected on a functioning, but intentionally inefficient television. The television monitor consists of a grid of light bulbs. Sengmüller materializes the transmission of each pixel sent by the camera and each pixel received by the monitor by connecting them with a thick copper wire. In this manner, all 2,500 pixels recorded by a camera are physically connected to 2,500 pixels projected on the television monitor. Additional wires supply electricity to each of the units that conduct this image transmission. The great tangle of copper wires that occupies the space between them represents energy being consumed.

By returning wires to wireless technologies, Sengmüller displays the material logic of the signal pathways that connect the photoconductors and the light bulbs. Such circuits are typically installed in the device's inner-workings, concealed from view. In Sengmüller's work, physical cables and lines transmit the immaterial coding signals, enabling visitors to see, and even touch them. Because this transmission

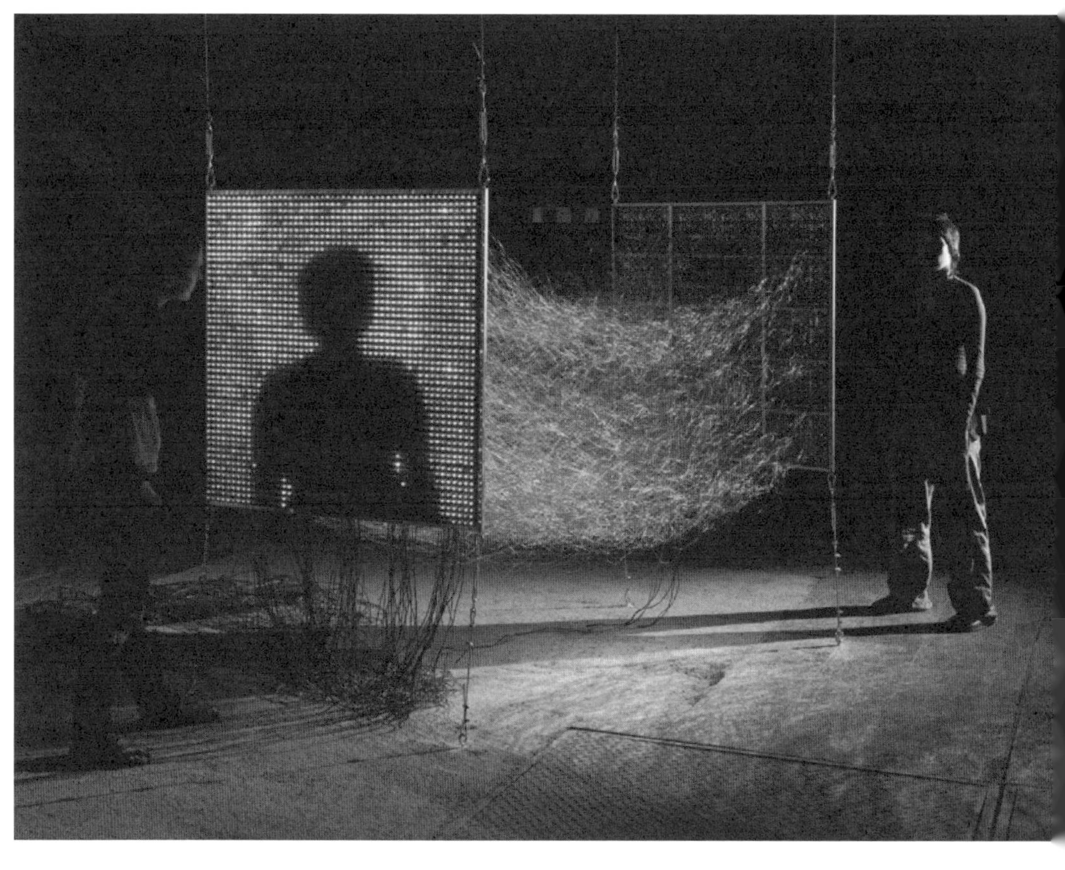

Figure 1a: Gebhard Sengmüller in collaboration with Franz Büchinger, *A Parallel Image* (2009–10), installation, copper wires, photo conductors, light bulbs, 100 × 199 × 180 cm. The Schmiede Hallein Festival, Hallein, Austria. Photo by Gebhard Sengmüller © 2008 by Gebhard Sengmüller.

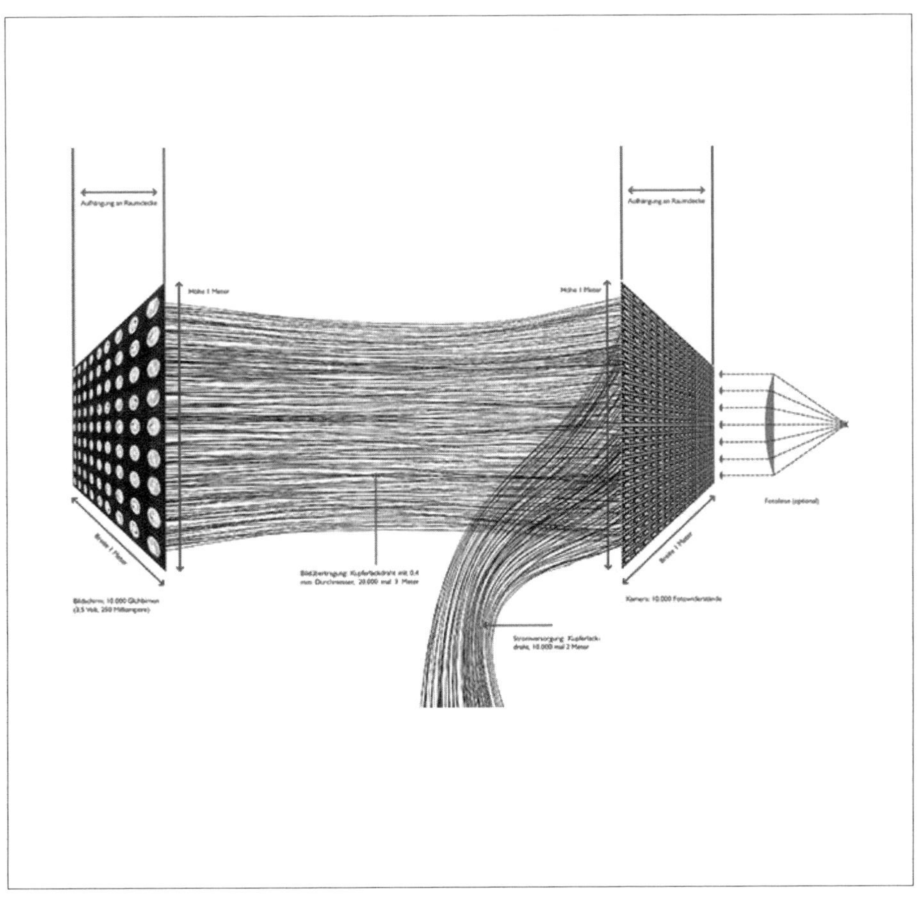

Figure 1b: Gebhard Sengmüller in collaboration with Franz Büchinger, *A Parallel Image* (2009–10), diagram of installation © 2008 by Gebhard Sengmüller.

from transmitter to receiver occurs in real time, it produces the 'parallel image' on the monitor that explains the title of this artwork.

Technically, the installation consists of two 1 × 1 m printed circuits hung from the ceiling about three meters apart. One is the camera. The other is the monitor. The camera circuit board is loaded with a grid of 50 × 50 in. light-sensitive photoconductors. The monitor circuit board is loaded with a grid of 50 × 50 in. small light bulbs. Sengmüller estimates that connecting them utilized about twenty-five thousand feet of glistening copper wire, which does not include the wires that lead to an external electricity supply.

Visitors who stand in front of the camera can observe the silhouettes of their bodies, which appear without delay, on the opposite monitor. They can manipulate this image by changing their distance from the camera, or by swiveling the photo lens to render bodies and objects in their gradations of brightness and plasticity. While the resolution of this image is necessarily minimal, the visual clarity is sacrificed to clearly reveal the energy transfer that serves as this work's theme. For this reason, Sengmüller refers to the 'techno-sculptural beauty of the installation' as a 'side effect'.[4] He explains:

> *Unlike familiar serial image transmission, the technology of A Parallel Image is completely transparent even to the lay viewer [...] there is a correspondence here between the real world and the transmission that can be sensually experienced. The television image is imbued with the directness of a film frame without the coding that normally takes place in the transmission of a television signal and does not allow for an easily comprehensible connection between the base image and the recorded signal (e.g. on video tape).[5]*

'In its directness, *A Parallel Image* is a radically new live medium that returns the visibility and comprehensibility of the process of transmitting an electronic image.'[6] The issues that *A Parallel Image* presents for the viewer's consideration exceed neglect among electronics users regarding electricity consumption. The prominence of copper in this installation, for example, demonstrates that this type of physical matter was not relegated to history. While contemporary users still value copper's durability, corrosion resistance, strength, and malleability, as those who lived during the Copper Age from approximately 6500 BC to 1700 BC, current technologies also exploit its high thermal and electrical conductivity. Copper greatly improves the efficiency of electricity generation, distribution, and use. Indeed, copper is essential to transmit power and information. It is the basis of contemporary lifestyles. The artwork also suggests that so much energy is lost during transportation from its source to its destinations in electrical appliances that only twenty-eight percent survives.[7] However, this number would be even more dismal if it were not for copper. *A Parallel Image* testifies to the fact that sixty percent of the copper produced worldwide is currently used in electrical motors, generators, transformers, printed circuit boards, microchips, wires, and cables.[8]

Sengmüller concludes:

> *You could call me a media artist, because I play with the medium and because I'm more concerned with the medium than the content [...] Yet I am specifically not interested in the so-called high-tech world, in which emphasis is placed on even more powerful computers, even more fantastic software. I'm interested in machines, but not in a way that means I'm concerned with newer, better machines or software that are even better tools for generating something even more spectacular. You could say that I prefer to manufacture tools other artists can use, or to create processes that are then capable of producing content independently.[9]*

**E-Omni
presence**

**Germaine
Koh**

*Germaine Koh (b. 1967, George Town, Malaysia) is a Vancouver-based
conceptual artist whose work draws on the use of everyday objects and
familiar concepts in order to examine the materiality of communications
among people throughout the world. She has exhibited her work at the BALTIC
Centre, Newcastle; De Appel, Amsterdam; Musée d'art contemporain de Montréal;
Para/Site Art Space, Hong Kong; Frankfurter Kunstverein; Bloomberg SPACE, London;
The Power Plant, Toronto. Koh was a recipient of the 2010 VIVA Award, and a finalist
for the 2004 Sobey Art Award. Formerly an assistant curator of contemporary art
at the National Gallery of Canada, she is also an independent curator and partner in
the independent record label '(weewerk)'. Koh studied fine arts and art history at the
University of Ottawa and received an MFA in 1993 from Hunter College in New York.*

People walking through the shopping area attached to an office tower heard
sounds that alerted them to the fact that something unusual was occurring. They
did not know that the strange noises were part of an installation created by the
artist Germaine Koh, working with the video game producer, Ian Verchere. *Broken
Arrow* (2009) was their contribution to Toronto's Luminato Festival of Arts +
Creativity in 2009. The sounds resembled insects chirping. While the volume of
the individual chirps and clicks was not loud, at times the layering was so dense
it became irritating. Thus, the insects' aural presence did not evoke pastoral
pleasantries; it was more likely to be experienced as an incursion into a space
claimed for humans.

Pausing to seek the source of these mysterious, misplaced sounds led to a visual
inspection of the area, revealing another alteration of the space. White, yellow,
and blue text was being projected on a nearby wall. The density of the words in
the projection correlated with the intensity of the sounds occupying the space at
that time, indicating that sound and image were responding to the same stimuli.

The text enabled people to surmise that the sounds were not related to insects.
Instead of biology, they were emanating from technology. The text identified
TV, Air Traffic Control, Search & Rescue, FM Radio, GPS, Montgomery Mobile,
GPS satellite emergency services, Bluetooth, Wi-Fi, cordless phones, radio, and
radio frequency identification as examples. Persons were surprised to find their
own names included in this projection alongside the words identifying the actual
devices they were currently carrying. This correlation indicated that surveillance
technologies were not only scrutinizing them; they were projecting the information
they gathered within that public space in real time.

Besides revealing the presence of digital technologies and the names of the
owners of these devices, Koh and Verchere applied visual and aural aesthetic

strategies to the presentation of the text to evoke the cluttered, but invisible domain of cyberspace that the projected words revealed. One visual strategy involved registering the intensity of the frequencies of the transmissions by presenting powerful signals from Bluetooth and nearby radio waves larger and brighter than those from weaker signals. Another strategy evoked the quantity of transmissions by varying the density of projected texts. The third strategy applied the aesthetic character of electronic transmissions in 'the cloud' to the formal presentation of the text. Words floated across the projected space without regard to gravity or spatial logic. The names of digital transmissions swelled in size and diminished. Colors intensified and faded. The letters ascended and descended. In all these ways, the text resembled the nebulous quality of clouds in the sky as within telecommunications.

The aural component of *Broken Arrow* was equally evocative. Various sensing hardware detected devices in the vicinity across a range of wireless frequencies. They generated a stream of Geiger-counter clicks for each device or frequency detected. Koh explains:

> *The project makes audible the constant networking performed by our devices (sometimes without our being conscious of it). It references and suggests relationships between different realms and eras of detection and locative technologies: the detection of radioactive particles by a Geiger counter, the acoustic pinging of sonar, the call-and-response of insects, and today's ubiquitous mobile devices. All are used as locative tools and, like a tuned sensor, are used to process or interpret a world of information around them.*[10]

In all these ways, *Broken Arrow* delivers visual and auditory indicators of the real but unseen realm of wireless transmissions.

The devices that intercepted the transmissions from people passing by were housed in a stack of boxes suspended from the ceiling in the middle of the space. It contained common electronic devices, such as a Mac Mini, that were easily converted into surveillance systems by software apps and additional hardware. *Broken Arrow* demonstrated that these simple technologies are capable of picking up specific waves and frequencies emitted by Wi-Fi, Bluetooth, radio and television broadcasting, cellular telephones, personal communications services (PCS), pagers, cordless telephones, business radio, radio communications for police and fire departments, amateur radio, microwave point-to-point links, and satellite communications. The immaterial signals, waves, and frequencies that comprise this chaotic realm consist of vibrating packets of electromagnetic energy travelling at the speed of light. They do not detour to avoid colliding with us. By configuring imperceptible but all-powerful cyber components as perceptible

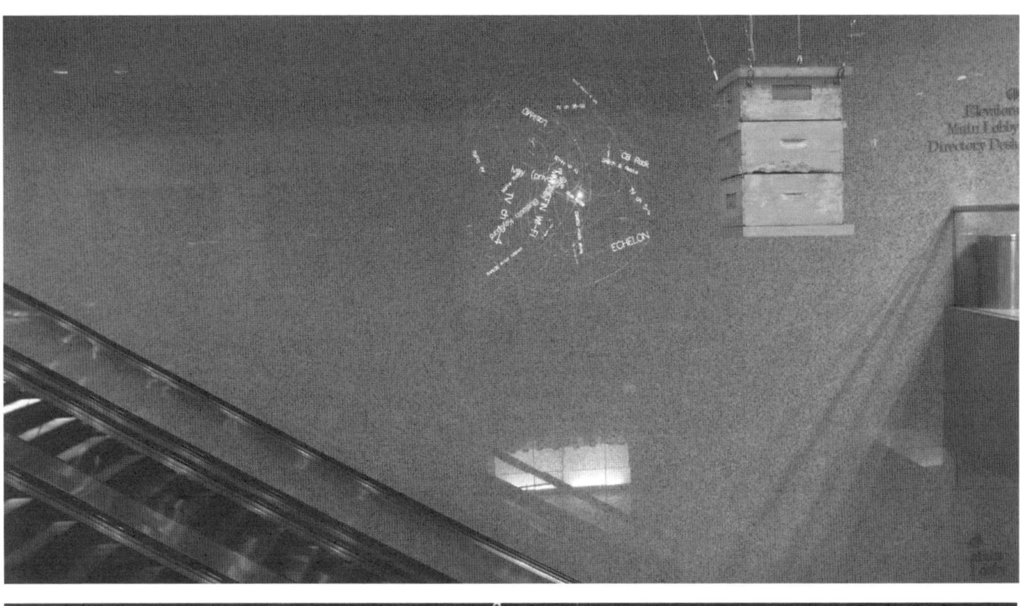

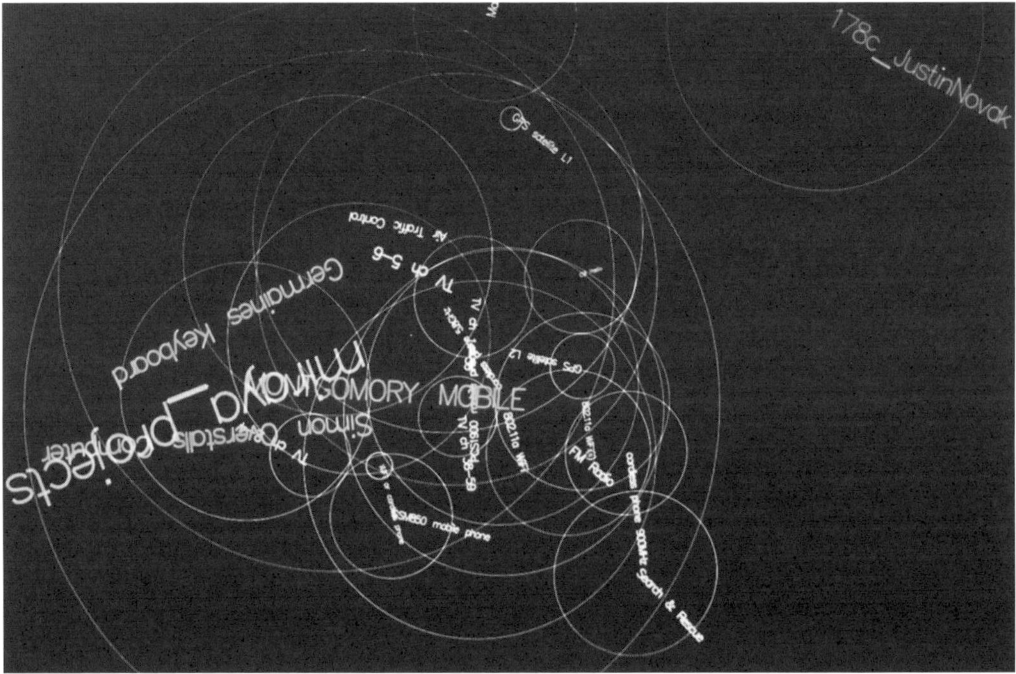

Figure 2a: Germaine Koh and Ian Verchere, *Broken Arrow* (2009). Photo by Germaine Koh. Courtesy of the artist.

Figure 2b: Germaine Koh and Ian Verchere, *Broken Arrow* (2009), detail of projection. Photo by Germaine Koh. Courtesy of the artist.

sensory experience, *Broken Arrow* exposes the bombardments that originate in these myriad telecommunications applications and their ability to intrude upon our personal space. Koh comments, 'I think that ideas often have to be experienced physically, and although some people think that my work is strictly conceptual, in fact it is always rooted in a physical encounter with the material of the world.'[11]

Besides the psychological ramifications of living in the midst of intense cyber traffic, *Broken Arrow* also introduces the possible health hazards associated with electromagnetic emissions from cell phones, wireless networking devices, and microwaves. The radio frequency waves they emit are suspected of causing cancer and brain tumors. This unsettling implication is conveyed by the work's title. According to Koh, 'broken arrow' is a military term for a missile that has gone astray, and cannot be controlled. This analogy suggests that the electronic communications we have unleashed into the atmosphere may be as dangerous as rogue missiles. Koh explains, 'The long-term effect of exposure to electromagnetic fields (EMFs) is still not clear, yet we willfully put those considerations aside in our rush to surround ourselves with radio frequencies.'[12]

Koh expanded these ethical and environmental considerations by choosing a three-tiered beehive to house the hardware that detects the wireless frequencies in the area. This biological reference, reinforced by the insect sounds, invites a comparison between biology and technology. The controlled complexity of relationships among bees within a crowded beehive offers one model of social structure. The boundless, frenetic, anonymous, chaotic relationships in cyberspace establish an opposite structure. Bees epitomize the productive advantages of distributing predetermined tasks among residents working toward a shared goal. While all these qualities are absent in cyberspace, both communities function with remarkable efficiency.

Broken Arrow exposes both the intended exchanges and the unauthorized surveillances that occupy the flip sides of current technologies. How did people respond to this unexpected encounter with their own susceptibility to surveillance tactics? Koh reports that some seemed uncomfortable and quickly shut their devices off. Others welcomed the opportunity, as if it was a new form of social media, by turning their devices on. Koh explains her response to this inconsistency:

> *I am interested in points of ambiguity as junctures at which viewers/ receivers/interlocutors are required to weigh for themselves not only various potential meanings but also their own predilections, and thus are points at which the processes of communicating, of making sense and of reckoning with things, are drawn out.*[13]

Significantly, Koh does not pass judgment on technologies for communication or surveillance. In the final quote, she refers to both as 'commonplace things', 'minor things', and part of the 'poetics of daily life'. This terminology reveals the ordinariness of cyber experiences precisely because they encroach upon so many aspects of daily life. She explains:

> I have great faith in the power of commonplace things to tell us about ourselves, how we live and how we relate to each other. I think that the minor things that mediate our everyday lives inevitably bear a residual meaningfulness, and much of my work has been an effort to allow these things to speak quietly back to us. I would characterize my work as a whole as an attempt to be attentive to the poetics of daily life by focusing on those phenomena that shape everyday experience, often slightly below the threshold of notice (and, yes, value). I would like to create moments in which the commonplace, the mundane and the ubiquitous are rendered remarkable again.[14]

E-Waste

Raphael
Perret

Raphael Perret (b. 1977, Uster, Switzerland) is a Zurich-based artist, exploring the interplay between physical and virtual spaces. He has taught at the Interaction Design Department of Zurich University of the Arts. Perret's projects have been shown across the globe from Gwangju, Delhi, Prague, and London to Rio de Janeiro. In 2014, he published Machines of Desire,[15] *documenting the* Recycling Yantra (2014) *project. Perret holds a BA in interactive media and a Master of Advanced Studies in scenography.*

Associating death with endings seems inevitable since departed loved ones vanish from our immediate environs. The same association is typically applied to digital technologies. These assumed terminations disregard the fact that all physical entities persist in the material realm even when they cease being present in our lives. Ironically, the effort expended by doctors and caregivers to postpone human deaths frequently matches efforts by manufacturers and retailers to accelerate commodity 'expirations'. Two very different afterlife scenarios emerge from this discrepancy. Organic molecules from deceased entities cycle back to restore Earth energies and matter. They perpetually renew ecosystems. The hereafter of electronic technologies, in contrast, often ends as accumulations of minerals and chemicals that persist as harmful contaminants. Efforts to recycle digital components are obstructed by the fact that computers contain a plethora of materials. The typical computer chip, alone, is composed of sixty different elements.[16] Comparable forms of matter are used in the central processing unit, hard drive, CD drive, RAM, and so forth. Some of these elements are common, such as chromium and cadmium. Some materials are unusual, like 7,7,8,8-tetracyanoquinodimethane. Some elements are dangerous, like lead, mercury, and bromine. Others are 'rare earth' minerals like lanthanum, lutetium, and dysprosium that deplete millions of years of formation to provide a human service for a few seasons. As a result, the temporal considerations regarding technologies expand from the time of their functionality for humans to occupy 'deep time'.[17] Deep time accounts for the depletion of the minerals mined to create technological media that originated in the far distant past, and the introduction of e-waste into the geological record of Earth that will endure eons into the future. The artist Raphael Perret tackled this environmental problem when he was invited by the Swiss company, Swico Recycling, to create an artwork to commemorate its fifteenth anniversary. Swico made headlines in 2013 when it safely recycled sixty thousand tons of discarded electrical and electronic equipment. The project Perret proposed to honor this accomplishment consisted of a large installation using electronic waste as his art medium. He intended to arrange the waste in the manner of symbolic visualizations of the cosmos, known as 'mandalas' in Buddhism and 'yantras' in Hindu religious practices. These sacred diagrams are traditionally created with colored sand. Formally, they appear as a circle with a point in the center, enclosed by a square with a 'gate' at each corner. Symbolically, they embody the celestial realm and the enlightened mind. Functionally, they facilitate meditation and opening channels of consciousness into a sacred space.

In addition to emulating the formal, symbolic, and functional characteristics of the mandala/yantra tradition, Perret also planned to replicate the Tibetan ceremony in which a mandala made of sand is swept up after it has served its purpose, and ritualistically released into a river to honor the cycle of life. Perret planned to enact this spiritual evocation of recycling to draw attention to a contemporary e-waste challenge. He proposed returning his e-waste mandala to a contemporary version of a 'stream' – the e-waste stream – to highlight the long-term effects of these toxic operations. Mountains of toxic e-waste testify to the scale of this problem. By conflating an ancient symbol of purity with a contemporary embodiment of blight, Perret highlighted the contradiction between the mindful reverence of the cosmos that mandalas and yantras cultivate, and the mindless disregard for unwanted electronic devices that consumer lifestyles allow. Within this sacred context, the degradation of ecosystems by electronic discards appears as a desecration of the eternal cosmos.

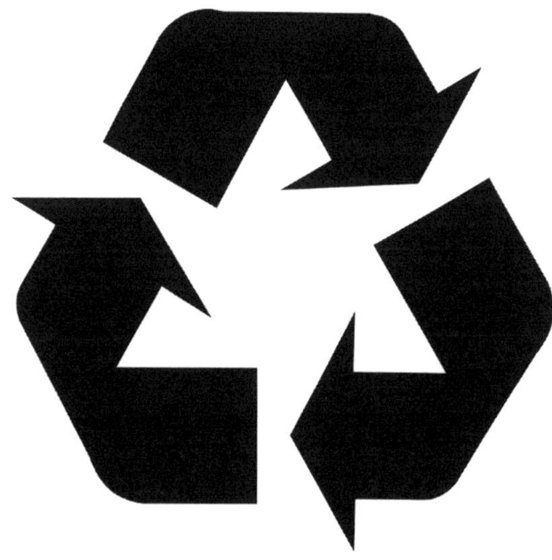

Figure 3: The universal recycling symbol. c. 2002.

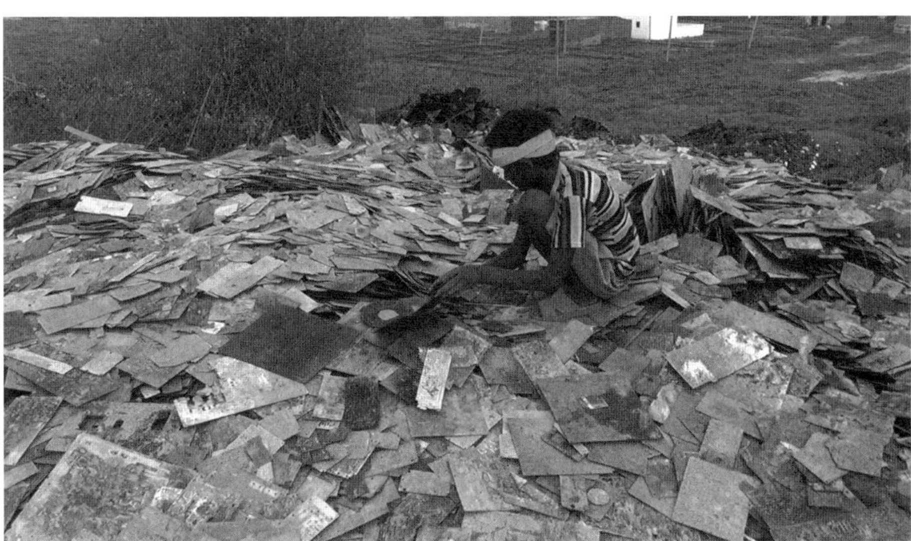

Figure 4a: Raphael Perret with Claude Hidber, *Recycling Yantra* (2014), video still of child rummaging for copper residue on etched PC boards . Photo by Raphael Perret. Courtesy of the artist.

Swico Recycling rejected Perret's proposal because, despite their commitment to recycling, they wanted a permanent artwork to commemorate their anniversary. Nonetheless, Perret launched an extensive study of the problem, explaining, 'After ten years of work in the electronic arts, we wanted to know more about what happens with old devices and what this means as a society and global entity.'[18] The work that resulted from this research did not merely present the quandary surrounding e-waste; it delivered a searing indictment of planned obsolescence as the source of the environmental blights and human hazards associated with e-waste. The digital media industry thrives on disposal and replacement. Evidence that obsolescence is an intentional strategy of designers and producers is apparent in electronic devices that are outfitted with difficult-to-replace batteries, proprietary cables, and chargers that are installed in plastic cases that are glued shut against repair. This strategy is shored up by marketing and advertising strategies that promote the purchase of upgrades. The twenty to fifty million metric tons of e-waste that are disposed of worldwide each year provides evidence of the success of these efforts to maximize sales.[19]

What happens to all this waste? Eighty to eighty-five percent is dumped into landfills or incinerated where it leaches toxic substances into water, soils, and the atmosphere.[20] The remaining fifteen to twenty percent is recycled, typically by travelling from Europe and America to Africa, China, and India. Perret admits that he, too, had subscribed to what he described as:

> *the never ending [sic] promise of a better life (or even a better world) with the newest feature/gadget/product/technology/upgrade/version/... you name it to have more, feel better and be faster. Yes, our future is bloated with promises, yet I never had the impression of having arrived there. Anyhow, on the way there, we leave a tremendous trail of data and outdated technology behind.[21]*

He then adds the wry truth that we shouldn't be bothered about dealing with the glut of non-biodegradable waste. 'It will most probably be somebody else's job, somewhere far away.'[22]

Perret and his collaborator, Claude Hidber, set about revealing the afterlife of discarded electronic devices.

> *We are living behind a veil in our western society. Artifacts appear in stores, we buy them, and after they [have] reached their lifetime [they] disappear under mysterious circumstances. The bigger part of a product's reality, its production and afterlife, remain hidden. Therefore we are interested in closing the circle by exposing the dark side of the trade as well as our society.[23]*

Examining this 'dark side' disclosed the electronics boom in the midst of an ecological bust. India is where Perret and Hidber launched their study. The artists travelled there in 2012 and again in 2013 to document the nation's massive e-waste recycling network. They immediately discovered how much it diverged from the meticulously regulated processes that earned Swico Recycling its reputation as a world leader in safe and efficient management of e-waste. Approximately ninety percent of India's e-waste was handled through 'the informal sector' that functions without labor protection or social welfare. Although electronic technologies are often touted as clean technologies in developed countries, in India, entire villages of men, women, and children are exposed to toxic mercury, lead, cadmium, PVCs, brominated flame retardants, and Chromium VI as they extract and hand-sort metals, toners, and plastics from electronic waste; or repair chips, transistors, and RAM. The artists

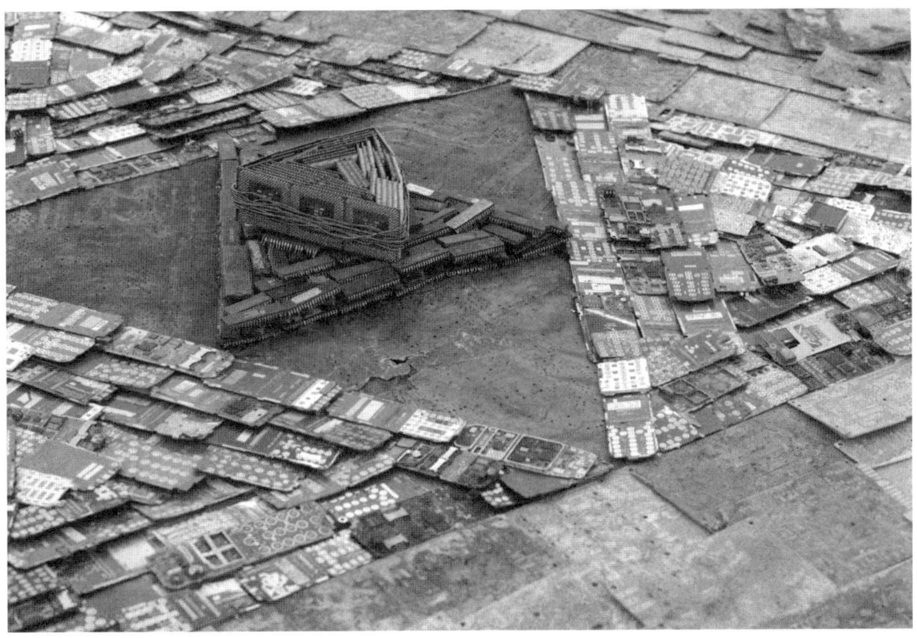

Figure 4b: Raphael Perret with Claude Hidber, *Recycling Yantra* (2014), detail of installation, sorted electronic waste, activated Yantra 5 × 5 m. Photo by Gion Pfander. Courtesy of the artist.

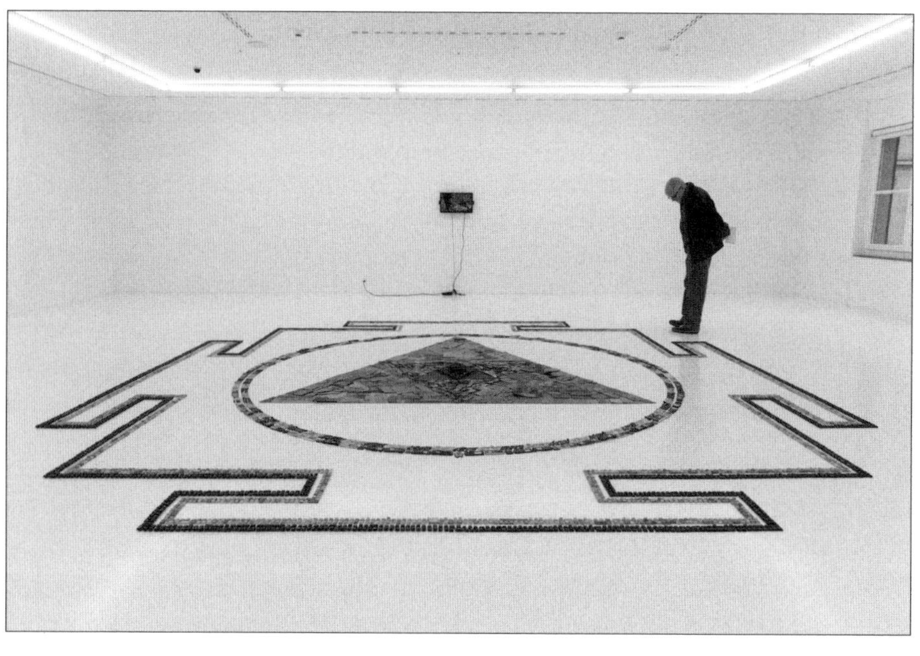

Figure 4c: Raphael Perret with Claude Hidber, *Recycling Yantra* (2014), installation view, sorted electronic waste, 5 × 5 m; activated *Yantra* video on four channels, 01.03.58. Photo by Gion Pfander. Courtesy of the artist.

documented India's e-waste recycling networks by producing photographs and a video recording how every bit of e-waste is retrieved and recycled without regard for the health of the workers or their surroundings.

The installation that represents this research occurred at Ars Electronica in Linz, Austria in 2016. In addition to the photographs, the artists installed a Hindu yantra comprised of collected e-waste that filled the floor of the exhibition space. The parts were carefully arranged to correspond with their material dismantling. Perret explains:

> If you look at the Yantra from edge to center, it starts with keys and plastic cases, the material we interact with, and the more to the center you go the more valuable the material gets, up to gold-plated pins.[24]

The form referenced a specific type of yantra known as the 'smara-hara' yantra. Within the tantric tradition, this yantra is used as an instrument to free people from desire. Because electronic devices rank high among contemporary desires, the smara-hara yantra presented itself as a counterpoint to consumer-culture.[25] The artists conclude, 'The particular thing that has created all the problems of life is the dissatisfied mind of desire, the mind clinging to this life.'[26] For this reason they promoted a form of 'enlightenment' that expanded awareness beyond personal desire to encompass an ecological and cultural awakening. They applied this traditional symbol to contemporary desires and fixations on digital media.

Recycling Yantra confronts the afterlife of wires, routers, switches, chargers, and batteries, as well as the heavy metals that are recovered as these parts are broken down. The lingering effects in the material environment constitute a 'fossil' record in terms of toxic water, contaminated air, depleted soils, and threatened life forms. Perret's artwork is an attempt to prevent this blight from becoming our enduring legacy to future generations. His prevention scheme is summarized in the title of the publication documenting this project – *Machines of Desire* (2014).[27] Yantras diminish the desires that drive electronic consumer-culture and produce a glut of toxic wastes. The book is described as a documentation of the *Recycling Yantra* project as well as an exploration of desire in electronic consumer-culture.

E-Systems

SEAD collective, Angelo Vermeulen

SEAD (Space Ecologies Art and Design) is an international transdisciplinary network of artists, scientists, engineers and activists founded by Angelo Vermeulen (b. 1971, Sin-Niklaas, Belgium). Vermeulen is a visual artist whose multidisciplinary practice merges the disciplines of biology, technology, and community. Vermeulen earned degrees in fine arts with a focus on photography, video, and installation art. He also earned a doctorate in biology and is working toward another degree in space systems design. In 2012, Vermeulen won the Witteveen+Bos Art+Technology Award. He was crew commander of HI-SEAS Mission I, a Mars simulation study on improving the nutritional value of space food, funded by NASA. As a Senior TED Fellow, he travels the world to share information about his art and scientific projects.

Despite his extensive training in photography and science, Angelo Vermeulen diverges from the camera and the microscope to create living computers that materialize the ability of biological processes and computer technologies to support each other. *Biomodd* is an ongoing art project by SEAD (Space Ecologies Art and Design), a collective co-founded by Vermeulen in 2009. The group's eligibility to represent 'What's next?' is apparent in its mission statement that includes such terms as 'interplanetary' and 'paradigm-shifting'. Referring to the artists, scientists, engineers and activists that comprise the SEAD collective, it states:

> *SEAD (Space Ecologies Art and Design) is an international transdisciplinary network of artists, scientists, engineers and activists. We reimagine and reshape the future through critical inquiry and hands-on experimentation. To achieve this, SEAD develops paradigm-shifting projects in which ecology, technology and community are radically integrated in unexpected ways.*[28]

SEAD's paradigm-shifting mission is to disprove the assumption that biological and technological systems are incommensurable. It not only merges these divergent systems into a unified, complementary, functional whole; the contributing systems are enhanced by this partnership. This premise is materialized in an ongoing project named *Biomodd*.

The prefix 'bio' in *Biomodd* consists of living single-celled algae, plants, and fish. The suffix 'modd' refers to a subculture of computer tinkerers who practice 'case modding'. These spirited rebels not only transform the standardized anonymous computer case into imaginative expressions of their personalities; they push the computer's intended levels of performance to heights unimagined by their designers. However, there is a practical limit to the ability of modders to test the extent of their ingenuity. Hyped-up home-built computers overheat! SEAD tackled this stubborn physical problem that stymies modders by offering a sustainable way to cool the computers, and thereby augment the speed of their functions – while simultaneously augmenting the growth of plants.

Biomodd is an evolving project that has been conducted in the United States, the Philippines, Belgium, Slovenia, New Zealand, the Netherlands, the United Kingdom, Chile, China, Germany, Kosovo, and Taiwan. Each iteration builds upon the accomplishments of previous innovations. Nonetheless, the fundamentals remain the same. Discarded, non-functioning computers are collected, dismantled, and their components are reassembled into functional devices. They are not, however, returned to their original conditions. Living ecosystems are installed among their parts. Some of the computer electronics are water-cooled using algae. The waste heat of the remaining computer electronics is captured and used to create a warm climate for the plants to boost their growth. In this manner the computer processors are cooled while algae are pumped through tubes to collect the excess heat, thereby cooling the computer processors so they can run faster. The heat they collect is circulated to augment the growth of a plant-based ecosystem that is installed with each computer. Vermeulen explains:

> *Part of the computer is cooled using a liquid just like in a car but instead of using a commercial coolant liquid we use a culture of single-celled algae. It's a living coolant liquid that runs through part of the computer, it's all connected.*[29]

This vibrant hybrid ecosystem is a perpetual work-in-process that expands and refines biological/engineered compatibility. Vermeulen explains, 'The challenge is to bring biological life as physically close to the electronics as possible, and allow them to communicate with each other through meaningful symbiotic relationships.'[30] These unlikely alliances of computers and ecosystems are installed in transparent cases so that viewers can observe a living ecosystem benefiting from technology (its excess waste heat helps plants grow) while technology is benefiting from a living system (the over-heated electronic device is cooled). Later versions of *Biomodd* make use of sensors and robotics that conduct 'intelligent' interactions between computers and plants. The promising results of this exchange are demonstrated by growing wheatgrass, oregano, basil, and tomatoes! How does this project qualify as an artwork, and not simply a clever contraption? Answers are provided by the six strategies that exceed material combinations and manipulations by offering compelling cultural insights.

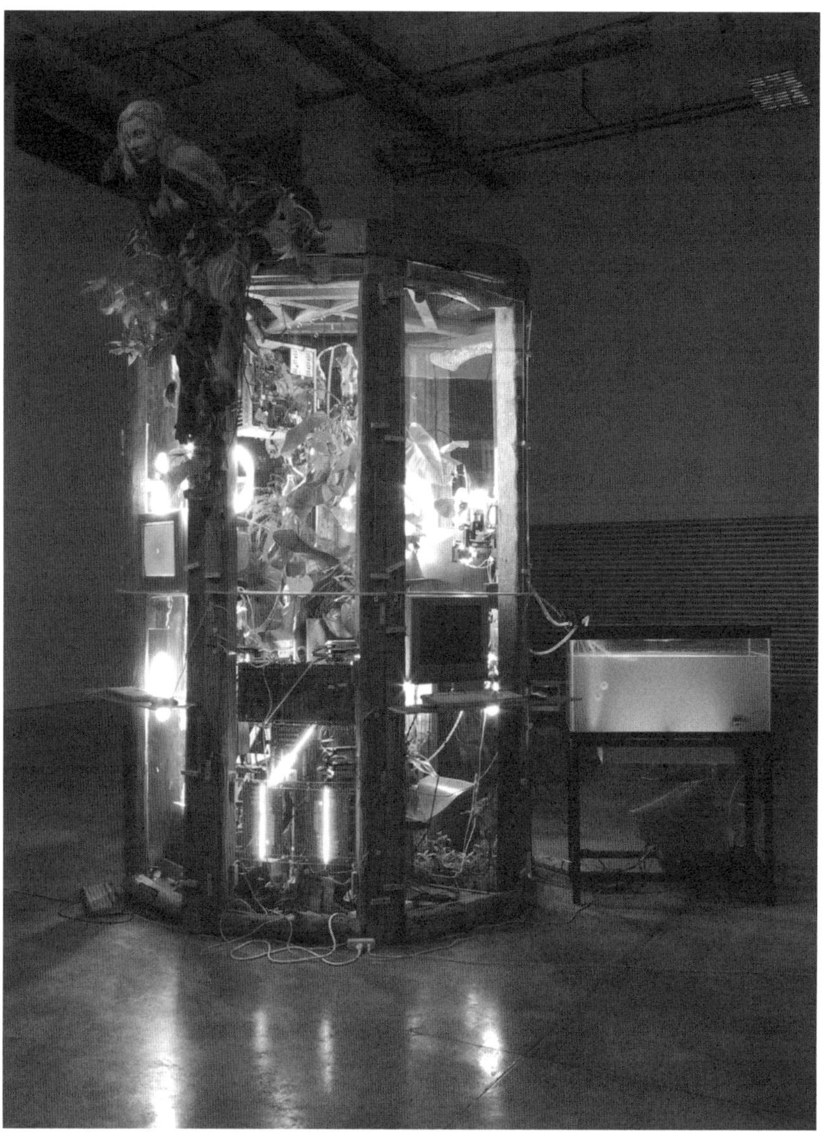

Figure 5a: SEAD (Space Ecologies Art and Design), *Biomodd [LBA]* (2009–10), installation view, upcycled computers with peripherals, network router and hub, personal computer (PC) water cooling systems, PC case modding lights, Arduino microcontroller with environmental sensors, various other electronics, hydroponics system, air and water pumps, PET (polycarbonate tubing, repurposed) bottles, aquariums, glass panels, reused coconut timber, wood carving and sculpture, lighting, various repurposed building materials and household items, various local plants, single-celled algae, goldfish, tailor-made multiplayer computer game, 3 × 3 × 3 m. Museum of Contemporary Art and Design, Pasay City, Metro Manila and the University of the Philippines Open University, Los Banos, Laguna, Philippines. Photo by Angelo Vermeulen. Courtesy of the artist.

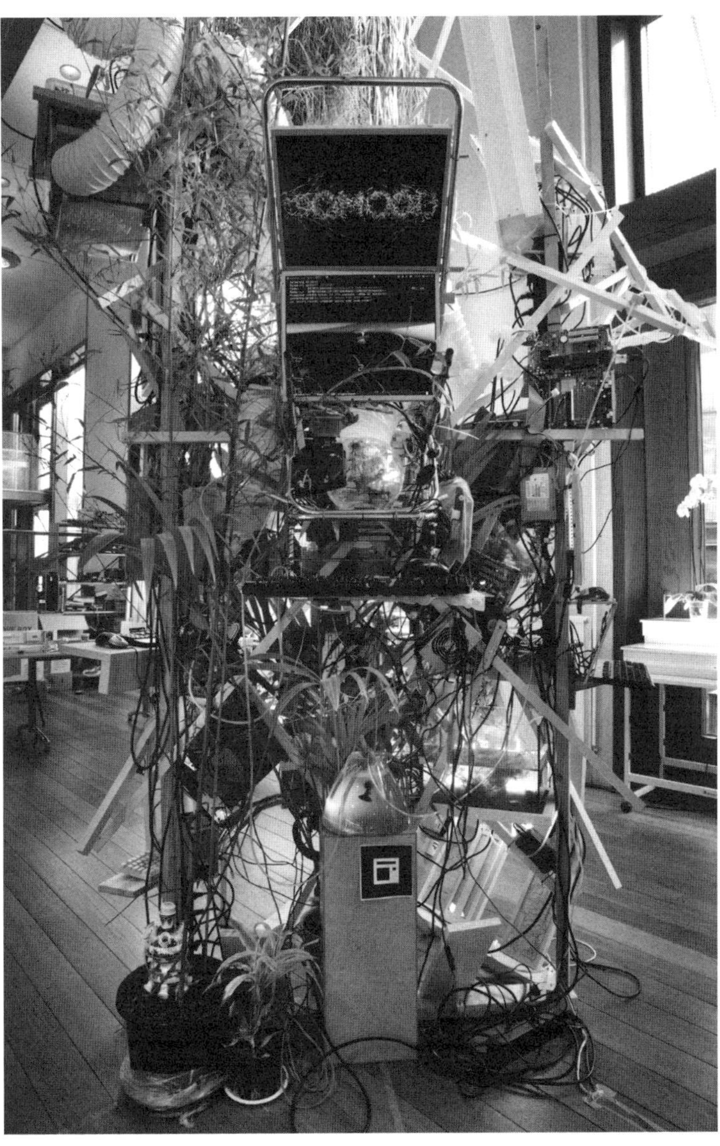

Figure 5b: SEAD (Space Ecologies Art and Design), *Biomodd [TUDelft3]* (2011), installation view, upcycled computers with peripherals, network router and hub, PC water cooling systems, various other electronics, steel utility cabinet, air and water pumps, aquariums, fluorescent lighting, white vinyl air duct, wood laths, various other reused building materials and household items, various house plants, single-celled algae, open-source multiplayer computer game, augmented reality system, tailor-made data visualization software, 4 × 3 × 4 m. Delft University of Technology, Delft, the Netherlands. Photo credit by Marijn De Reuse. Courtesy of the artist.

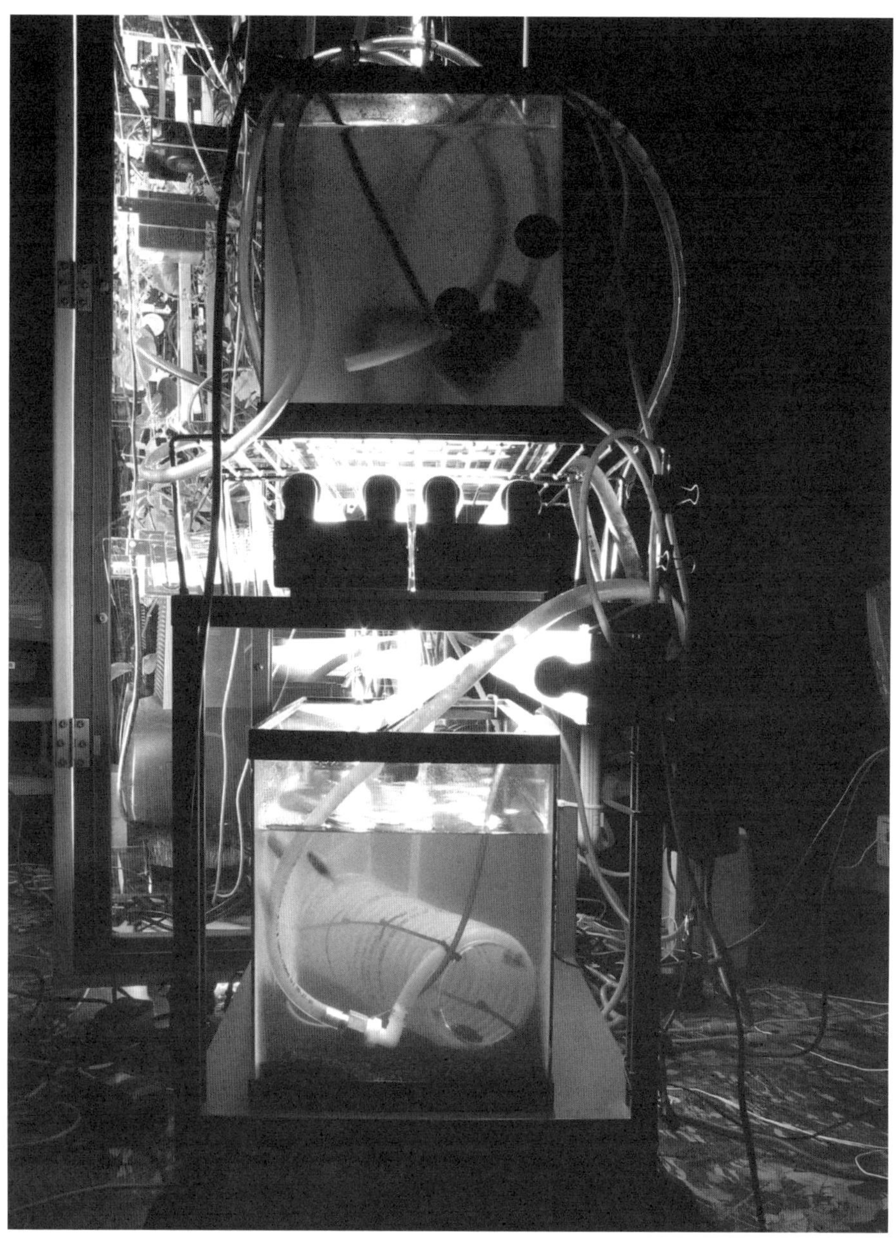

Figure 5c: SEAD (Space Ecologies Art and Design), *Biomodd [ATH1]* (2007–08), installation view, large computer case installed inside transparent structure that contains a living ecosystem of plants, transforming the computer into a greenhouse cyber-sculpture and gaming stations stations for a multiplayer computer game. Union Arts & Aesthetic Technologies Lab, Ohio University, Athens OH, US. Photo by Angelo Vermeulen. Courtesy of the artist.

***Biomodd* is communal:** The work provides a creative catalyst to visitors because the computers in each installation are not merely 'on'; they function as gaming stations for a fully operational multiplayer digital game that is run on a LAN. The game involves growing a virtual organism whose life-sustaining needs are triggered by sensors embedded among living plants. Sometimes a separate row of vegetables grows by the windows of the exhibiting institution. These planters are outfitted with light, humidity, and temperature sensors that connect back to the computer system through a LAN. Playing the game deploys robotic caretakers to water, feed, and trim the plants when needed. Thus, through game-playing, the public not only competes or collaborates to play the games; they contribute to the techno/eco system's vitality. This multiplayer game thereby establishes a feedback loop that allows biological life and technology to coexist. Vermeulen calls this intertwining of the virtual and the real 'entangled reality'. He notes, 'The more the network is used, the more heat is generated and the more the ecosystem is being boosted.'[31]

***Biomodd* is global:** The communal component of *Biomodd* is not limited to visitors at exhibition sites. By posting *Biomodd* instructions online as an open-source manual, and also by developing open-source game software, SEAD invites off-site co-creation. In this manner the project serves as an international, cross-cultural platform for interdisciplinary collaborations between artists, scientists, and other social change makers. Various online applications to accomplish tasks outside the construction of the art installation are coursed through several avenues, including the university's official learning management system, text messaging, mailing lists, online photo albums, personal blogs and mass media.

***Biomodd* is sustainable:** *Biomodd* introduces a new creative reuse of e-waste because its systems function with obsolete hardware that has been refurbished. Vermeulen comments, 'I'm not interested in trash art. I'm interested in finding ways to recycle electronic waste.'[32] He is so committed to this endeavor that he has conducted classes in safe dismantling techniques of e-waste computer parts for youngsters in neighboring communities. In addition to inventing such creative reuse of e-waste components and cases, the rebuilt computers are fully disassembled when an exhibition closes. Most of the components are reused by the project participants. The rest are sustainably recycled. Optimizing sustainability by recycling also applies to the living entities in each installation. The plants, fish, and other biological components are either adopted by people who will care for them, or they are returned to their natural habitats.

***Biomodd* is nomadic:** *Biomodd* travels lightly. While selected parts of previous versions are integrated into each new installation, the installation basically arrives in the form of a strategy associated with a concept. New 'modds' are configured at each venue according to the skills and interests of local collaborators, local cultural traditions, and local materials. Besides performing prescribed functions, these techno-bio systems perform experiments on such diverse topics as aquaponics, edible plants, tropical plants, plant communication, and robotics.

***Biomodd* is co-created:** *Biomodd* creation is always a collaborative effort. Each version is formed through the exchange of ideas between numerous participants representing a wide range of skills and talents. While SEAD provides the creative spark by establishing the principles of the work, the design is conceived and fabricated by an intentionally unstructured team of local contributors. Top-down and bottom-up organizational structures are avoided to optimize co-creation. Vermeulen states, 'This is why the project begins with a solely conceptual focus, and grows organically from there.'[33]

***Biomodd* is evolutionary:** As a developmental biologist, Vermeulen is aware that future prospects are contained within present possibilities, but they can never be predicted. *Biomodd* applies this theory to an 'evolvable' technological/biological system. As in biological evolution, *Biomodd* develops without a master plan by adapting to ever-changing conditions as they occur. Its hybrid systems emerge and are continually altered by random collections of people, electronics, and living entities.

***Biomodd* is visionary:** *Biomodd* models an 'entangled reality' between humans, organic materials, and machines in which the virtual and real are mutually reliant instead of mutually endangering. In this manner, this project imagines 'a new ecology of relations'[34] and 'a critical awareness of connectedness.'[35] Maranan & Vermeulen explain, 'By this we mean a shared understanding of how the flourishing of human societies relies on our interaction with the natural ecosystems on which we depend, and on a critical engagement with the artificial systems that we create.'[36] In sum, *Biomodd* has multiple identities. It belongs, simultaneously, to art, gaming culture, horticulture, environmentalism, technology, and community building. It adds equal measures of virtuality and materiality to this mix. In all these ways, *Biomodd* offers hope that the current limits of technological progress can be extended in a manner that increases the prospects for a survivable future. This is another way of describing Eco Materialism.

Alice Anderson (b. 1972, London, United Kingdom) works primarily in post-digital materials such as copper. Her work has been shown at the following exhibitions: in Vivo *at the Centre Pompidou in Paris (2017) and* Champagne Life *at the Saatchi Gallery in London (2016), and the Wellcome Collection in London (2015). She received her MA in fine art from Goldsmiths College, London.*

When computers are whizzes at conducting mathematical calculations, organizing text, tracking coordinates, translating texts, and storing formulas, should people bother to add numbers, or find a square root, or alphabetize, or find our way home, or learn a foreign language, or memorize a formula? These intelligent machines are not only capable of conducting such tasks with uncanny speed and accuracy; their portability enables them to provide these mental services wherever and whenever the need arises. Furthermore, computer memory trumps human memory because, once pieces of information are recorded on a computer's hard drive, they never change. Memories stored in the brain, in contrast, are unreliable because they continuously erode and are reconstructed in response to emotions and life experiences. Few of us can match Kim Peek, the savant who inspired the movie *Rain Man*. He purportedly memorized nearly twelve thousand books![37] Alas, for the rest of us, the brain does a notoriously wretched job as a recorder. How much time is lost just trying to remember the name of an acquaintance or the location of our car keys? Michael Anderson, a memory researcher at the University of Oregon in Eugene, decided to find out. He concluded that people squander more than a month of every year dealing with the inconveniences caused by forgetting.[38] Are we smarter, better informed, and more competent if we outsource memorization to electronic devices?

While it is generally assumed that our lives are greatly enhanced by the memorizing capacities of cyber technologies, Alice Anderson's artistic practice is devoted to reclaiming the aspects of remembering that are sacrificed when computers perform this function for us. Her efforts may seem out of place considering the availability of hard disk drives, random access memory, read-only memory, cache memory, and flash memory that can be accessed through cell phones, personal digital assistants, game consoles, car radios, VCRs, toys, ovens, and watches. Anderson provides an Eco Material explanation for why she denies herself the advantages of electronic memory aids by stating:

The digital world gives more freedom, information, and creativity, but how are we meant to cope with it? This revolution is just beginning and it's already affecting the whole of society and all its current models – economic, of course, but also social and ideological. I don't know if it's good or bad yet, but our everyday life already sources a lot of its basic reflexes from automated information or service-sharing through the likes of Google, Wikipedia, Uber, and so on, so I have to find my own methods of slowing down, of keeping a sort of intimacy with the world surrounding me, of understanding – learning – and memorizing differently. It's a paradox, but the more my everyday existence fills up with digital data about the things around me, the greater my need to get to grips with their material, physical data.[39]

Anderson identifies a specific and crucial aspect of memory that the automatic recording of data by digital technologies precludes: a personal relationship with the objects being memorized. She pursues a double strategy to cultivate such personalization. First, she chooses to remember only objects that she has personally selected, used, stored, and cared for. These objects become eligible for her ritualized efforts to remember them once their functionality and appeal diminish, and they are destined to either be stored or abandoned. Thus, all the objects that Anderson commits to memory are slated to be removed from the present tense of her biography, and relegated to the past tense where they are likely to be forgotten.

An original sculptural process marks the second strategy Anderson employs to personalize her memories. She meticulously winds thin copper wire around an object she wants to remember, covering its surface row by row, until it is completely enclosed in a glistening copper coating. In this manner, Anderson methodically interacts with every fraction of the object's surface, gradually constructing a memory of information received by her fingers, hands, and muscles. This process digresses from the detached observations of technological memorization and from the dematerialized memory procedures of the human brain. By converting these digital and mental exercises into intimate sensory interactions, mentality becomes integral to physicality. She explains:

I always worry to break or lose an object, therefore I have established rules: When one of the objects around me is likely to become obsolete or is lost in stream of our lives [sic], I 'memorize' it with thread before it happens.[40]

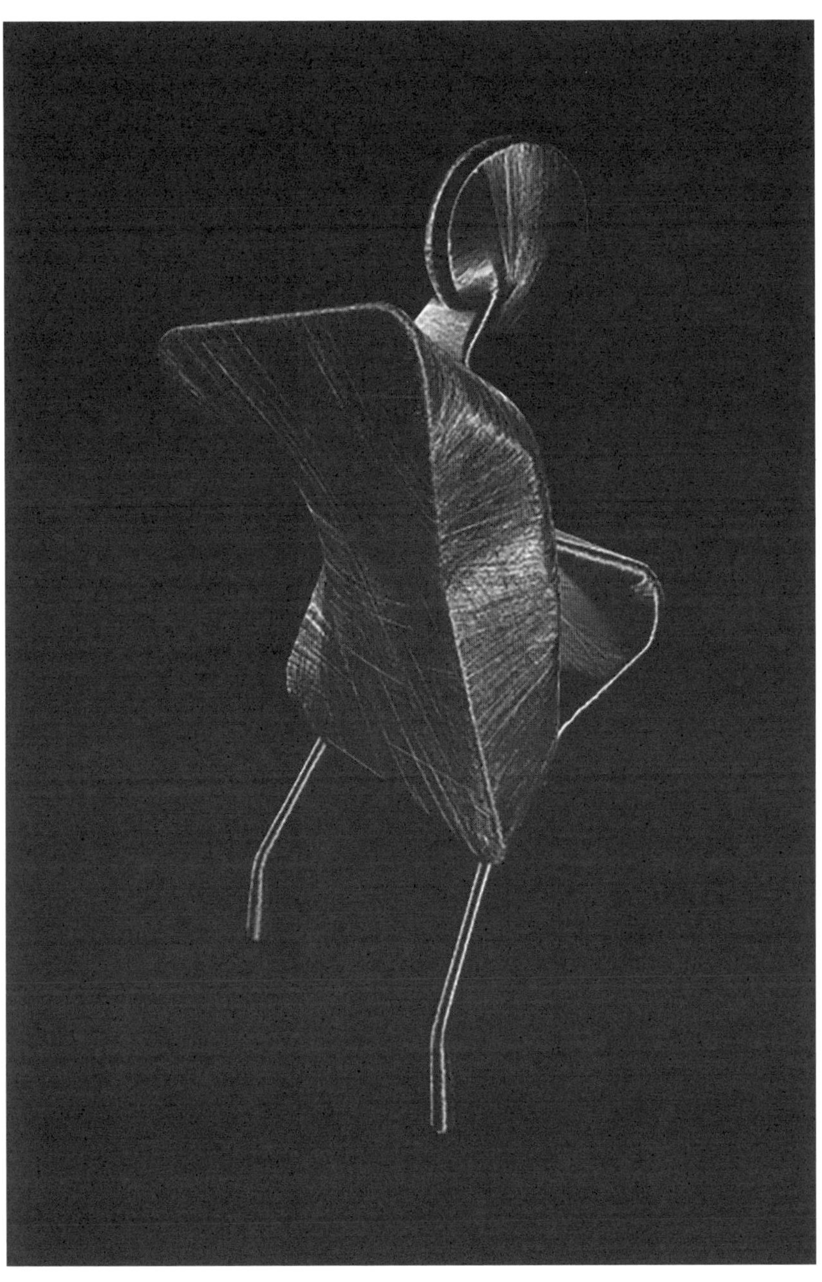

Figure 6a: Alice Anderson, *Memorized Objects series (distorted object)* (2013), wheelbarrow, copper-colored wire © 2013 by Alice Anderson Studio. Courtesy of the artist.

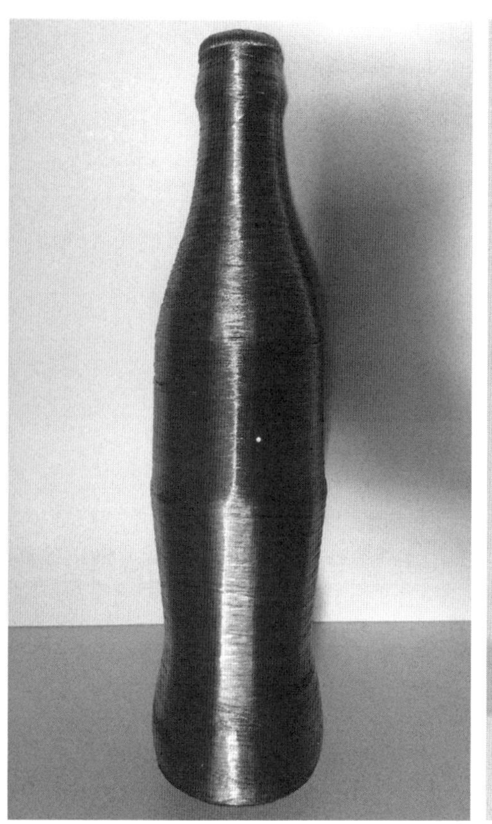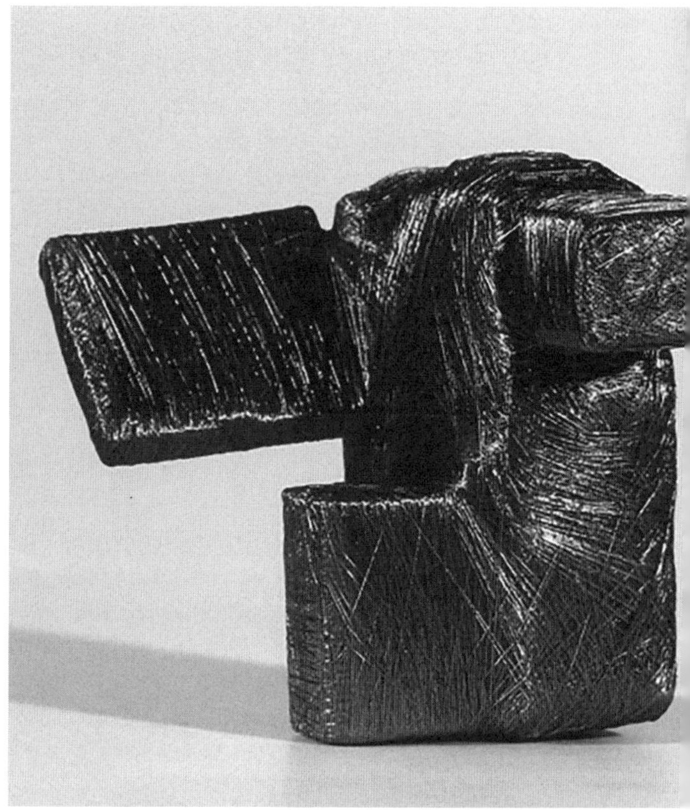

Figure 6b: Alice Anderson, *Memorized Object series (recognizable object)* (2013), Coca-Cola bottle, copper-colored wire © 2013 by Alice Anderson Studio. Courtesy of the artist.

Figure 6c: Alice Anderson, *Memorized Objects series (recognizable object)* (2010), camera video, copper-colored wire © 2010 by Alice Anderson Studio. Courtesy of the artist.

For this reason, she corrects anyone who refers to her process as 'wrapping' instead of 'recording'. She states:

> *I don't work to produce an artwork, but simply to live this digital transformation. I have developed my own way of learning and memorizing with thread. It corresponds to a necessity. I don't refuse this world. I just need to explore alternatives to live with it today.*[41]

It is significant that the first object Anderson treated in this manner was her video camera. The act was inspired by the arrangement of many threads on a loom on display at the Freud Museum, where she was scheduled to mount an exhibition in 2011. This spontaneous action was prophetic because it anticipated her decision to shift her career focus from making filmed representations of the material environment, to connecting with tangible objects without digital intermediaries. Anderson explains:

> *I felt a need to interact physically with these 'objects subjects', but my films offered little scope for this. I realized I was no longer interested in making images dance, and started thinking about working differently with these objects [...] This was also when I decided to immerse myself in the memory processes that are being changed irreversibly by the digital era. Our everyday memory is being externalized, and shared, and part of it is becoming instantly collective.*[42]

Anderson has been memorizing objects by recording them in copper wire ever since. She has applied this copper technique to old TVs, loud speakers, turntables, charger plugs, coins, a telescope, a kayak, spectacles, transistor radio, obsolete telephone, toaster, a Coke bottle, and other products from everyday life. These physical engagements in memorizing occur at varying scales. The winding process for small items such as keys and eyeglasses involves delicate and deliberate manual operations. However, her entire body is activated when she wraps large objects, such as a staircase. In these instances, Anderson repeatedly walks around the object, gradually unwinding the copper wire from a large spindle. At both scales, her process reverses the effortless, disembodied nature of digital memorization and retrieval. Unlike the speed and ease of automatic operations tethered to digital data collecting, memories formed from personal encounters with tangible objects are slow, concentrated, and meditative. These physical actions match the mental processes required to memorize them. Consider, for example, the difference between registering the details of a scene through mindful concentration, and accomplishing this act with a click of a camera.

The public is offered two points of entry into Anderson's realm of heightened somatic memorizing. The 'retrieval' half of remembering is activated when viewers confront the enclosed objects. It is stimulated as brains search for signs of recognition among the uncertain identities of the camouflaged objects. The public can also participate in the 'formation' half of memorizing by participating in the tactile version of digital memorization. Anderson installed the *Studio* in 2015 as part of a large exhibition of her work.[43] By converting the isolation of an artist's studio into the public space of a gallery, visitors could observe the creative process, and also activate the process by producing their own copper-encased sculptures. Visitors either brought their own unwanted objects, or they selected one from the choices that lined the shelves in the *Studio* installation. Old CDs, VHSes, cassette tapes, a typewriter, and even a 1967 Ford Mustang were available for rescue from the destinies that await unwanted commodities: discarded and forgotten, given away and forgotten, or recycled and forgotten. Once the objects were entwined in copper wire, however, they were preserved in the minds of the people who helped enclose them.

Anderson's creative process ensured that the unwanted objects would be preserved in her life, despite the fact that their physical presence had been obscured by copper that also prevented them from functioning. Nonetheless, these items survive in the form of vivid and detailed memories, as opposed to the inattention paid to many mass-produced objects that are visible and functional. Anderson guarantees the permanence of their status as a memory by pledging to never replace the objects that have been encircled by wire. She explains:

> *I select an object when I have the feeling that it is starting to fade in my memory and when I feel that I have to 'reactivate' it. The selection is a strong commitment because I never undo what I have done and I never buy a replacement object. If I do a scanner, for example, I won't get another one, so I am committing to living without one.*[44]

How long will these individual time capsules survive as material evidence of her memory? Anderson ensures a long duration by treating the copper to prevent its corrosion and decay. Nonetheless, these sculptures will eventually succumb to the same fate as every object and every memory by fading away. Perhaps the most enduring aspect of these sculptures is unrelated to their physicality and their mentality. They exemplify the Eco Material principle of paying attention to all forms of materiality, even when they were damaged or obsolete. Anderson concludes, 'These charged works are markers of time and affirm for me the value of physical records and the power of human memory in the fascinating digital age we live in.'[45] Her copper entwined objects carry the aura of a new kind of funeral rite, one that reveals remorse regarding our casual disregard for the products of the material world, mourns the loss of sensory intimacy, and seeks emotional affinity with our memories.

E-Prayer

Kal Spelletich

Kal Spelletich (b. 1960, Davenport, Iowa, United States) is a San Francisco-based artist working at the intersection of art and technology. In 1988, Spelletich founded Seemen, an interactive machine art performance collective that allows the audience to operate dangerous machinery. Spelletich's work has been included in numerous museum exhibitions, including the De Young Museum, San Francisco, CA; the San Francisco Museum of Modern Art (SFMoMA); the Exploratorium Museum and the Yerba Buena Center for the Arts, San Francisco; California Folk Art Museum, Los Angeles; Museum of Art and History, Santa Cruz; Oakland Museum of California, Oakland; and Headlands Center for the Arts, Marin. Recent international exhibits have been in Namibia, Ljubljana, Berlin, Vienna, New York, and India. Spelletich received an MFA from the University of Texas at Austin in media art.

In a series of computerized and mechanized sculptures, Kal Spelletich scrutinizes a current version of faith by testing 'the alarming promise that new advances in information and communication technologies will solve social and global problems and fulfill our private and public dreams.'[46] He satirizes these presumptions by wondering if technology may be humanity's best chance to achieve spiritual fulfillment. Spelletich's logic follows the widespread conviction that engineered products and services offer superior speed, power, efficiency, and reliability. Such claims have been proven regarding brushing teeth and guiding spacecraft. Why not summon technology to conduct our spiritual pursuits as well? Thus, Spelletich created *Praying Robots* (2015) to allocate praying, a quintessential function of the human spirit, to technology.

The ungainly clumsiness of the seven human-scaled, headless, handless robots in this series confirm his belief that 'we may think all we need to do is update, upgrade and replace our devices, but we can't just upgrade ourselves to enlightenment, right?'[47] The robots' twitches and jerks imitate the repertoire of physical movements that humans enact during prayer: genuflecting, prostrating, whirling, clasping hands, bowing, and wailing. Instead of positioning each praying robot sitting on church pews or facing an altar, they conduct their appeals to the gods upon a sled, a coffee table, a plank floor, and other platforms. The robots' clothing is equally banal. Each robot is dressed in a casual outfit that was owned and worn by artists and poets whom Spelletich admires. Presumably, the mystical powers endowed upon these robots does not depend upon sacred paraphernalia, or the interventions of priests or shamans, or such conduits to the divine as candles, incense, icons, and choirs.

The individuals who lent their clothes also lent their names to the sculptures. Individualizing the robots in this manner parodies the goal of constructing mechanics and electronics that behave and think like humans, which means granting them

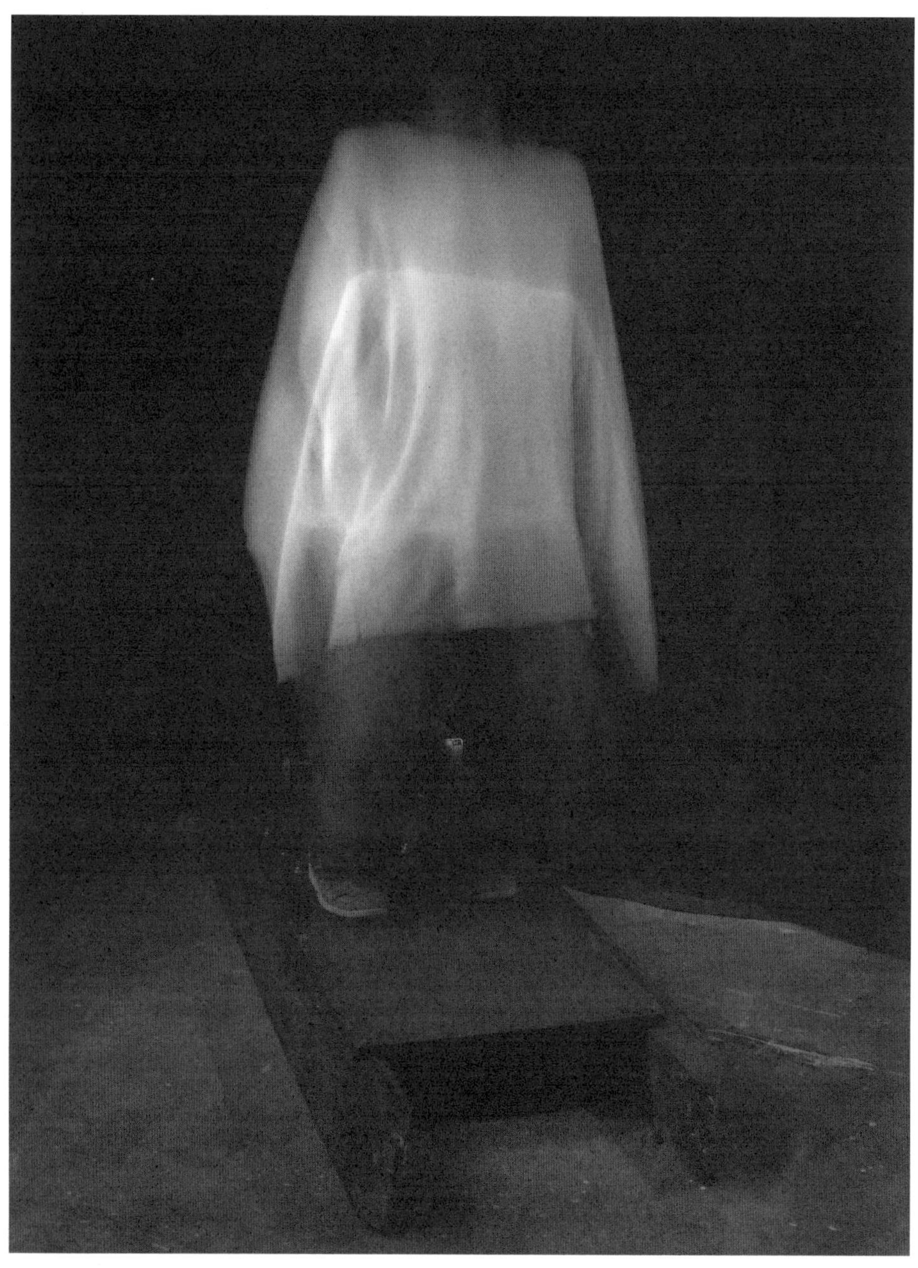

Figure 7a: Kal Spelletich, *Praying Robots, In Bot We Trust* (2015), robots dressed in Emory Douglas clothing, electronics, steel, H: 5 ft. × 7 in.; W: 2 ft. × 1 in.: L: 5 in. × 2 in. *Intention Machines* exhibition, Catherine Clark Gallery, San Francisco, 2015. Courtesy of the artist.

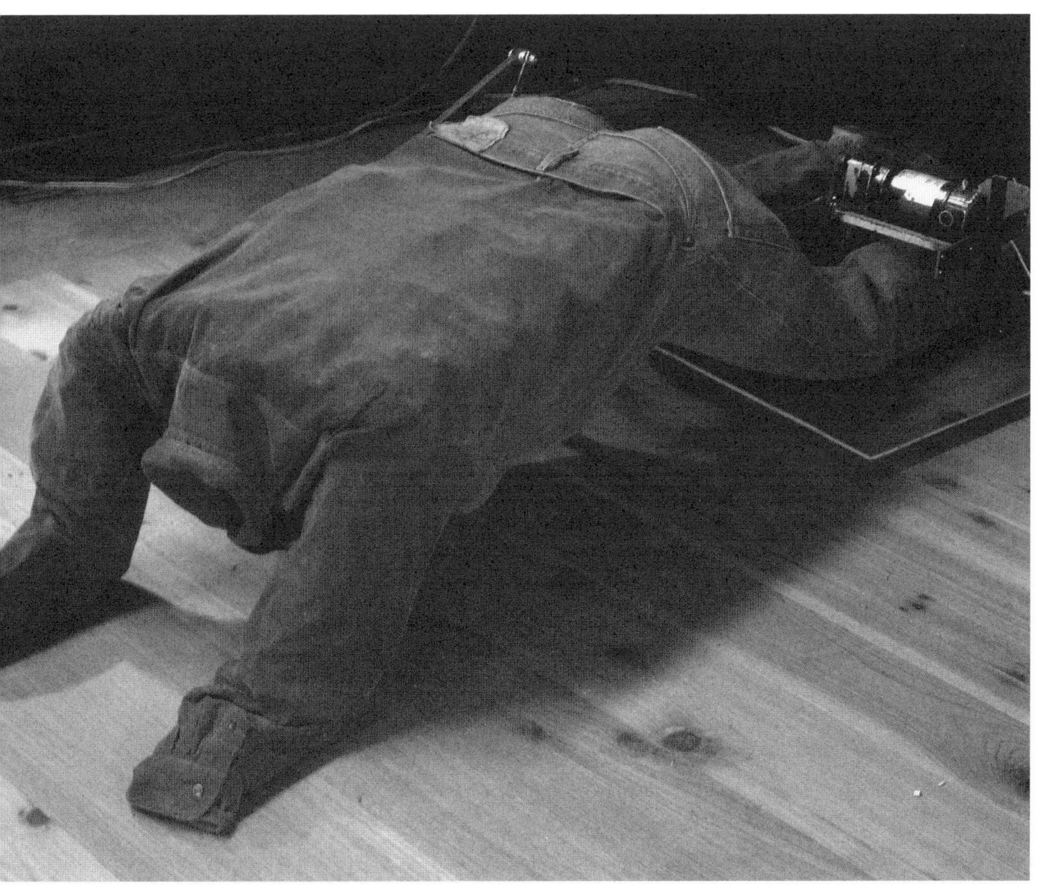

Figure 7b: Kal Spelletich, *Praying Robots, In Bot We Trust* (2015), robots dressed in Les Blank clothing, rug, electronics, steel, H: 1 ft. × 10 in.; W: 1 ft. × 7 in.; L: 7 ft. × 5 in. *Intention Machines* exhibition, Catherine Clark Gallery, San Francisco, 2015. Courtesy of the artist.

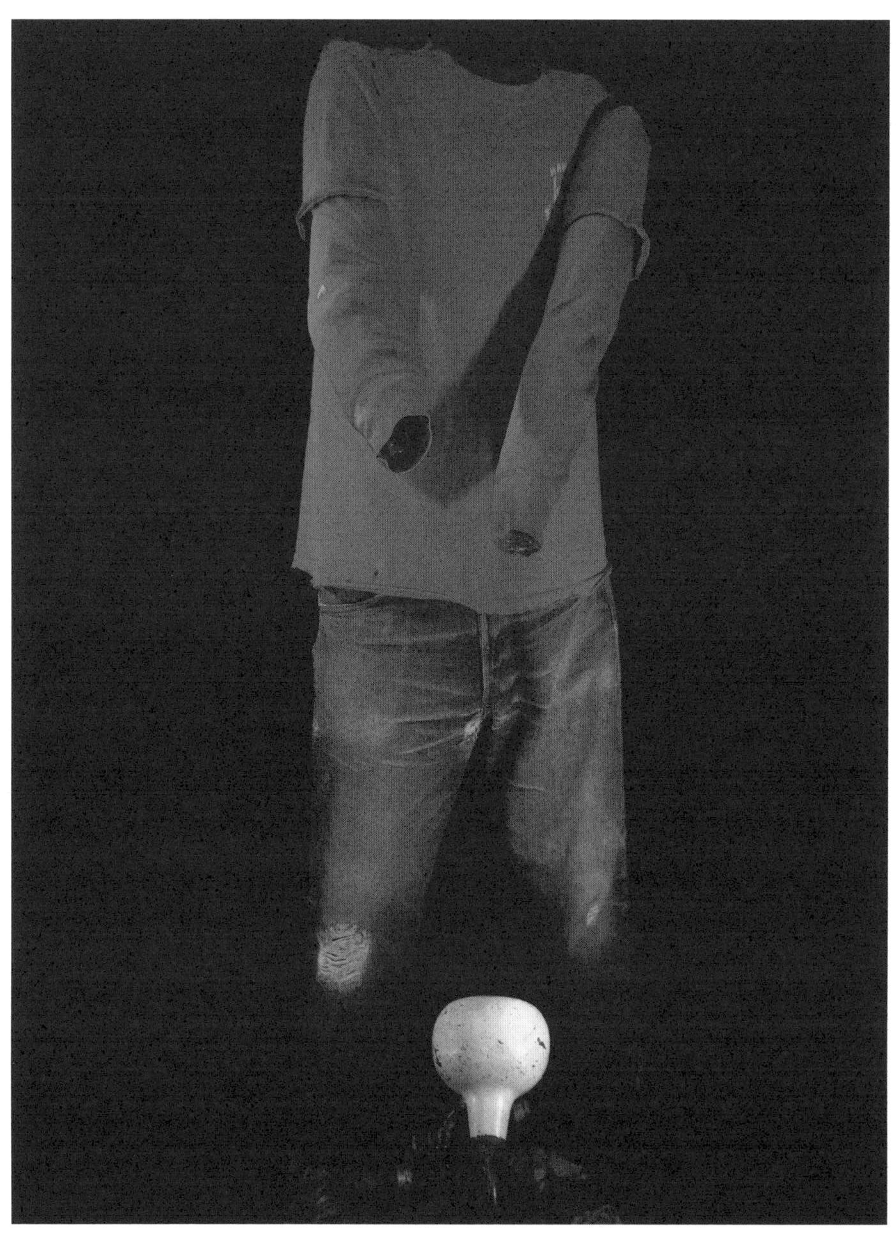

Figure 7c: Kal Spelletich, *Praying Robots, In Bot We Trust* (2015), robot dressed in Chris Johanson clothing, life-sized. *Intention Machines* exhibition, Catherine Clark Gallery, San Francisco, 2015. Courtesy of the artist.

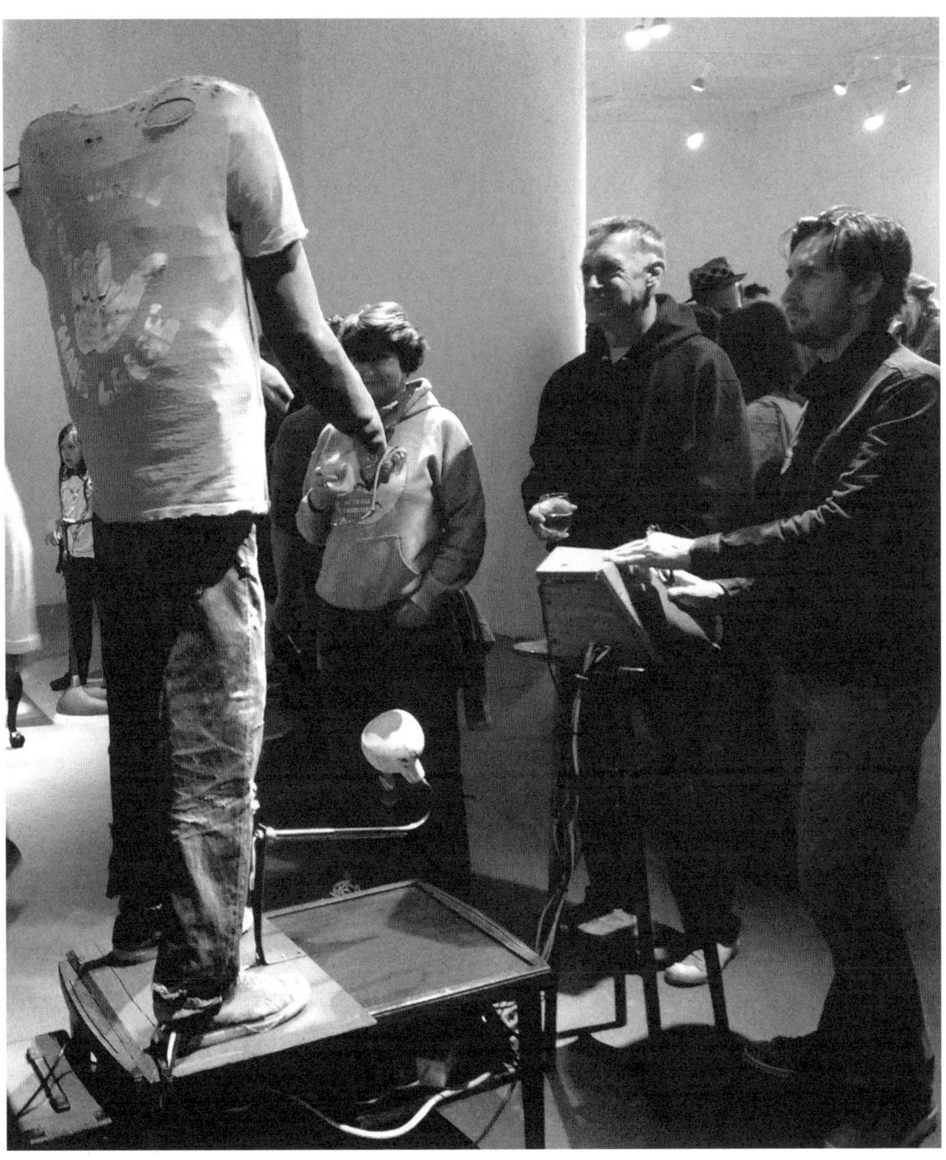

Figure 7d: Kal Spelletich, *Praying Robots, In Bot We Trust* (2015), installation view, robots, furniture, clothing. Exhibition included photographs of the sun taken with a digital camera modified with various apparati and inspired by the individuals that inspired the robots. *Intention Machines* exhibition, Catherine Clark Gallery, San Francisco, 2015. Courtesy of the artist.

individuality. Spelletich even individualized their manner of conducting prayer. Kay Miller, an artist, spins like a whirling dervish; Martha Wilson, a feminist artist, slowly raises and lowers her hands; Emory Douglas, Minister of Culture for the Black Panther Party, genuflects; Les Blank, documentary filmmaker, prostrates himself; and so forth.

The individuality of the people who solicit the robots' praying services is also factored into this artwork because each is programmed to detect proximity, touch, sound, and even the alcohol content of breath. Gallery-goers activate these transmissions by touching a round lever wired to an array of devices that enable the robot to detect the person's physical state (thermometers, pulse measuring devices, etc.) and mental state (lie detectors and electrocardiogram [EKG] machines). Within this context, the process of analyzing personal data ceases to be a routine calculation. Instead, it resembles detecting the invisible emanations of a person's 'aura'. The robot then responds by performing the prayer gesture it deems most suitable for that individual. In this manner, Spelletich introduces an incongruous conduit to spiritual enlightenment.

While the robots' bumbling movements are comical reversals of the efficiency we associate with advanced technologies, neither chuckles nor spirituality dilute the probing question posed by this artwork – is it reasonable or preposterous to imagine robots adding prayer to the innumerable services they are depended upon to provide? As in muscle power, sensory power, and mental power, would humanity's soul power be augmented if it were allocated to technology? Robot prayer offers many advantages. Its appeals to an almighty being would not only be more fervent, it would never be diminished by such human foibles as distractions, fatigue, and doubt. Alternatively, are interventions with divinity the sole remaining territory where robot assistance is not welcome? Spelletich has countered faith in technology with faith in religion to satirize the mania for allocating human tasks to non-human devices. Spelletich doubles the absurdity of his strategy by creating robots that certainly do not embody the solemnity of someone communing with God or the gods.

Absurdity may not be the end of this story. Each robot is accompanied by photographs, sound, and video of the sun and moon during peak moments (such as sunspots, solar flares, blue moons, and eclipses). The artist explains in earnest, 'I tie nature to spiritualism. This work is inspired by my eighteen-year study of Zen and Taoist thought.'[48] He then comments, 'Why can't technology have that mystical energy – or does it? This is my experiment to see if I can make machines or robots that help the planet.'[49] These are curious and thought-provoking revelations from an artist who is known for the robotic antics devised by a mad scientist in which terror, humor, and technology coalesce.

Your opinion: What's next?

Reader Interaction

Unplug: Readers are invited to conduct a private action or public performance in which you unplug a device that routinely contributes to your everyday activities. Set this device aside and accomplish the task the device was designed to perform. Please suppress the expectation that this action will be time-consuming, arduous, tedious, or any other negative quality. Instead, undertake this task as a search for a positive experience you have been missing. Did unplugging provide sensory pleasure, pragmatic advantage, environmental conservation, social opportunity, community consolidation, long-term advantage, interspecies alliance, or any other benefit?

Examples:

Speak to a friend without texting, phoning, or writing.

Brush teeth without an electric toothbrush.

Add numbers without a calculator.

Go someplace unfamiliar without a GPS.

Record a visual scene without a camera.

Write a letter instead of texting.

Play a board game instead of a computer game.

Ask someone a question instead of searching the Web.

Walk upstairs instead of taking an elevator.

Dry your hair without a hairdryer.

Spend 24 hours without artificial illumination.

Change television channels without a remote.

Write a story instead of going to the movies.

Take a walk outdoors instead of using a treadmill.

Build a fire instead of turning up the thermostat.

Purchase something with cash instead of using a credit card or electronic banking.

Tell time by studying the location of the sun and shadows instead of reading a clock.

Your comments: What's next?

Endnotes

1. Bryan Walsh, 'The Surprisingly Large Energy Footprint of the Digital Economy,' *Time*, August 14, 2013, accessed September 21, 2017, http://science.time.com/2013/08/14/power-drain-the-digital-cloud-is-using-more-energy-than-you-think/.

2. Mark P. Mills, 'The Cloud Begins with Coal: Big Data, Big Networks, Big Infrastructure, and Big Power,' Digital Power Group, August 2013, accessed April 10, 2016, http://www.tech-pundit.com/wp-content/uploads/2013/07/Cloud_Begins_With_Coal.pdf?c761ac.

3. Joshua Melvin, 'What's the Carbon Footprint of an Email?' November 26, 2015, accessed March 10, 2018, https://phys.org/news/2015-11-carbon-footprint-email.html#jCp.

4. Copper Development Association Inc., 'Electricity Generation and Supply,' 60 *Centuries of Copper: Electricity Generation and Supply*, accessed September 21, 2017, https://www.copper.org/education/history/60centuries/electrical/electricity.html.

5. Gebhard Sengmüller and Franz Buchinger, *A Parallel Image*, exhibition catalog, 2009–10. Supported by Fels-Multiprint,Vienna; Land Salzburg Kunst und Kulturforderung; Bundesministerium fur Unterricht, Kunst und Kulture, Kunstsektion; Ku Ituramt der Stadt Wien. Schmiede Hallein Festival, Hallein, Austria.

6. Sengmüller and Buchinger, *A Parallel Image*.

7. SchoolScience.co.uk, 'Electrical Efficiency,' accessed September 21, 2017, http://resources.schoolscience.co.uk/CDA/16plus/copelech1pg1.html.

8. SchoolScience.co.uk, 'Electrical Efficiency.'

9. Gebhard Sengmuller and Dominik Landwehr, 'Fictive Media Archeology' in Dieter Daniels and Barbara U. Schmidt (eds), *Artists as Inventors* (Linz: Ludwig Boltzmann Institute Media. Art. Research, 2007), accessed April 10, 2016, http://www.hatjecantz.de/files/9783775721530_06.pdf.

10. Germaine Koh, 'Broken Arrow,' accessed September 12, 2016, http://germainekoh.com/ma/projects_detail.cfm?pg=projects&projectID=5.

11. Germaine Koh, interview by Mathew Kabatoff, *Rhizome*, January 30, 2001, accessed September 12, 2016, http://germainekoh.com/content/press/kabatoff-rhizome.html.

12. Koh, interview.

13. Koh, interview.

14. Koh, interview.

15. Raphael Perret (ed.), *Machines of Desire* (Zurich, Switzerland, Amsel Verlag, 2014).

16. Jussi Parikka, *The Anthrobscene* (Minneapolis, MN: Minnesota Press, 2015), 15.

17. Deep time is geologic time. It differs from historic time. The concept was developed by the Scottish geologist, James Hutton (1726–97).

18. Raphael Perret, *Works by Raphael Perret: Recycling Yantra*, accessed November 22, 2017, http://raphaelperret.ch/recycling-mandala/.

19. Electronic Takeback Coalition, 'Facts and Figures on E-Waste and Recycling', 2014, accessed December 12, 2016, http://www.electronicstakeback.com/wp-content/uploads/Facts_and_Figures_on_EWaste_and_Recycling.pdf.

20. EPA (United States Environmental Protection Agency), 'Reduce, Reuse, Recycle,' April 25, 2017, accessed September 21, 2017, http://www.epa.gov/epawaste/conserve/materials/ecycling/manage.htm.

21. Raphael Perret, *Works by Raphael Perret: Recycling Yantra*.

22. Raphael Perret, *Works by Raphael Perret: Recycling Yantra*.

23. Raphael Perret, *Works by Raphael Perret: Recycling Yantra*.

24. Raphael Perret, email correspondence with the author, November 21, 2017.

25. Perret, *Machines of Desire*.

26. Perret, *Machines of Desire*.

27. For a full discussion, see Richard Maxwell and Toby Miller, 'Cultural Materialism, Media and the Environment,' *Key Words* 11 (2013), 86–102, accessed March 3, 2018, http://tobymiller.org/images/techenviro/Cultural%20materialism%20Media%20Environment.pdf.

28. SEAD (Space Ecologies Art and Design), accessed March 23, 2016, http://www.sead.network/.

29. Felicity Ross, 'Computer-Plant Connection Sci-Fi Comes to Life,' *Stuff*, January 24, 2011, accessed September 21, 2017, http://www.stuff.co.nz/taranaki-daily-news/news/4572703/Computer-plant-connection-sci-fi-comes-to-life.

30. Angelo C. J. Vermeulen and Diego S. Maranan, *Biomodd as a Paradox*, July 2010, accessed March 23, 2016, https://www.researchgate.net/publication/263844675_Biomodd_as_a_Paradox.

31. Angelo Vermeulen, email correspondence with the author, November 30, 2017.

32. Danone, 'Biotech Art Garden: Is This How We'll Grow Oregano on Mars?,' *Down to Earth*, accessed September 21, 2017, http://downtoearth.danone.com/2012/11/14/biotech-art-garden-is-this-how-well-grow-oregano-on-mars/.

33. Angelo Vermeulen, 'Artist Statement,' *Biomodd*, accessed September 21, 2017, http://www.biomodd.net/overview/artist-statement.

34. Angelo C. J. Vermeulen and Diego S. Maranan, *When Ideas Migrate: A Postcolonial Perspective on Biomodd [LBA2]*, August 2015, accessed December 21, 2017, https://www.researchgate.net/publication/275256076_When_Ideas_Migrate_A_Postcolonial_Perspective_on_Biomodd_LBA2.

35. Vermeulen and Maranan, *When Ideas Migrate*.

36. Vermeulen and Maranan, *When Ideas Migrate*.

37. Joshua Foer, 'Extremes of Human Memory,' *Mind Power News*, accessed September 21, 2017, http://www.mindpowernews.com/ExtremeMemory.htm.

38. Joshua Foer, 'Extremes of Human Memory.'

39. Anna McNay, 'Alice Anderson: "I Don't Refuse This World, I Just Need to Explore Alternatives to Live with It Today",' *Studio International*, February 5, 2016, accessed September 21, 2017, http://www.studiointernational.com/index.php/alice-anderson-interview-i-dont-refuse-this-world-champagne-life-saatchi-gallery-london.

40. Paul Carey-Kent, 'Alice Anderson: Post-Digital – 11 Nov 2016–18 Feb 2017,' *Paul's Art World*, December 20, 2016, accessed September 21, 2017, http://paulsartworld.blogspot.com/2016/11/alice-anderson-post-digital-11-nov-2016_4.html.

41. McNay, 'Alice Anderson: "I Don't Refuse This World".'

42. Saatchi Gallery, 'Alice Anderson,' accessed September 21, 2017, http://www.saatchigallery.com/artists/alice_anderson_articles.htm.

43. *Memory Movement Memory Objects*, July 22–October 18, 2015, Wellcome Collection, London, United Kingdom.

44. McNay, 'Alice Anderson: "I Don't Refuse This World".'

45. McNay, 'Alice Anderson: "I Don't Refuse This World".'

46. Natasha Boas, 'Poetic Kinetics or Intentional Imperfection,' *NYAQ/ LXAQ/ SFAQ*, accessed September 21, 2017, http://sfaq.us/2015/03/poetic-kinetics-or-intentional-imperfection/.

47. Boas, 'Poetic Kinetics.'

48. 'Intention Machines/Praying Robots,' *Kaltek*, June 22, 2017, accessed September 21, 2017, https://kaltek.wordpress.com/intention-machinespraying-robots/.

49. Kimberly Chun, 'Kal Spelletich's Mystical Robots Intended to Help the World,' *SFGate*, April 16, 2015, accessed September 21, 2017, http://www.sfgate.com/art/article/Kal-Spelletich-s-mystical-robots-intended-to-6202910.php.

WHAT'S NEXT?

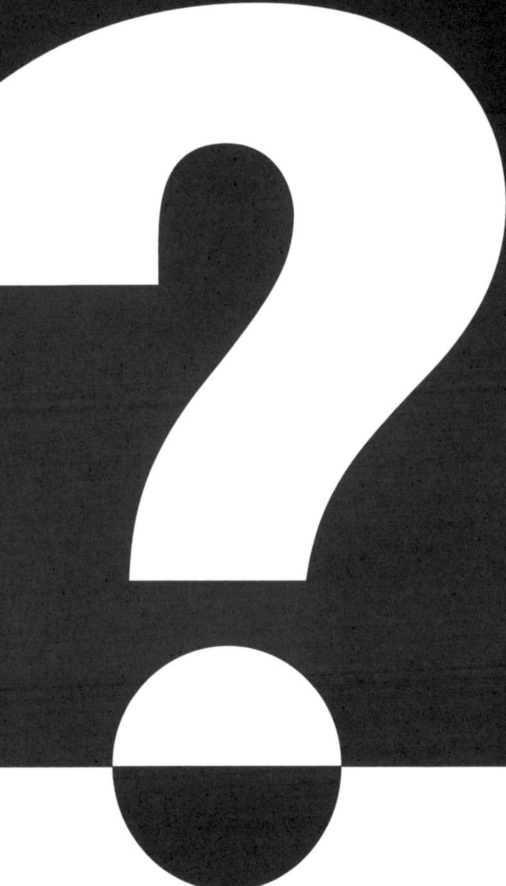

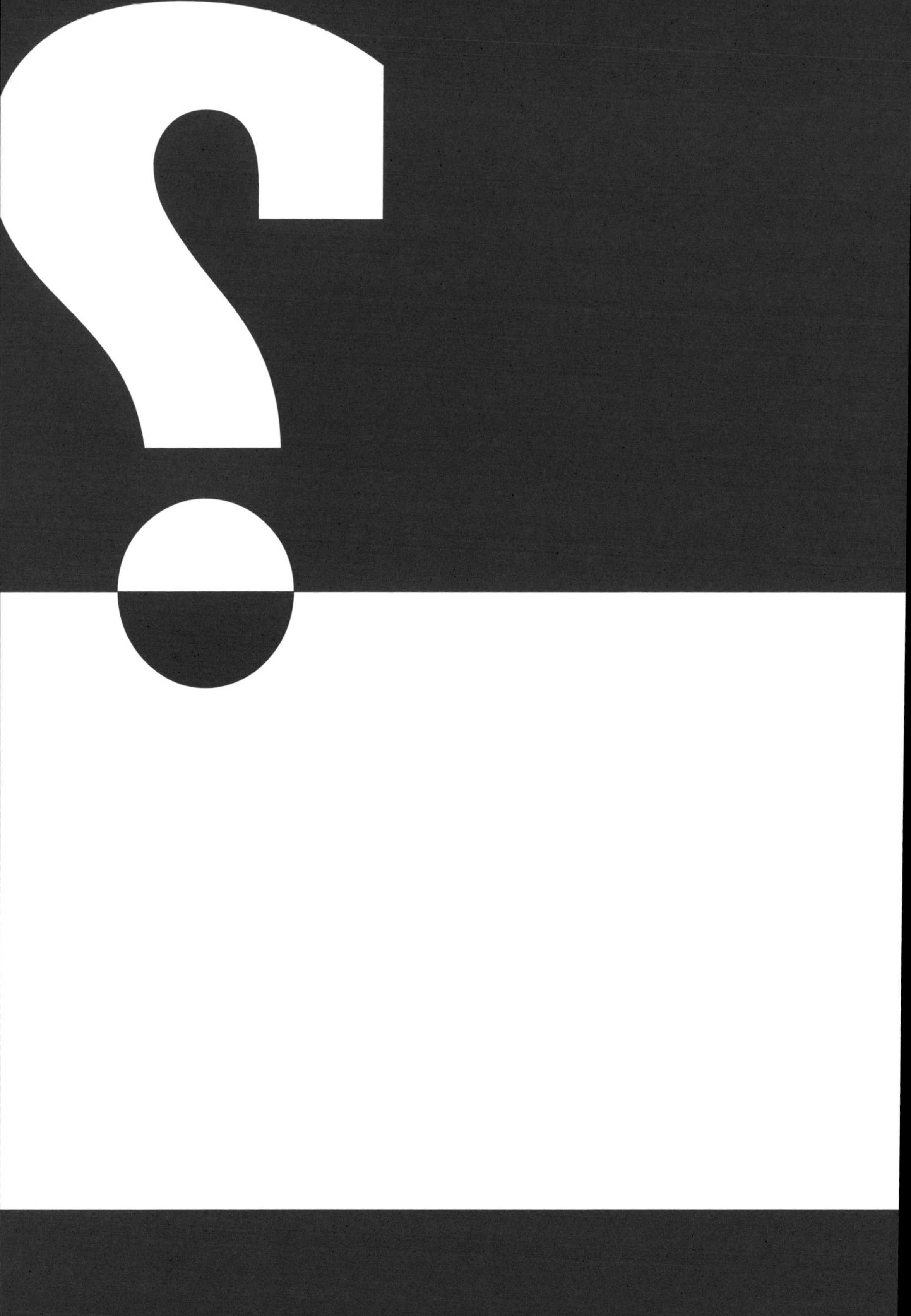

WHAT'S NEXT?

'What's next?' for planet Earth and its inhabitants? Three possible variants of this question are envisioned by the artists presented in the book's concluding chapter:

Are we, as humans, poised to suffer the fatal consequences of our previous indiscretions? Chim↑Pom stages an act of desperation to dramatize the dire deprivation the group foresees. But even the dismal circumstances it constructs are not cause for despair. Although the artwork suggests that the disruptions are greater than humanity's ability to craft solutions, humans may yet survive by adapting.

Will humanity's creative ingenuity unleash previously unimagined strategies for survival? Zbigniew Oksiuta's audacious, speculative, and ingenious initiative occupies the visionary zone of Eco Materialism. His projection into the future is located at the frontier of innovation where survival is accomplished by discarding and replacing familiar strategies for acquiring shelter and resources.

Will the public activate its ability to integrate responsible stewardship into daily encounters with the materials of the planet, ensuring the well-being of the planet through the incremental accumulation of many small actions? Natalie Jeremijenko's clever strategies represent the congenial zone of small actions designed for immediate use and adoption. They offer the reassurance that humans can attain resourceful and respectful material interactions that allow ecosystems to thrive.

**Option 1
Desperation**

Chim↑Pom

Chim↑Pom is a Tokyo-based artist collective formed in 2005. Its six members are Ellie, Ryuta Ushiro, Yasutaka Hayashi, Masataka Okada, Toshinori Mizuno, and Motomu Inaoka. Their practice responds to contemporary social issues. In addition to developing international projects, in 2015 the collective organized an international exhibition/ project Don't Follow the Wind *in Fukushima inside the area restricted after the 2011 nuclear disaster. Their intervention is scheduled to continue until the ban is lifted and public access is again allowed. Chim↑Pom has been the subject of many solo and group exhibitions, including at the Saatchi Gallery (2015); MoMA (Museum of Modern Art) PS1 (2011, 2014); the 9th Shanghai Biennale (2012); and the 29th São Paulo Biennial (2010).*

'What's next for life on Earth?' Dismal answers to this question were encapsulated in a three-week 'performance' orchestrated by Chim↑Pom, a group of six subversive Japanese artists. They created desperate conditions for survival, and then stepped back to see what would happen. Would there be more casualties than survivors? Would fights ensue? Or would restraining rules be imposed? Or kindly efforts to ensure co-existence? The real life/death contest the artists staged sabotages reassuring prospects for the future.

Long-standing feuds were deeply rooted among the members of the interspecies society that Chim↑Pom assembled to create *Becoming Friends, Eating Each Other or Falling Down Together* in 2008. Fights over claims of turf and contested provisions had been bitter and were ongoing. Chim↑Pom intervened by creating conditions to intensify these conflicts. First, they reduced the members of this society to three living entities: one human (Toshinori Mizuno, a Chim↑Pom member), one crow, and one rat. Then they shrunk the territory this trio occupied by confining them in an enclosed space that was only 10 × 6 1/2 × 6 1/2 ft. Next, they restricted their access to food to the contents of one bag of garbage each day, plus water. Each day's diet, therefore, might be abundant or meager, gross or palatable. The survival of the three confined occupants depended upon how much discarded food they could claim for themselves for the day's offerings. Why? Ryuta Ushiro, a member of Chim↑Pom observes, 'The only thing humans, rats, and crows have in common is garbage.'[1]

Not even the artists could anticipate how this real-life drama for survival, with real-death consequences, would develop. But this much was known – once the 'flight' response was eliminated from the 'fight or flight' alternatives to danger that are ingrained in living entities of all kinds, the three inmates would be forced

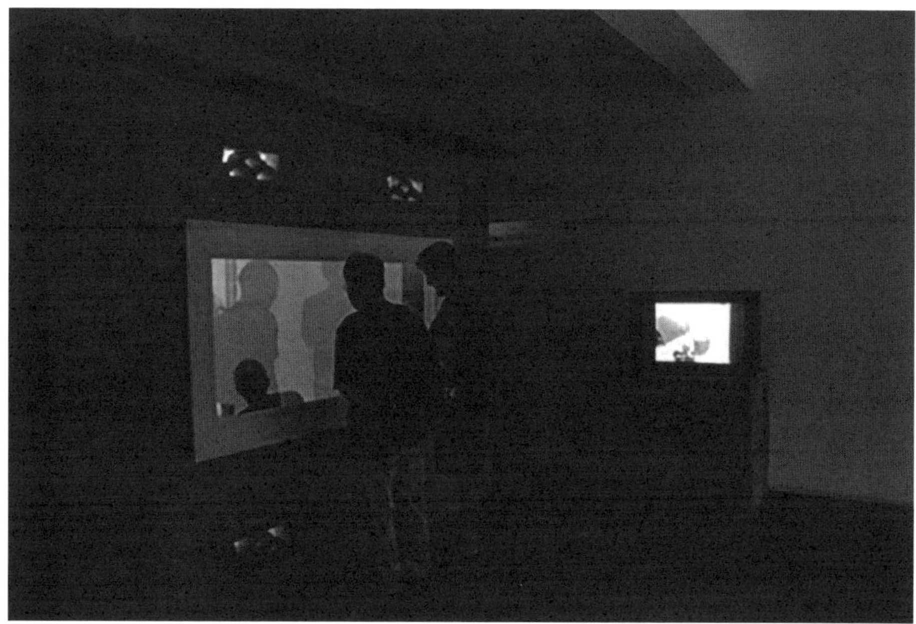

Figure 1a: Chim↑Pom, *Becoming Friends, Eating Each Other or Falling Down Together* (2008)
© by Chim↑Pom 2008. Courtesy of the artist and MUJIN-TO Production, Tokyo.

to confront each other. The society that emerged from this grueling, three-week experiment in interspecies habitation could resemble a harmonious ensemble, or a well-matched competition, or a brutal massacre.

When entering Tokyo's Hiromi Yoshii Gallery in 2008, visitors of Chim↑Pom's exhibition could observe the exterior of a small room the artists had constructed toward the back of the gallery space. It was made of concrete blocks. Visitors might also hear the sounds of a scuffle and then, perhaps, a screech, or howl, or caw, or grunt. There was a one-way mirror that allowed people to observe the interior and the events transpiring within. The inside of the space resembled a typical garbage collection shed that is a common adjunct to apartment buildings in the city of Tokyo. The absence of doors prevented exiting and entering, heightening the drama of confinement, and its attendant requirement for the occupants to negotiate their options among themselves.

The highly regulated system of separating garbage that is maintained throughout Japan to facilitate recycling provides a municipal context for this art event. Garbage is typically classified into four types: bottles and cans, combustible, non-combustible, and oversized. Meat, vegetables, fish, bones, fruit, and other

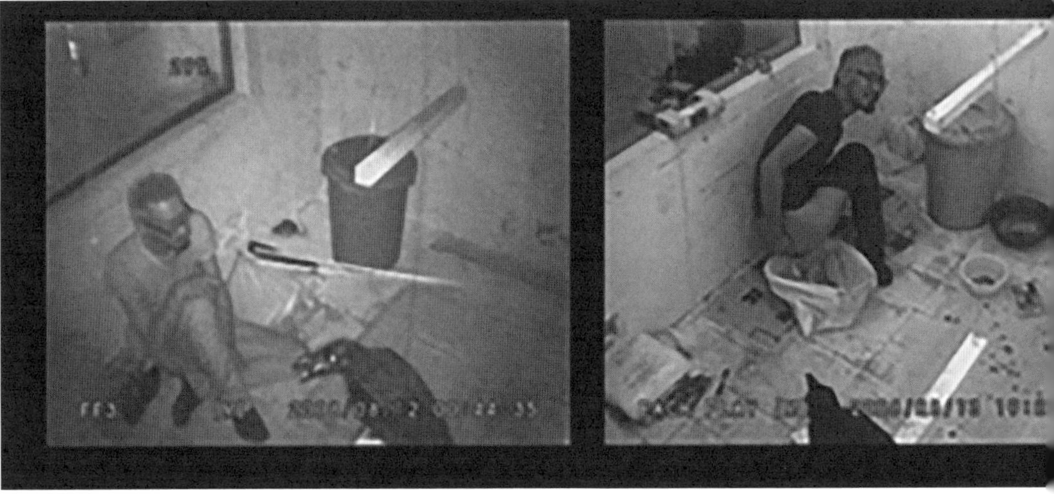

Figure 1b: Chim↑Pom, *Becoming Friends, Eating Each Other or Falling Down Together* (2008) © by Chim↑Pom 2008. Courtesy of the artist and MUJIN-TO Production, Tokyo.

edibles are considered 'combustible', along with paper, wood, clothing, burnable plastics, leather, rubber, and disposable diapers. This unsavory assembly is placed in transparent or semi-transparent bags. Such rules are not only monitored by municipal authorities, they are also overseen by individuals who scrutinize their neighbors. Nonetheless, garbage remains an extremely troublesome problem throughout Japan. Despite its strictness, the system of managing waste collapses every time the garbage is viewed as 'food' by competing populations of scavengers: crows, rats, and homeless people! That is when bags are torn open and the contents are strewn about.

Long before the artists enlisted these species into their art installation, they were embroiled in bitter rivalries for food. Rats and crows are well-known scavengers. Ushiro justifies the inclusion of humans by noting that homeless people are already engaged in a food-grab in the city, particularly in Yoyogi Park, where the crow was acquired.[2] The huge size of Japanese crows matches their appetites. They not only feast on the garbage left out for collection; they defend these feeding stations by aggressively dive-bombing intruders. Some of these intruders are trash-producing residents, but others are homeless people who also depend on garbage for their food. The Tokyo government has retaliated by waging an assault against the crow by allocating $5.3 million each year to conduct a war on crows.[3] One offensive strategy consists of erecting large, two-hundred-square-foot crow traps. The crow in this artwork by Chim↑Pom was snatched from such

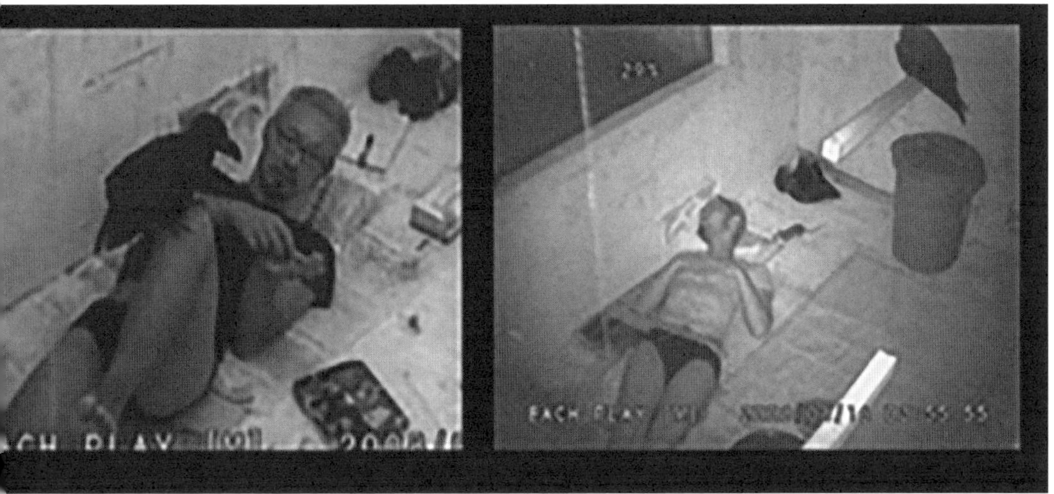

a trap in Yoyogi Park. But even these deadly contraptions have not succeeded in limiting their numbers. Toshimasa Uno, Tokyo's crow czar at the Bureau of the Environment, blames the relentless rise in crow populations on commuter crows that fly into the city from the suburbs to feast on the capital's plentiful garbage.[4]

The rat in *Becoming Friends, Eating Each Other or Falling Down Together*, was captured in Shibuya's Senta-Gai, a popular district where youths converge to enjoy shopping, fast food, and dancing in nightclubs. The area is infested with rats. Crows congregate to hunt the rodents and devour them with relish, a classic example of a 'predator/prey' relationship[5] that may determine the outcome of this artwork. But it is equally possible that the rat will prevail because of its remarkable ability to withstand assaults that would decimate other populations. Indeed, rats not only survive chemical poisons, they are known to endure nuclear contamination. Ushiro explains, 'We identify with the super rats. We have some sympathy for them and respect them as peers.'[6]

Ushiro explains:

> *Garbage always reflects the paradox of our society. At night, rats roam around in the garbage but, in the early morning, crows gather to pick at it as well, so this garbage bag becomes the symbolic medium that connects human beings to these animals; it is the link between mass consumerism and disposal.*[7]

He summarizes his diagnosis of the problem by stating simply, 'As we grow more affluent and produce more waste in Tokyo, we have more rats and crows.'[8]

Despite their aggressive intrusions, rats and crows are not the cause of these escalating urban pressures. Humans are guilty because they produce the garbage. The artists introduce this truism into their work by referring to 'Crow Tengu', a mythical creature popular in traditional Japanese lore. Tengu are elevated beings known for their skills in sword fighting and weapon smithing. Mythic crows use these powers to protect the Buddhist law against transgressors by playing tricks on arrogant priests and punishing vain samurai. Chim↑Pom updates this traditional symbol of power and morality by directing the traditional legend to current waste-producing consumers.

Might this drama, carried out within the confined space of an art gallery, foretell a time when humans grovel among animals over scraps in order to survive? Will we learn to magnanimously share provisions? Or will devastation be so extensive that it will even engulf the 'last being' standing? These options are all accounted for by the work's title, *Becoming Friends, Eating Each Other or Falling Down Together.*

What actually happened? The artists' description in an email correspondence with the author is presented in its entirety:

> *In a small shelter set up in the gallery, Mizuno spent twenty-one days of the exhibition period together with Chu, a rat caught by Chim↑Pom in Shibuya, and Katan, a crow captured in a trap laid by the Tokyo Metropolitan Government, while being watched from outside through a one-way mirror, and visually shut out from inside under the surveillance of security cameras. Katan was quite tame and became friendly with Mizuno, sometimes resting on his legs, but suddenly died on the tenth day, due to stress from being trapped, according to a veterinarian. As stated in the exhibition title, Mizuno mourned over the death by eating Katan then having its body stuffed. While he described his sorrow in his diary, 'the body of Katan was very light. Too light, even if it's made to fly [...] Katan was to be killed by the public organization. We tried to help him but made him die anyway.' He also wrote his calm feeling 'stabbing a knife into the belly of Katan. No negative impression like feeling pity. Strange even to me.' Mizuno spent the remaining time alone with Chu until the last day of the exhibition and had intended to keep Chu as his pet, but he finally released Chu on the street to let him go back to the wild.[9]*

The 'What's next?' scenario offered by this artwork is a complicated amalgam of possible conclusions. Besides dramatizing the possibility that survival may require groveling for scraps of garbage, the artwork introduces several compelling insights. It demonstrates that interspecies cooperation can overcome greed; that confinement can be as life-threatening as starvation; that adaptation is a viable survival strategy; that practical considerations can overcome sentiment; that affection can exist alongside adversity; that people can overcome biases, and animals can subvert predatory instincts; that altruism is real.

**Option 2
Relocation**

**Zbigniew
Oksiuta**

Zbigniew Oksiuta (b. 1951, Mulawicze, Poland) is an architect and artist experimenting with the possibility of designing biological structures. Oksiuta sets aside social and aesthetic factors to concentrate on verifiable physical and chemical possibilities. Oksiuta's work has been shown at venues worldwide including the Venice Biennial (2004); the ArchiLab d'Orleans (2004); Ars Electronica Linz (2007), the Biennale of Electronic Arts Perth (2007); the Center for Contemporary Art, Warsaw (2007); FACT Foundation for Arts and Creative Technology, Liverpool (2008); the Biennial for Electronic and Unstable Art, Stavanger (2008); Casino Luxembourg (2009); the Kapelica Gallery, Ljubljana (2010); and the Science Gallery, Dublin (2011). He studied at the Faculty of Architecture, Warsaw University of Technology. He currently teaches in the architecture department at Rensselaer Polytechnic Institute in Troy, New York.

If Zbigniew Oksiuta succeeds in fulfilling his monumental ambitions, he could earn the status of the hero who triumphs over adversity to rescue humanity from the dismal fate they would otherwise suffer. This artist/architect/scholar/scientist is conducting a bold experiment to ensure humanity's survivability by enabling colonization in the outer reaches of the cosmos. Those who consider his work 'far out' might be referring to outer space, the location he explores for future human habitation after Earth has become too corrupted to support life. Alternatively, 'far out' might refer to his insistence that 'science is magic,' which has led him to pursue 'a new fairytale for now.'[10] Fairytales are where states of beauty, perfection, luck, and happiness are imagined. By aspiring to generate such fairytale conditions with actual matter, Oksiuta's pursuit may appear as implausible as a fairytale, but his otherworldly investigations are based on scientific research conducted in collaboration with international scientific institutions. He explains:

> *We have entered the era of synthetic biology. The discoveries that are occurring before our eyes reveal the unity of the biological and the digital language. This opens up unimaginable possibilities. We will no longer copy nature, we will create it. It seems that everything can be possible and attainable. Very brutally speaking, by constructing new DNA in a computer and then translating this information to several bottles of chemicals,[11] we will be able to create living organisms about which evolution never dreamed. They will be born and will grow in laboratories. Their development will not be decided by natural selection, it will be decided by you and me.[12]*

Oksiuta is not engaging in science fiction entertainment. His efforts address actual threats to humanity's capacity to secure its own survival. Such concerns date back to 1700 BC when the citizens of Sumer and Akkad in the Tigris and Euphrates River systems were forced to abandon their homelands because their intemperate use of irrigation degraded the soil, rendering it infertile. This event foreshadowed recent threats to human survivability. The litany of human-induced environmental ills includes desertification, acidified oceans, toxic waste dumps, extinctions, depleted rainforests, among many other hazardous conditions. Meanwhile, headlines regularly announce such cataclysmic events as the recent Fukushima nuclear disaster that required 470,000 people to abandon their homes.[13] Projections extend this worrisome scenario into the near future. It is estimated that 250 million people in Africa may suffer from water and food insecurity in the twenty-first century.[14] Furthermore, a recent Environmental Justice Foundation report predicts that there could be as many as 150 million climate change refugees by the year 2050.[15]

How will provisions and services be generated within compromised ecosystems to sustain these human populations? Oksiuta is not waiting for the disasters that these conditions foretell. Since the end of the 1970s, he has been devising practical ways to establish human colonies in outer space as well as on Earth! While creating enclosures is his primary focus, the enclosures he is developing bear little resemblance to the kind of architecture he was taught as a student at the Faculty of Architecture at the Warsaw University of Technology. Oksiuta's maverick approaches are based upon an overarching conviction that it is folly to attempt to colonize the universe by exporting architectural systems that accommodate Earth gravity and Euclidean geometry. Instead of loading rockets with Earth materials and tools and sending them throughout the cosmos, Oksiuta set about inventing an architecture specifically suited to the conditions of weightlessness and the absence of 'up' and 'down' orientations that he envisions at the distant frontier of human habitation.

The basic unit of matter that Oksiuta employs is also the basic unit of life – the cell. His goal is to create a new model of human habitation that utilizes the cell's structure and functioning. Cosmic structures modeled on a living cell would consist of elastic membranes that enclose air or fluids. They would assume a spherical form, and they would function like a chemical factory, not a machine. Like a cell, the surfaces of these structures would be amorphous, moist, and elastic. Because they are soft, they would resemble the clothes we wear to protect us, and the mattress we lie upon to sleep. Furthermore, they would not be tied to any defined location; and they would have no top or bottom. Because these living enclosures would continuously form themselves as they interact with matter and energy, they could not be predetermined by a human designer. This means their dimensions and shape perpetually change in unpredictable ways. Oksiuta explains:

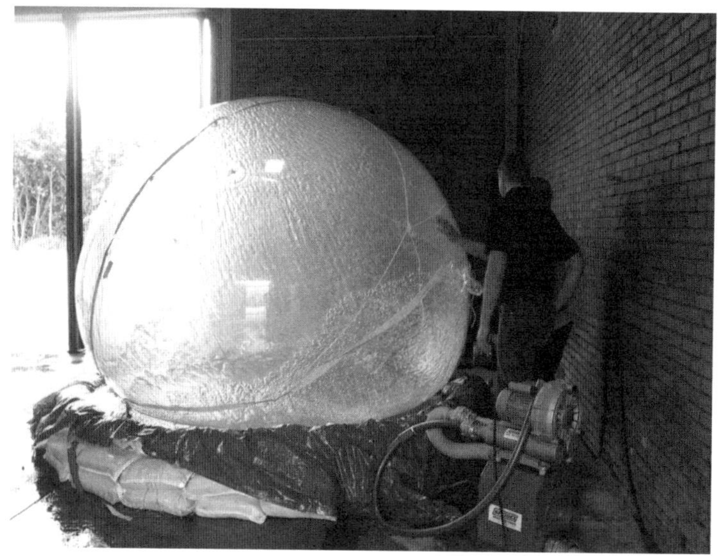

Figure 2a: Zbigniew Oksiuta, *Spatium Gelatum* (2003), form 090704, Biological Habitat. Gelatine 270 degrees Bloom. Physical properties: viscosity 31.8mP, transparency 93.4%, conductivity 248 uS. Chemical properties: pH 5.65, ashes 0.5%, water before drying 70%, metals 40 ppm Bacteriological properties: aerobe germs 1,000/g. 2.5 m. Courtesy of the artist.

Matter itself is creative. We should give matter a chance to trust us. We need to create not objects, but the conditions for life processes. My aim is not novelty of form. Instead, I want to learn what a material can do. My architecture is like a three-dimensional Petri dish or biological reactor where structures can grow.[16]

Oddly, Oksiuta traces the inspiration for this futuristic vision to the quaint farmstead in Poland where he was born. Because the farm supplied all the needs of his family of eight, this historic model of local productivity and efficiency inspires the version of 'What's next?' he is striving to realize. He explains that his futuristic interpretation of his family farm expands human engagements in two cardinal directions: 'The first one is directed inwards, into the micro cosmos of Life, the world of molecules, genes and chromosomes. The second one is leading us into the macro cosmic space, into the ocean of planets and stars.'[17]

By stating, 'Modernity does not mean using complicated high-tech which has been transferred from institutes to the exploration of outer space,'[18] Oksiuta reveals that his visionary enterprise not only repeals and replaces Earth-bound architecture; it

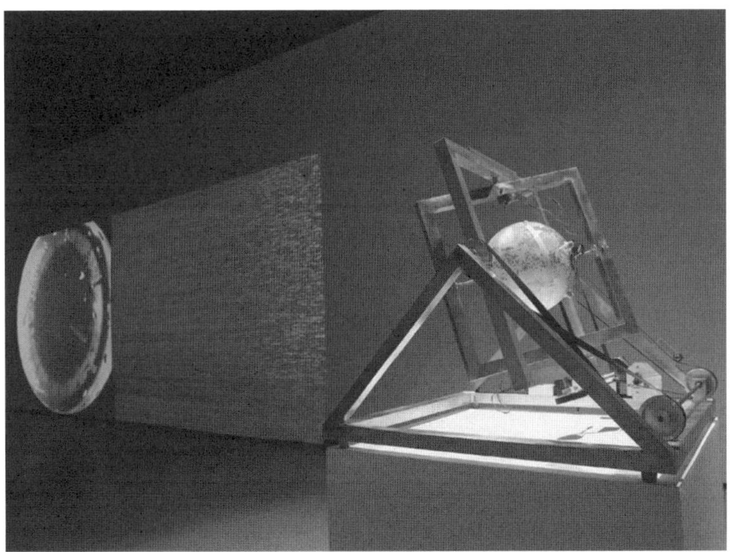

Figure 2b: Zbigniew Oksiuta, *Cosmic Garden* (2017), model on a 3-D clinostat, a Random Positioning Machine that recreates conditions of quasi-zero-gravity for research into growing plants in the cosmos. Seeds are sewn into the agar surface inside the continually moving transparent, introverted biosphere. Photo by Nick Kozak, Toronto. Courtesy of the artist.

also subjects the tools, machines, and instruments that comprise the contemporary 'techno-economic world' to a comparable overhaul. Such artifacts of civilization are not only unsuited for outer-space occupation; Oksiuta believes they are responsible for Earth's impending crisis. By identifying their defect as 'they are dead,'[19] he exposes the radical nature of his propositions to resolve humanity's current predicament. What is wrong with dead buildings and dead tools? He explains their inherent shortcoming – lifeless things do not die. Things that don't die also don't biodegrade. Dying and biodegrading are essential for the self-regulating, self-perpetuating economy of living systems. They enable life to function. As such, current modes of surviving and constructing disobey the planet's overarching systems that ensure the continuation of life. Relying on dead materials like steel and fossil fuels disrupts the continuity of these systems. Oksiuta affirms, 'Existence of the biosphere and existence of life on Earth, both depend on whether we shall be able to move from the over-exploiting economy of linear systems to the cyclical economy of Nature.'[20]

Thus, instead of depending upon the 'dead technologies' of traditional terrestrial settlements, Oksiuta enlists 'life' as his powerful ally for installing humans in cosmic settlements. His strategy exercises the creative and adaptive capabilities of living systems to provide the materials and methods to create structure and provide resources in outer space. This 'fairytale' concept is grounded in the biological fact that living processes, because of their inherent complexity, precision, and economy, far surpass the inert procedures of production thus far devised by humans. Thus, the research Oksiuta conducts focuses on the newly discovered principles of biological transformation of energy into form.

Oksiuta refers to his ongoing explorations of space habitats formed with biological materials and processes as *Breeding Spaces* (2003-ongoing)[21] to distinguish them from inert 'living spaces' designed by architects. *Breeding Spaces* are alive. They take the form of isolated spatial entities, created out of vegetable matter that not only provide habitat; they actively absorb, transform, and synthesize matter and energy from their surroundings. These living enclosures 'breed' by conducting chemical transformation of solar energy through photosynthesis.

What will it be like to inhabit this new domain of living structures, if it should provide the answer to 'What's next?' for humanity? Oksiuta offers an appealing answer:

> *This space will be born with you, as it will be part of your body. It will grow, develop and bloom with you. It is beautiful, when you are young, but it will not survive a thousand years as the stone would. It will become old like the owner and eventually dry, wrinkle. Even the tools will die!* [22]

Oksiuta defines the challenge of inhabiting the far reaches of space in scientific terms by asking:

> *Can life processes, which normally take place at the nano-scale, in proteins, acids and saccharides, happen on the macro-scale? We understand that a stretching of a living cell to the size of a house is not possible. It is necessary to develop new ways, until now not existing in nature, which would make it possible for the biological processes to occur at an architectural scale.*[23]

He also acknowledges the challenges of adopting living conditions beyond the known structures, forms and norms. It requires re-thinking 'thinking', and re-doing 'doing'. Both are essential for relocating fairytales from the imagination to the material realm. Regarding 'thinking', Oksiuta comments, 'This is not a matter of computing. My aim is not novelty of form.'[24] Regarding 'doing', he insists, 'We need to touch, to get our hands dirty. Only computing won't solve this challenge. It is about materiality.'[25]

Oksiuta manifested innovative thinking and doing at the Max Planck Institute for Plant Breeding Research in Cologne, and the University of Cologne, where he tested biological membranes to enclose space structures. These structures are mostly created out of polymers like gelatin (an animal product) or agar (a vegetable product). Oksiuta explains how they conduct worldly functions in otherworldly contexts, 'The compartments acting as "bio fabricators" will produce on the spot and in a decentralized way, all that the system needs: food, tools, and energy.'[26]

In some experiments, plant and animal cells are injected in this surface medium. Other experiments inoculate surfaces with strands of DNA (deoxyribonucleic acid). These are particularly promising for outer-space occupation because DNA is not only capable of growth (increasing mass and volume); it is also capable of differentiation (changing form and function). Sending microscopic strands of DNA into the universe to multiply, differentiate, and grow into a new cosmic civilization may seem daunting, but Oksiuta is confident because DNA strands have already amassed millions of years of evolutionary experience on Earth. Somewhere in this long past they might have configured a strategy suitable for the conditions of space. Thus DNA might introduce an entirely new kind of organism that is perfectly adapted to weightless conditions. Such autonomous self-creation means that the *Breeding Space* structures will generate resources. Furthermore, because they are alive, they will perpetuate a cyclical self-regulating economy. The artist envisions a future in which everything is subjected to the cycle of birth, life, death, and decay. Oksiuta envisions humanity's future in outer space in the following way:

> *In this maximally condensed introversive garden, a variety of organisms, similar to the plants, animals and people on my grandfather's farm, would be living together. Warm-blooded flowers that heat the sphere are growing there. Gene fabricators, equipped with universal sets of DNA, 'produce' things and tools that are needed on the spot. In the bladders life boils. These are bioreactors producing plant and animal proteins. Maybe the cultivation of plant and animal products for our consumption could occur within these new systems, without victimizing living organisms.*[27]

Oksiuta's bio-based initiative seems elegant when it is compared to alternatives for outer-space exploration that are currently being studied. New ventures in mineral extraction, for example, envision exotic minerals being mined on distant planets and returned to Earth.[28] Researchers are already developing industrial equipment, such as space backhoes, to conduct operations off planet Earth, taking advantage of the efficiencies afforded by near-zero gravity.[29] Mercury, Venus, Moon, Mars, Jupiter, Saturn, Titan, Uranus, Neptune, Pluto, Ceres, asteroids, and comets have all been destinations of such space explorations. Oksiuta is confident that 'biological phenomenon will be the main topic of this century.'[30]

Oksiuta expands Eco Materialism's considerations of agents of material change beyond the creativity of humanity, beyond the impact of living species, and beyond the self-organizing capacities of matter. He also assigns material creativity to the smallest units of living matter, believing they not only ensure human survival; they provide unimagined opportunities. He explains:

> The discoveries that are occurring before our eyes reveal the unity of the biological and the digital language. This opens up unimaginable possibilities. We will no longer copy nature, we will create it. It seems that everything can be possible and attainable.[31]

**Option 3
Recuperation**

Natalie Jeremijenko

Natalie Jeremijenko (b. 1966, Mackay, Queensland, Australia) is a New York-based artist and engineer whose background includes studies in biochemistry, physics, neuroscience, and precision engineering. She has alternatively described her work as 'X-Design' (short for experimental design) and herself as a 'thingker', a combination of thing-maker and thinker. In 1999, the Technology Review *named her among the 'Top 100 Young Innovators'[32]. Since then she has received awards from* i.D. *magazine,* Fast Company, *and* Creative Capital. *She is currently an associate professor at New York University in the Visual Arts Department, and has affiliated faculty appointments in the school's Computer Science and Environmental Studies Departments. She holds a Ph.D. in Computer Science and Electrical Engineering from the University of Queensland, Australia.*

A 'happy ending' is not the optimal answer to the nagging question, 'What's next?' The condition that is desirable, as apparent in vital ecosystems, resists endings of all kinds, even happy ones. Finalities are alien to material existences on Earth because they are ever-beginning and never-ending. Environmentalists bring eternity down to Earth, both literally and figuratively, by demonstrating that discarded goods are never 'departed'. They contribute to a shared legacy that is inherited from the past, and bestowed upon the future. This favored condition involves perpetuity constructed out of a balance between evolutionary change and cyclic return. Thus, whether or not an afterlife of heavenly abodes exists, the material cycling and mutating of life relies upon avoiding system malfunctions here on Earth.

Regrettably, the legacy left by many contemporary material interactions threatens future prospects. This ill-fated bequest reflects the boom mentality of contemporary consumer cultures that takes abundance for granted, tolerates waste, and disregards contamination. For this reason, the finale of this book presents Natalie Jeremijenko, an artist whose professional credentials also include biochemistry, physics, neuroscience, and precision engineering. As an artist, she goes beyond fretting about the future of this everlasting principle by originating material interactions designed to perpetuate ecosystem functions, and that can be performed by non-professionals. They are conceptually innovative, engagingly humorous, and pragmatically effective. All three qualities converge to avoid the tiresome reminders to recycle; to choose 'green' commercial products; and to source materials from the waste stream. Jeremijenko's strategies are conceived to reduce toxins, produce less waste, and minimize energy consumption. In 2007, she established the X-Design Environmental Health Clinic (EHC) at the Steinhardt School of Culture, Education, and Human Development at New York University, where she teaches. She featured the word 'clinic' in the name of her new art facility to announce its shared mission with fertility clinics, cancer clinics, and methadone clinics. Like them, the Environmental Health Clinic treats illness and promotes health. It differs only in terms of the patient being treated. This clinic ministers to the 'environment', that dynamic and amorphous territory where atmosphere, water, earth, microbes, insects, fungi, wildlife, and minerals intersect. The mission of this X-Design Environmental Health Clinic, therefore, is to devise ways that art can serve as a healthcare provider for an ailing planet.

When art studios double as clinics, artistic creativity is couched in terms of the diagnosis of maladies and treatments. The patient being treated in the Environmental Health Clinic is the Earth. The artist's 'medium' consists of the intricately formulated and delicately balanced assemblies of water, air, soil, minerals, plants, and animals. When patient Earth is healthy, these components resolve into functioning systems that perpetuate themselves by metabolizing inputs, eliminating wastes, accommodating fluctuations, and defending against threats. However, Earth's medical chart is currently alarming. It reveals that Earth's systems are disrupted and obstructed. Entries on such charts include maladies that resemble familiar medical afflictions: invasives (cancer), algae blooms (immune deficiency disease), overstuffed landfills (obesity), smog (emphysema), traffic jams (arterial sclerosis), desertification (malnutrition), soil fatigue (infertility), global warming (fever), coral reef loss (osteoporosis), and nitrogen fertilizer dependency (opiate addiction).

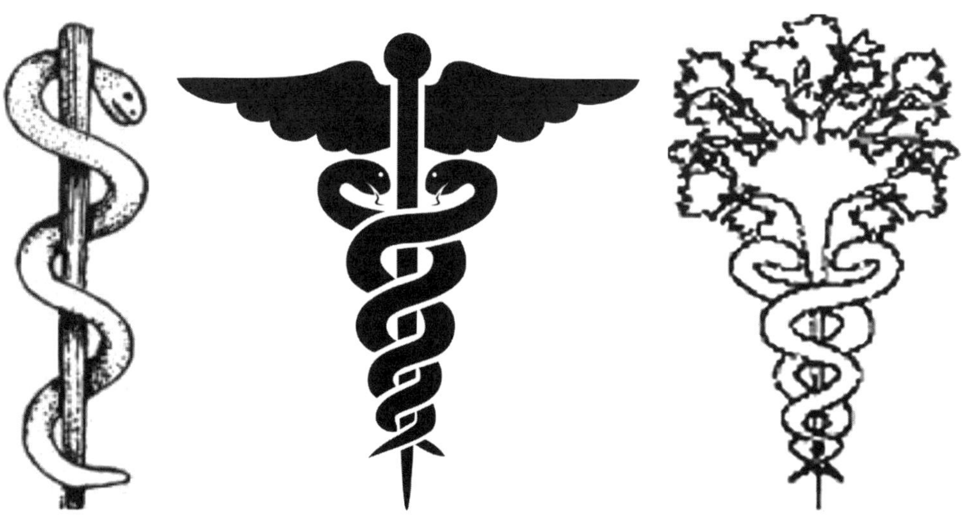

Figure 3a: Symbols of healing: Staff of Asclepius (symbol of medical organizations); Caduceus of Mercury (Roman); Environmental Health Clinic sign developed by Natalie Jeremijenko.

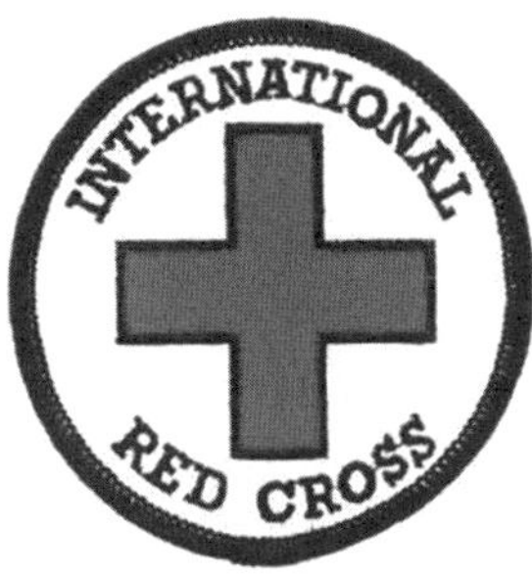

Figure 3b: International Red Cross symbol.

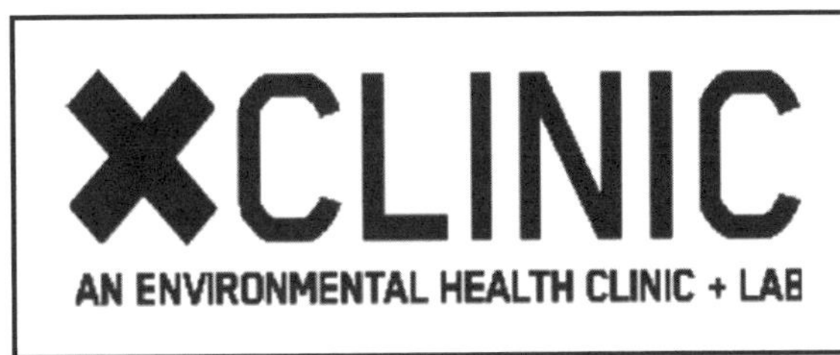

Figure 3c: Environmental Health Clinic symbol.

Are its maladies life-threatening? Earth's medical chart might look like this:

Patient = Earth	
Age:	4.54 billion years old
Size:	24.901.45 miles diameter
Weight:	13,000,000,000,000,000,000,000,000 pounds
Pressure:	High
Symptoms:	Erratic behavior, chills and sweats, wheezing, infertility, fatigue, indigestion, blocked arteries

Jeremijenko developed a pair of visual symbols to encapsulate her special enterprise. One symbol is that of Asclepius, the Greek god associated with healing and medicine. The 'Staff of Asclepius' is depicted as a serpent-entwined rod. This symbol of medical care morphed in the seventh century AD when a second snake and wings were added, shifting the medical reference from the god Asclepius to Hermes, whose magic wand identifies him as the inventor of supernatural incantations, and whose winged sandals depict him as the divine messenger. Jeremijenko retained the staff's association with healing, but she altered its alliance with myth and magic by replacing Hermes's magic wand and wings with the Tree of Life. The resulting variant conveys her intention to dispense pragmatic healing regimens that treat environmental ills.

The second visual symbol of the Clinic's mission is the Red Cross, which was first introduced at an international conference in 1862. It was adopted to protect medical personnel from military attack on the battlefield. By rotating the Red Cross ninety degrees, the revised symbol takes the form of a bold red X that marks the site of each X-Design Environmental Health Clinic initiative.

Jeremijenko activates this medical metaphor by designing actual health interventions for ailing ecosystems. Her diagnosis and treatment of environmental maladies extend the goal of medical care beyond relieving symptoms to include treating the causes of the maladies. In this manner, the Environmental Health Clinic both perpetuates and expands the Hippocratic Oath that is still required of medical students upon graduation in most medical schools. One passage from the oath is particularly relevant to the X-Design Clinic. It states, 'Treatment of the inner requires treatment of the outer.'[33] Hippocrates, who lived in the fourth century BC, could never have predicted the significance of this statement in the twenty-first century AD. The current harmed and harmful conditions of the air, waters, soils, and microbiome are implicated in asthma, cancer, obesity, diabetes, and numerous other common maladies. Rather than despairing, Jeremijenko is confident that environmental impairment can be treated, stating, 'It's external, it's shared, we can do something about it.'[34] Thus, Jeremijenko invents healing initiatives to enhance ecosystem functions.

How are trees choking on smog; birds tormented by urban clamor; and soils suffocated by asphalt treated at the Clinic? Since they cannot visit the Clinic, people are invited to make an appointment on their behalf. Jeremijenko refers to these volunteer caregivers as 'impatients' because they are too anxious about environmental ills to wait for government agencies to act. Impatients explain their environmental concerns to the Clinic's medical providers who are dressed in white lab coats with a rotated red cross on the pocket. These clinicians review the ailments, diagnose the problem, and write a non-pharmaceutical prescription detailing how to act on behalf of the afflicted patient. 'Referrals' consist of leads to local environmental organizations, government departments, and civic groups that might assist in carrying out the prescription. The Clinic thereby empowers citizens to tackle environmental issues that might otherwise seem daunting. Many people harbor environmental concerns, few know how to address them. 'Coming to the clinic doesn't require that you are an environmental activist or a community organizer,' explains Jeremijenko. 'It's a way of coordinating diverse, local actions and making them amount to something significant [...] People *can* actually do something.'[35]

The following four Environmental Health Clinic prescriptions reflect the range of environmental concerns and the innovative nature of the recommended prescriptions:

NoPark (2010–11)

Impatient concern: Contaminated storm water in urban areas. Hard surfaces prevent rainwater from being absorbed into the ground, and pollutants from

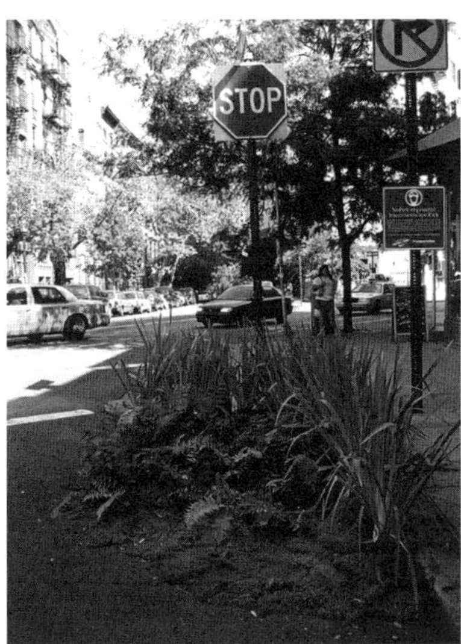

Figure 3d: Natalie Jeremijenko, *NoPark* (2010–11), plants, moss, grasses. Size variable. Lent courtesy of the artist and Postmasters Gallery, New York.

being filtered out by plants. This runoff collects dog urine, oil residues from roads, sewage, and other contaminants and deposits them in rivers, loading them with an unsavory mix of sediment, nutrients, metals, chloride, bacteria, and hydrocarbons.

Symptoms: Waters laden with pathogens, excess nutrients, heavy metals, and other toxins kill aquatic life and create algal blooms that can suffocate fisheries. Because the pathogens can reappear in drinking water and swimming areas, symptoms of contaminated river water are apparent in humans too. Indicators are diarrhea, vomiting, respiratory infections, hepatitis, dysentery, and other diseases.[36]

Prescription: Convert 'No Parking Zones' at fire hydrants throughout a city into *NoPark* zones by digging up the asphalt and planting low-growing, low-maintenance mosses and grasses. A *NoPark* is the opposite of a No Parking Zone. The latter are reserved for fires or other emergencies and are, therefore, unused most of the time. Because they are paved, they are environmentally damaging all of the time. Each *NoPark* transforms these vacant and lifeless spaces into a mini park that contributes to urban ecosystem health by capturing and filtering contaminated storm water, preventing it from discharging into rivers. Other advantages include recharging and replenishing soil moisture for nearby trees and other foliage, which improve urban air quality, beautify the city, and invite bird habitation. In addition, each *NoPark* prevents puddles that inconvenience pedestrians and serve as breeding sites for mosquitoes. Jeremijenko explains:

By creating engineered micro landscapes to infiltrate in them, we don't prevent them from being used as emergency vehicle parking spaces, because, of course, a fire truck can come and park there. They flatten a few plants. No big deal, they'll regenerate. But if we did this in every single... every fire hydrant, we could redefine the emergency. That ninety-nine percent of the time when a fire truck is not parking there, it's infiltrating pollutants. It's also increasing fixing CO2s, sequestering some of the airborne pollutants. And aggregated, these smaller interceptions could actually infiltrate all the road-borne pollution that now runs into the estuary system, up to a seven-inch rain event, up to a hundred-year storm.[37]

Carbon Pencils (2013)

Impatient concern: Smog and soot-laden air from cars and trucks, factories, power plants, incinerators, engines – anything that combusts fossil fuels.

Symptoms: Air quality can be measured according to concentrations of particulate matter (PM). Particles derived from dust, dirt, soot, or smoke are large enough to see with the naked eye. Those that result from complex chemical reactions emitted from power plants and automobiles are miniscule, thirty times smaller than a single hair![38] When these particles are inhaled, they can lodge in lungs and enter bloodstreams. Smog and soot affect plants by interfering with their ability to photosynthesize, reproduce, and store food, making them susceptible to diseases and pests. The maladies associated with polluted air are not confined to the environment. Humans and other life forms suffer the ill-effects of reduced lung function, irregular heartbeats, and irritated eyes, noses, and throats.[39] Jeremijenko notes that carbon black, with ozone, is responsible for about half of global warming's effects. 'It settles on the snow, it changes the reflectors, it changes the transmission qualities of the atmosphere.'[40]

Prescription: Air quality alerts issued by government agencies typically include recommendations to reduce exertion and remain indoors. The EHC replaces these reactive responses with a proactive initiative. Instead of retreating until a favorable wind disperses the contaminants, 'impatients' help remedy the situation by filtering particles out of the air they are breathing. The prescription involves removing and collecting unwanted carbon particulates that are trapped in the air, and transforming them into a useful product. Jeremijenko explains, 'In this project our aim is to mine the black carbon found in the polluted air we breathe and transform it into pencils. We explore different ways to make tangible what could be imprinted in our lungs.'[41]

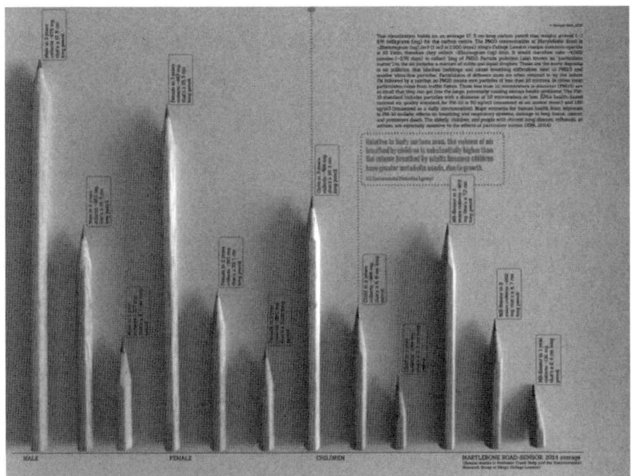

Figure 3e: Natalie Jeremijenko, *Carbon Pencils* (2013), ultra-fine particles (UFPs) and black carbon from soot in the air, mixed media, size variable. Natalie Jeremijenko, NYU with Professor Frank Kelly, King's College London), and John Nussey, Ding Labs. Lent courtesy of the artist and Postmasters Gallery, New York.

This prescription for decontaminating air combines two simple devices. The first is a solar chimney, which is featured in many contemporary energy-efficient building designs to control interior temperatures. The chimney takes advantage of the fact that hot air naturally rises. The EHC prescription recommends that black plastic be placed on the side of a building to augment the upward flow of the polluted air through the chimney.

The second component of the prescription consists of installing standard HVAC (heating, ventilation and air conditioning) filters on top of the chimney. These filters are capable of removing ninety-five percent of the carbon black contained in the air that passes through them. Jeremijenko describes the air contaminant that is collected in the filter as 'that grime that otherwise lodges in your pretty pink lungs.'[42]

This prescription was first tested on Marylebone Road, a heavily trafficked thoroughfare in London where the particulate matter concentration was already being monitored. Jeremijenko and her collaborator, Gyorgyi Galik, then calculated the quantity of pollution that would provide the amount of carbon in a pencil. They compared this number to the PM on Marylebone Road and determined that it would require four thousand minutes (2.78 days) to collect enough carbon to make a seven-inch pencil. The first phase of this project was exhibited in the London Design Festival in 2013. In subsequent exhibitions, the length of the pencils fabricated with the carbon black varied depending upon the amount of grime pulled out of the air. Each one removes offending particles from the air while simultaneously creating a useful pencil that materializes a serious environmental malady.

Lure (2010)

Impatient concern: Waterways polluted with PCBs (polychlorinated biphenyls) do not exist naturally. PCBs are manufactured for use in industrial products and chemicals. Because of their low volatility, PCBs typically accumulate in sediments where they continue to be released over long periods of time. While PCBs have low solubility in water, they are highly soluble in fat, which is why they accumulate in the bodies of aquatic organisms.

Symptoms: PCB poisoning is evident in multiple species of birds, fish, reptiles, and mammals. Their symptoms include developmental toxicity, reproductive impairment, compromised immunologic function, and other adverse effects. People absorb PCBs primarily by eating contaminated fish. The illnesses associated with PCB consumption include cancer, liver dysfunction, digestive disorders, chloracne, headaches, nausea, and fatigue. PCBs can also affect the respiratory, immune, and nervous systems. Recently, scientific research has linked PCB exposure with a variety of reproductive disorders, including male sterility, developmental abnormalities, learning disorders, and birth defects.[43]

Prescription: Conventional water pollution treatments dredge up contaminated sludge and dump it somewhere else. Conventional protective strategies recommend avoiding consumption of contaminated fish. The prescription offered by the EHC replaces reactive and inactive responses with a proactive strategy that removes the toxins from the fishes' bodies. This strategy transforms two fishing conventions. The first involves the use of lures without hooks to attract fish for the purpose of snagging them. In contrast, the EHC prescription uses lures to attract fish for the purpose of feeding them. The second reverses conventional approaches to protecting aquatic life by announcing, 'Don't feed the fish.' This prohibition is necessary to protect fish from an unwholesome diet that Jeremijenko describes as 'cigarette butts or Doritos or chewing gum.'[44] The EHC prescription, entitled *Lure*, encourages people to attract the fish with the lure for the purpose of feeding them a specially concocted food that enables the fish's digestive system to neutralize the toxins it has ingested. This fish food contains 'chitinase', a chelating agent that grabs minerals and metals as it travels through the fish's body. In this manner, calcium, lead, mercury, cadmium, copper, aluminum, and iron are removed from the bloodstream, delivered to the kidneys, and excreted as urine. Once these toxins and minerals are bound by a chelating agent, they cease being dangers. Jeremijenko explains the function of a chelating agent by stating:

> *When either people or fish ingest it, it binds to the bio-accumulated heavy metals and PCBs. There's no other way to get them out – you can't slice mercury out of a fish, or prepare it in a way that removes the PCBs [...] By*

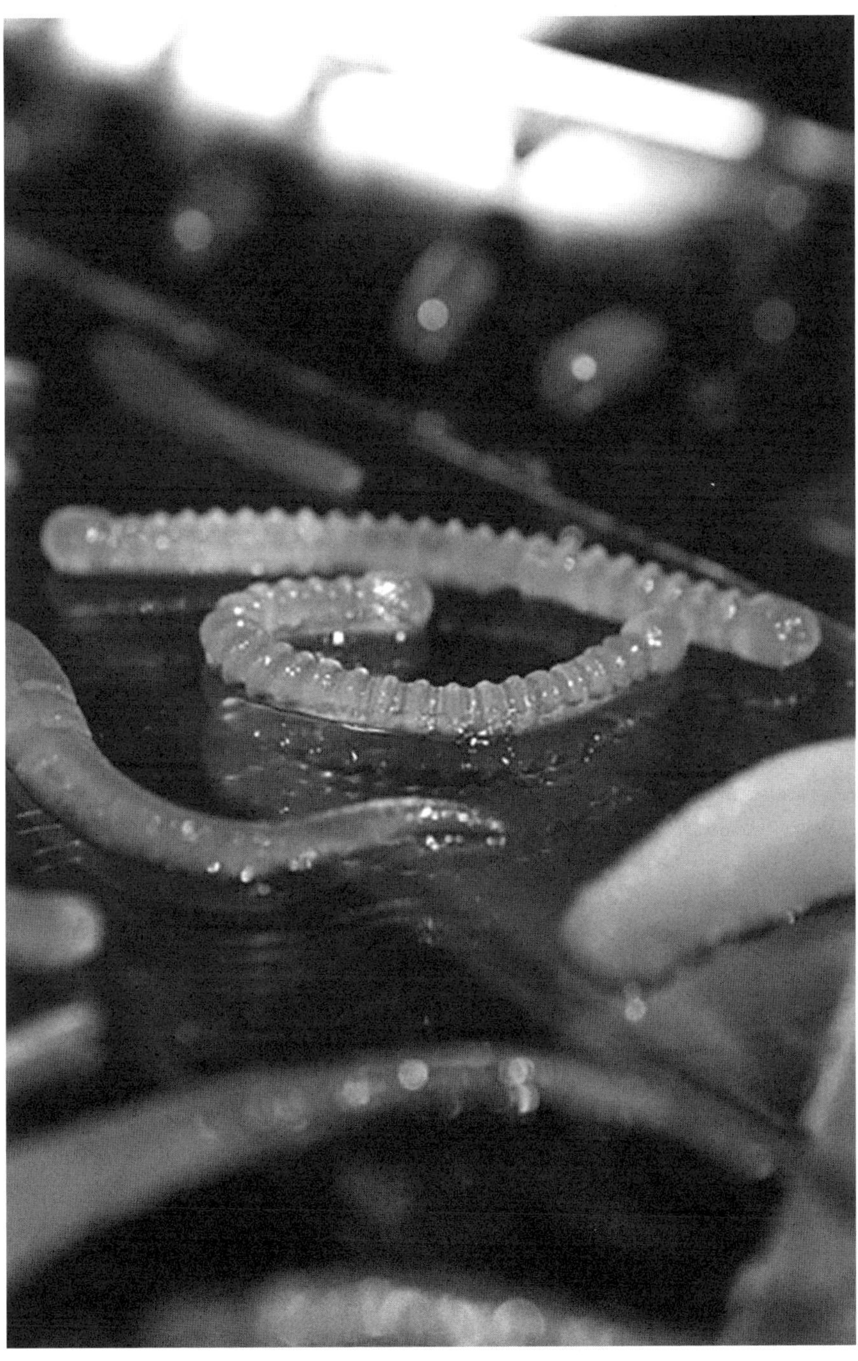

Figure 3f: Natalie Jeremijenko, *Lure* (2010), gellan with chelating agent (chitanase) to bind heavy metals and PCBs, size variable. Lent courtesy of the artist and Postmasters Gallery, New York.

feeding fish the lure with the chitinase in it, it binds to the bio-accumulated mercury and PCBs, and passes out as a harmless salt. It's much less reactive and much heavier, so it settles into the silt and is effectively removed from bio-availability.[45]

Thus, this prescription introduces several opportunities for healing and health. First the food that is fed to the fish purges the mercury and other metals that have accumulated in their bodies. Second, the amount of accumulated toxins in the waterways decreases. Finally, those who eat the fish are no longer endangering themselves by ingesting heavy metals. Jeremijenko comments:

Our approach is rather like targeted drug delivery, rather than treating the whole system as they're doing in the Hudson by dredging all the sludge and shipping it off to Texas or to Pennsylvania or to the nearest Third World country where it continues to be toxic sludge. Instead, we're using the bio-amplification that happens through the food web.[46]

Tadpole Bureaucrats (2009)

Impatient concern: Water that is muddy contains little dissolved oxygen, contains disease-causing pathogens, and is contaminated with sewage, pesticides, phosphorus, or another chemical toxicity. Temperature, acidity, dissolved oxygen, excess nutrients, and electrical conductance all factor into water quality.

Symptoms: While some indicators of poor water quality are perceptible, most are not. They include damaged plumbing equipment, detergent smell, gritty feel, murky color, and metallic taste. Impurities in lake and river waters can be discerned from the absence of aquatic organisms, such as water pennies, riffle beetles, and gilled snails that only tolerate clean water. Other symptoms can only be observed through a microscope. These tests are typically conducted in a laboratory, but invisible indicators of water quality can also be tested by observing the health of the aquatic occupants. Jeremijenko explains that tadpoles are highly sensitive to industrial contaminants and hormone-mimicking substances in water. She comments, 'Frogs are the canaries in the industrial coal mine. Animals always tell us how to redesign and improve their health.'[47] Such direct bio-monitoring is the basis of the prescription issued by the EHC.

Prescription: Impatients are instructed to acquire a tadpole from the body of water that the impatient has identified as a concern, and give the tadpole the name of a bureaucrat whose decisions affect water quality. For example, one tadpole was named James Gennaro, the chairman of New York City's Council Committee on Environmental Protection. Another was named Pete Grannism, commissioner of the New York State Department of Environmental Conservation.

Impatients place their named tadpole in a sample of the waterway and monitor the amphibian's condition and behavior. Jeremijenko refers to this process as 'keeping tabs' by 'keeping tads.'[48] Since people and tadpoles share the same water, using actual organisms to conduct bio-monitoring personalizes the results, as opposed to the technical data generated by sensors. Once the tadpole monitoring has produced results, an impatient packs it up and takes it to the office of the influential bureaucrat who serves as the tadpole's namesake. As they are introduced, the bureaucrat confronts the condition of the tadpole that bears his or her name. If its eyes are cloudy, or it is listless, skinny, or deformed, it is not well. All these conditions can be symptoms of poor water quality.

This cross-species strategy is amusing and plucky, but it is not frivolous because it is calculated to cure an ailing waterway. First, the shared name establishes a personal connection between responsible officials and victims of their neglect. The officials' sympathy may be evoked by confronting a pitiful specimen. Sympathy translates into concern. Concern leads to guilt. A sympathetic, concerned, guilty bureaucrat is primed to promote protective legislation and initiate remedial action for the waterway.

In order to expand personalized communications between people and the tadpole, the prescription also recommends that the impatient take the tadpole out for a walk in the evening. Jeremijenko explains why:

> *Your neighbors are likely to say, 'What are you doing?' And then you have to introduce your tadpole and who it's named after. You have to explain what you're doing and how the developmental events of a tadpole are, of course, very observable and they use the same T3-mediated hormones that we do. And so next time your neighbor sees you they'll say, 'How is that tadpole doing?' And you can let them social network with your tadpole, because the Environmental Health Clinic has a social networking site for not only impatients, humans, but non-humans, social networking for humans and nonhumans.*[49]

The relationship between ecosystem health and human health is commonly referred to as 'the health paradox'. It is paradoxical because human health often benefits by sacrificing the health of wild ecosystems. Examples include damming wild valleys, destroying mosquito-bearing wetlands, diverting water for irrigation, converting wilderness to farmland, logging forests, and eradicating wolves. Another paradox emerges when these protective actions are interpreted as destructive acts. The

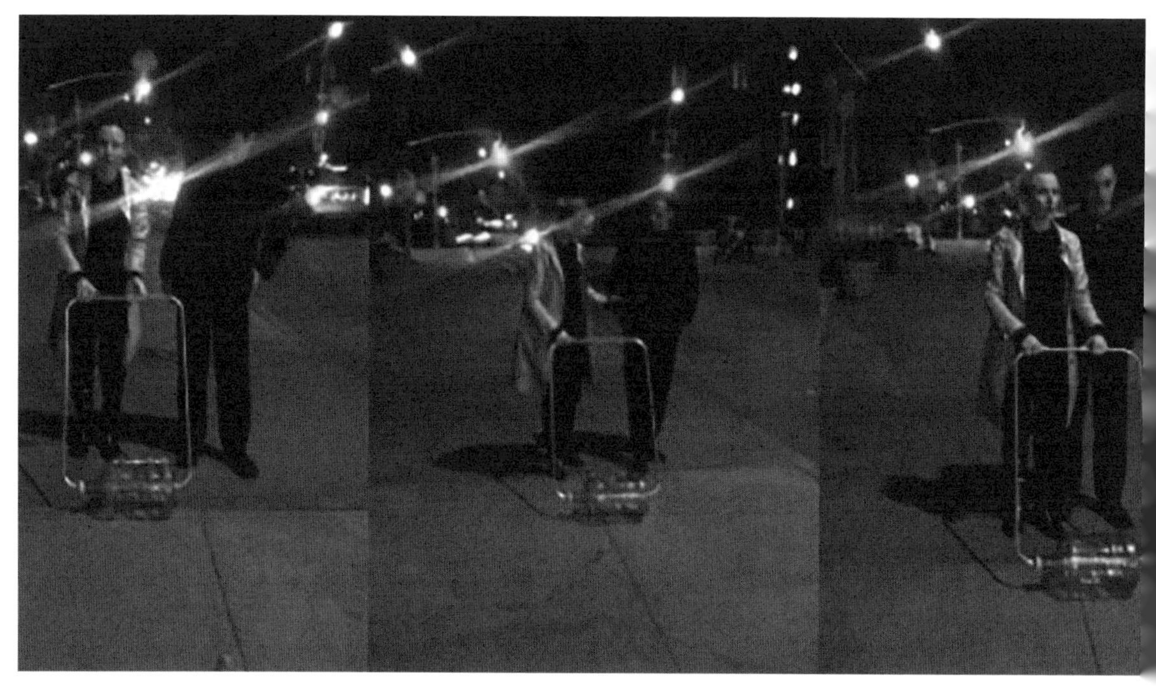

Figure 3g: Natalie Jeremijenko, *Tadpole Bureaucrats* (2009), tadpole; pond, lake or river water; mixed media. Lent courtesy of the artist and Postmasters Gallery, New York.

initial paradox assumes that the environment must be sacrificed to protect humans. The second paradox insists that human preferences must be sacrificed for the sake of the environment. Jeremijenko critiques the typical course of the self-sacrificing route to environmental health. She notes that eco-conscientious individuals tell her,

> *'I print on both sides of the paper.' 'I catch public transport.' 'I'm a vegetarian'. 'I'm doing what I can, but isn't suicide the best thing that I can do as an environmentalist? Then I'll use less gas and eat less food.'*[50]

While this conclusion might be the logical extension of the fixation on reducing the size of footprints, left by energy and material consumption, the Environmental Health Clinic encourages people to do more (that is productive), rather than less (that is destructive). By focusing on effecting measurable improvements in human health alongside environmental health, the Clinic's positive interventions offer uplifting alternatives to the disagreeable aspects of both environmental paradoxes. The nutrient cycles and energy exchanges it prescribes fortify social and ecological networks. Such mutual benefits introduce the possibility of circumventing the conventional 'cost–benefit' analysis that assumes there will be costs associated with every gain. Jeremijenko has invented the term 'benefit-benefit' analysis to indicate that interactions with the ecosystem can become 'a form of cultivation rather than extraction or damage.'[51]

To date, none of the Clinic's innovative prescriptions has actually cured a chronic environmental malady. Such an outcome would depend upon greatly increasing the scale of remediation efforts to match the scale of remedial needs. Recruiting environmental health practitioners by empowering concerned citizens represents one approach utilized by the Clinic to amplify its interventions. In addition, Environmental Health Clinic advisory boards are being established throughout the world. Jeremijenko refers to board members as 'Doctors without Disciplinary Borders' (DR.X*), to contrast them with 'Doctors without Borders'. Unlike this well-known medical humanitarian organization that sends doctors anywhere medical care is lacking, the 'Doctors without Disciplinary Borders' provide care where they reside. The creative focus of its initiatives is thereby developed with regard to the specific expertise of its members, and knowledge of local resources greatly enhances the effectiveness of its medical interventions. Thus, by recruiting 'impatients' and establishing an organizational structure for the material interactions of the Environmental Health Clinic, the efforts of 'Doctors without Discilinary Borders' might eventually match the scale and urgency of patient Earth's maladies.

Daily headlines identify situations where the dedicated efforts of Earth stewards are in dire need. The following list compares such environmental ills to familiar categories of healthcare providers. This tactic translates environmental problems into career opportunities. Readers are urged to consider each entry as a way to practice Eco Material art. Please join this initiative. By signing up for active Eco Material service, you will join a community of individuals who are devoted to positive change and beneficial outcomes.

Urgent care specialists to save species from extinction.

Geriatric doctors to prolong the life of old-growth forests.

Surgeons to remove invasive species.

Gastrointestinal specialists to process inputs and manage wastes.

Primary care doctors to diagnose maladies.

Preventive care providers to maintain healthy systems.

Physical therapists to rehabilitate damaged systems.

Trauma doctors to intervene in emergencies.

Nurses to facilitate recovery.

Chronic disease managers to oversee persistent ailments.

Infectious disease specialists to balance microorganisms.

Cancer physicians to halt the spread of algae blooms and other scourges.

Cardiovascular specialists to maintain the passageways through which energy, nutrients, carbon, water, and nitrogen circulate.

Dietitians, occupational therapists, technicians, instructors, acupuncturists, naturopaths, and researchers – lots of researchers!

Conclusion

The finale of this book marks a beginning, not an ending. Like a commencement exercise that situates past achievements within the context of future accomplishments, the pioneering initiatives undertaken by the contributing artists in this book are presented as prototypes for the coming Eco Material era. Fulfilling this potential ultimately depends upon whether the behaviors they envision become normalized, actualized, and stabilized, not just by artists, but also by businesses, government agencies, educational institutions, religious organizations, and individuals.

Art, which is systemically committed to inspiring, innovating, and informing, is poised to play a leadership role in establishing a culture of Eco Material values. All the artworks compiled in this volume manifest a core belief that optimizes healthful outcomes. They demonstrate that in ecosystems, everything matters! There is no time-out, no reruns, no backups, no substitutes, no escapes, no vacations, and no exemptions. Humans matter exceptionally because of the powerful tools and technologies they have amassed to satisfy their desires. Eco Material artworks address this 'matter' by demonstrating that humans do not need to endanger the mutually dependent, self-correcting functions of ecosystems in order to survive. They can quell net environmental losses and they can encourage net environmental gains. The artists who create these congenial artworks consciously welcome non-humans and non-living entities as creative partners, not resources to consume, situations to avoid, or obstacles to overcome. In all these ways, these artists manifest a holistic consciousness that expands the 'creativity' factor of their art practice beyond self-expression and personal predilection.

Eco Material artists are, therefore, primed to retire the question mark of inquiry following the phrase 'What's next', and replace it with an exclamation point of certainty.

Your opinion: What's next?

Reader Interaction

What Do I Believe?

A Personal Environmental Self-Assessment

A planet in peril has convinced many environmentalists to call for a complete overhaul of humanity's current means of acquiring, using, and discarding resources. They share a widespread conviction that our material interactions with the planet are more likely to originate in attitudes and assumptions than knowledge about objects, entities, and substances. Since no authority exists to define and enforce the cultural values that generate sustainable actions, the following self-assessment has been prepared to help readers fashion a personalized environ-mentality out of their responses. These responses may help you construct a blueprint of your individual environmental beliefs and the course of action they suggest. A variety of responses are provided for each of the ten environmental beliefs and conditions that follow. Rate their relevance to you on a scale of 1 to 10. Ten is your strongest belief. One is your weakest belief. Add items that are missing from the choices provided.

1. I believe the current state of planet Earth is:

Ailing

Robust

Deteriorating

Recuperating

Resilient

Fragile

Other

2. When I envision 'nature', I imagine:

Pure air, clean water, blossoms, trees, rainbows, butterflies

Germs, rotting carcasses, blight, weeds, mold, mutants, slime

Clear-cut forests, strip mining, landfills, toxic waste dumps,
oil spills, smog

Dams, highways, industrial parks, high-rise buildings, nuclear reactors,
pipelines

Parks, zoos, aquariums, planetariums, botanical gardens, game
farms, theme parks

Residential homes, yards, driveways, barbecues, air conditioners,
lawns

Shopping malls, e-commerce, brand names, advertisements,
credit cards

Websites, blogs, instant messaging, iPods, emails, YouTube,
MySpace, Facebook...

GM seeds, fertilizers, pesticides, herbicides, tractors, columbines

The human microbiome: bacteria, fungi, archaea

Other

3. The geological, botanical, zoological, and energetic components of the environment where I conduct my daily activities are my:

Refuge

Responsibility

Liability

Resource

Source of shame

Source of reverence

Obstacle

Delight

Recreation

Nothingness

Other

4. The non-visible parts of the environment I pay attention to are:

Molecules, microbes, algae, fungi, spores, nematodes

Contaminants in water, air, soil

Electromagnetic fields, ultrasonic waves, infrared waves,
radio waves

Conditions such as air pressure, wind, temperature, humidity

Spirits that reside in land, rocks, waters, plants, animals

Souls of the deceased

Climate change

Other

5. My physical organism is primarily sustained by:

My physical strength

The physical strength of personal acquaintances

The physical strength of people I do not know

Machines

Pharmaceuticals

Money to purchase things

Electronic technologies

My skills and knowledge

The skills and knowledge of personal acquaintances

The skills and knowledge of strangers

Animals and plants I cultivate

Animals and plants I forage

Other

6. **The groups that have the greatest impact upon the wildlife, air, water, and soil within the region where I live are:**

Environmentalists

Corporate managers

Government officials

Religious leaders

Financial authorities

Farmers

Engineers

Consumers

Merchandisers

Other

7. **My familiarity with the following entities is:**

Pets

Farm animals

Wild animals

Ornamental plants

Utilitarian plants

Weeds

Soil

Microbes

Wilderness

Suburbia

Metropolises

Electronics

Mechanics

Media

Other

**8. The factors that most influence my opinions about the current state of
planet Earth are:**

Personal observation

Research

Academic studies

Art

Music

Literature

Dance

Religious teachings

Media news

Science fiction

History

Nostalgia

Instincts

Popular films

Social media

Politics

Other

9. Environmental protocols I practice include:

Recycling

Retooling

Downsizing

Energy conservation

Composting

Minimizing consumption

Creating habitat

Removing litter

Creating soil

Planting native fauna

Political action

Other

10. I anticipate the following within the next one hundred years:

Life as we know it will continue

Debilitated ecosystems will be vitalized

New species will evolve to replace extinct species

Humanity will adopt sustainable practices and avert disaster

Shortages of food, water, top soil, and energy will cause
worldwide turmoil

Homo sapiens will become extinct but other life forms will survive

Worldwide extinctions will end life on Earth

Other

Endnotes

1   Edan Corkill, 'Art Collective Chim↑Pom Provokes Tokyo with a Crow and a Rat,' *Japan Times*, August 8, 2008, accessed February 12, 2018,https://www.japantimes.co.jp/culture/2008/08/14/arts/shock-tactics-return/#.WqbZeXyQy70.

2   Louisa Lim, 'As the Crow Flies, Tokyo Battles Avian Pest,' NPR, January 12, 2010, accessed September 29, 2017, http://www.npr.org/templates/story/story.php?storyId=122291084.

3   Lester Haines, 'Tokyo Battles Monstrous Murder of Crows,' *The Register*, accessed September 29, 2017, http://www.theregister.co.uk/2009/07/27/murder_of_crows/.

4   Lim, 'As the Crow Flies.'

5   'Crows in Japan: Menace or Guide of the Gods?,' *The Japan Chronicles*, February 26, 2012, accessed September 29, 2017, http://thejapanchronicles.blogspot.com/2012/02/crows-in-japan-menace-or-guide-of-gods.html#.WVG6yIG2xSE.

6   AND, '"Six Members Is Already a Society",' *Frieze*, 21 October, 2015, accessed September 29, 2017, https://frieze.com/article/%E2%80%98six-members-already-a%C2%A0society%E2%80%99.

7   AND, '"Six Members".'

8   PBS, 'Art Cannot Be Powerless: An interview with Ryuta Ushiro,' accessed September 29, 2017, http://www.pbs.org/wgbh/pages/frontline/the-atomic-artists/art-cannot-be-powerless/.

9   Chim↑Pom, email correspondence with the author, December 5, 2017.

10   Zbigniew Oksiuta, interview with the author in Troy, NY, September 8, 2017.

11   'Several bottles' refers to four nucleotides: A-Adenine, G-Uanine, C-Cytosine, and T-Tymine. These components of DNA (deoxyribonucleic acid) carry the genetic information of all living organisms.

12   Zbigniew Oksiuta, email correspondence with the author, September 20, 2017.

13   Bevan Hamilton, 'Fukushima 5 Years Later: 2011 Disaster by the Numbers,' March 10, 2016, accessed June 23, 2016, http://www.cbc.ca/news/world/5-years-after-fukushima-by-the-numbers-1.3480914.

14   Michael Werz, 'Cause Behind African Migrant Flood Has Terrifying Implications for the World,' *Reuters*, April 21, 2015, accessed June 23, 2016, http://blogs.reuters.com/great-debate/2015/04/21/not-science-fiction-changing-climate-pushing-migrants-out-of-africa-and-into-tragedy/.

15   John Vidal, 'Global Warming Could Create 150 Million "Climate Refugees" by 2050,' *The Guardian*, accessed November 3, 2104, https://www.theguardian.com/environment/2009/nov/03/global-warming-climate-refugees.

16   Oksiuta, interview.

17   Zbigniew Oksiuta, 'Breeding Spaces', accessed November 5, 2014, http://www.stillliving.symbiotica.uwa.edu.au/pages/artists/zbigniew.htm.

18   Oksiuta, interview.

19   Oksiuta, interview.

20   Video – Zbigniew Oksiuta – *Biological Habitat*, AA School of Architecture, Lecture November 5, 2009, Published September 3, 2015, accessed November 12, 2017, https://www.youtube.com/watch?v=jTI3y2oZKKc.

21   The *Biological Habitat: Breeding Spaces Technology, Made in Space* project was exhibited at the OK Centrum, Linz in 2007.

22   Oksiuta interview.

23   'Zbigniew Oksiuta,' *Article*, accessed November 12, 2017, http://article.no/en/artists/2008/zbigniew-oksiuta-pl.

24   Oksiuta, interview.

25   Oksiuta, interview.

26   'Zbigniew Oksiuta,' *Article.*

27   Zbigniew Oksiuta, *Your Personal Biosphere*, 2009, published 2011, Science Gallery, Trinity College, Dublin, https://www.youtube.com/watch?v=hOEpmzOly2A.

28   Brian Wang, 'New Space Age Materials for a New Space Age,' *Next Big Future*, accessed February 24, 2011, http://nextbigfuture.com/2011/02/new-space-age-materials-for-new-space.html.

29   Wang, 'New Space Age Materials.'

30   Oksiuta, interview.

31   Oksiuta, interview.

32   MIT Technology Review: Innovators Under 35. Year honored - 1999. Region - global. Organization - 'Bureau of Inverse Technology'.

33   Natalie Jeremijenko, 'Transcript: The Art of the Eco-Mindshift,' TED, October 2009, accessed September 29, 2017, https://www.ted.com/talks/natalie_jeremijenko_the_art_of_the_eco_mindshift/transcript?language=en.

34   Jeremijenko, 'Transcript.'

35   Sabine Heinlein, 'Avant-Garde Rx,' *NYU Alumni Magazine: Avant-Garde Rx*, accessed September 29, 2017, https://www.nyu.edu/alumni.magazine/issue10/10_nyc_environmental.html.

36   American Rivers, 'How Stormwater Affects Your Rivers,' *American Rivers*, May 31, 2016, accessed September 29, 2017, https://www.americanrivers.org/threats-solutions/clean-water/stormwater-runoff/.

37   Jeremijenko, 'Transcript.'

38   EPA (United States Environmental Protection Agency), 'Particulate Matter (PM) Basics,' September 12, 2016, accessed September 29, 2017, https://www.epa.gov/pm-pollution/particulate-matter-pm-basics.

39   Jillian Mackenzie, 'Air Pollution: Everything You Need to Know,' accessed November 1, 2016, https://www.nrdc.org/stories/air-pollution-everything-you-need-know#sec3.

40   Jeremijenko, 'Transcript.'

41   Gyorgyi Galik, 'Experiments with Invisible Pollutants I: Carbon Pencils, London,' accessed September 29, 2017, http://gyorgyigalik.com/Experiments-with-Invisible-Pollutants-I-Carbon-Pencils-London.

42   Jeremijenko, 'Transcript.'

43   Rensselaer Polytechnic Institute, 'PCB Contamination of the Hudson: Is Dredging an Appropriate Cleanup Strategy?,' accessed September 29, 2017, https://www.rpi.edu/dept/environ/orgs/Clearwater/dredging.html.

44   Nicola, 'Cross-Species Dining: An Interview with Natalie Jeremijenko and Mihir Desai,' *Edible Geography*, October 27, 2010, accessed September 29, 2017, http://www.ediblegeography.com/cross-species-dining-an-interview-with-natalie-jeremijenko-and-mihir-desai/.

45   Nicola, 'Cross-Species Dining.'

46   Nicola, 'Cross-Species Dining.'

47   Natalie Jeremijenko, interview with the author in Manhattan, August 2013.

48   Jeremijenko, interview.

49   Jeremijenko, 'Transcript.'

50   Natalie Pompilio, 'Robot Dogs and Other Weird Creatures Bring Nature to the City,' *Yes! Magazine*, January 19, 2015, accessed September 29, 2017, http://www.yesmagazine.org/issues/what-would-nature-do/shocking-the-big-city-with-a-little-green-grass.

51   Nicola, 'Cross-Species Dining.'

SUGGESTED READINGS

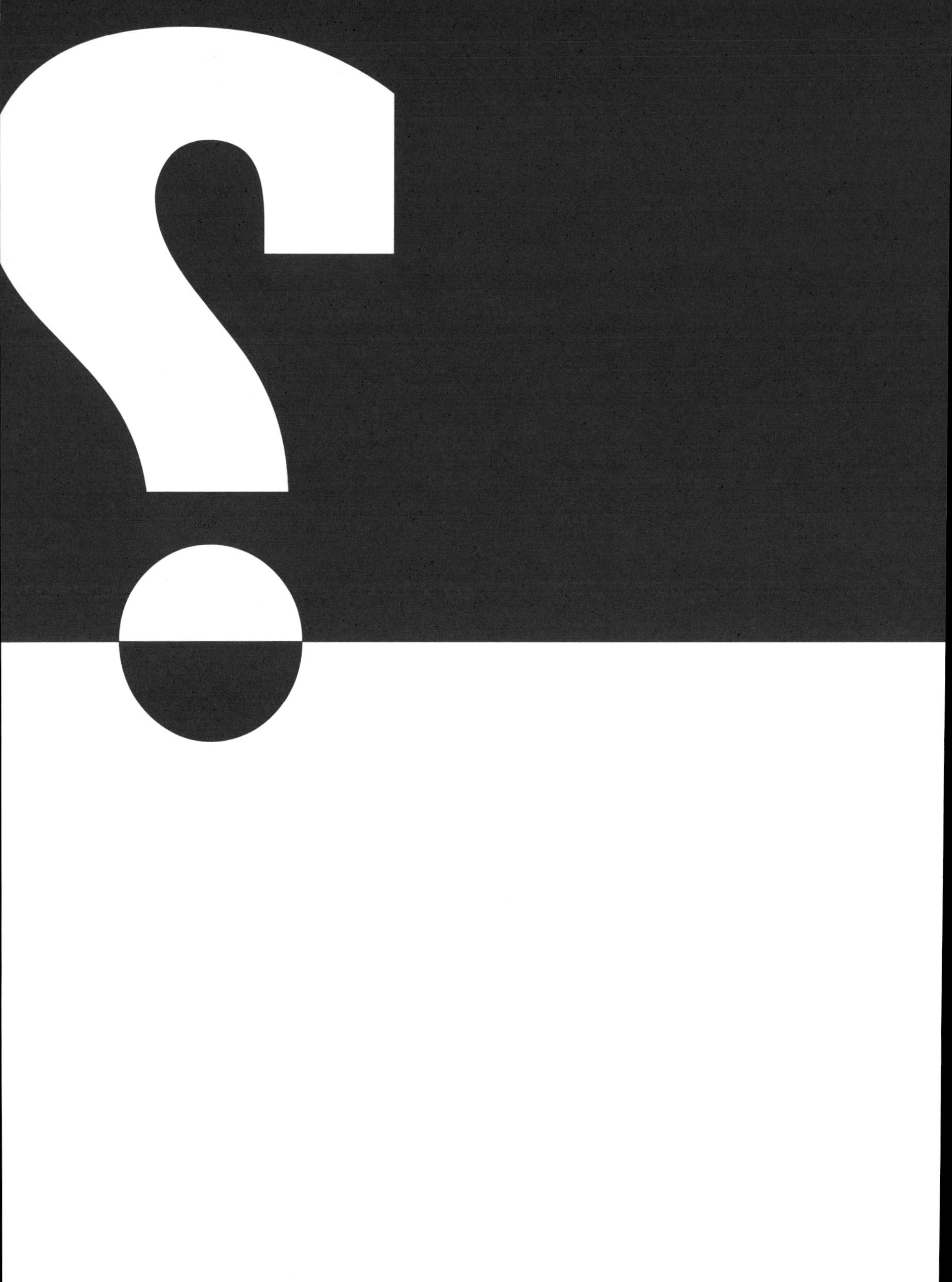

New Materialism

Alaimo, Stacy and Susan Hekman, *Material Feminisms* (Bloomington, IN: Indiana University Press, 2009). *Material Feminisms* assembles theories generated by an international group of feminist thinkers that are grounded in recent reconsiderations of the human body, the natural world, and the material world. 'By insisting on the importance of materiality, this volume breaks new ground in philosophy, feminist theory, cultural studies, science studies, and other fields where the body and nature collide.'[1]

Bennett, Jane, *Vibrant Matter: A Political Ecology of Things* (Durham, NC: Duke University Press, 2010). Bennett shifts focus from the human experience of things to things themselves, arguing that political theory needs to recognize the active participation of non-human forces in events.

Bogost, Ian, *Alien Phenomenology, or What It's Like to Be a Thing* (Minneapolis, MN: University of Minnesota Press, 2012). Bogost develops an object-oriented ontology that puts things at the center of being. He encourages professional thinkers to become makers as well.

Bryant, Levi, Nick Srnicek, and Graham Harman, *The Speculative Turn: Continental Materialism and Realism* (Melbourne: re.press, 2011). This book offers a sweeping critique of the self-enclosed Cartesian subject referred to as 'anti-realist'. It introduces a new era of 'reality'.

Coole, Diana and Samantha Frost, *New Materialisms: Ontology, Agency, and Politics* (Durham, NC: Duke University Press, 2010). *New Materialisms* explores the innovative materialist critiques that are emerging across the social sciences and humanities. This collection documents the reworking of older materialist traditions, contemporary theoretical debates, and advances in scientific knowledge.

Crockett, Clayton and Jeffrey Robbins, *Religion, Politics, and the Earth: The New Materialism* (London: Palgrave Macmillan, 2015). This book proposes that western culture is approaching its limits, which is apparent economically, politically, culturally, ecologically, and philosophically.

DeLanda, Manuel, *Intensive Science and Virtual Philosophy* (London: Bloomsbury Academic, [2002] 2013). This book tackles new developments in geometry, complexity theory, and chaos theory, which generate dynamic processes that constitute the identity of objects through time.

Dolphijn, Rick and Iris van der Tuin, *New Materialism: Interviews & Cartographies*, New Metaphysics (London: Open Humanities Press, 2012). The first part of this book contains elaborate interviews with some of the most prominent New Material scholars of today. The second part situates the New Material tradition in contemporary thought by developing the ethical and political consequences of New Materialism.

Eagleton, Terry, *Materialism* (New Haven and London: Yale University Press, 2016). Eagleton argues that materialism is at the center of today's important scientific and cultural as well as philosophical debates. The text demonstrates how it is our bodies and corporeal activities that make thought and consciousness possible.

Fox, Nick J. and Pam Alldred, *Sociology and the New Materialism: Theory, Research, Action* (London: Sage Publishing, 2016). This book applies New Materialism to creativity, sexuality, emotions, health, research, and change.

Harman, Graham, *Guerilla Metaphysics: Phenomenology and the Carpentry of Things* (Chicago: Open Court Publishing, 2005). Professor Harman argues for a radical shift in the phenomenological attitude to objects, and explains how phenomenology can be reunified with the physical world.

Harman, Graham, *Towards Speculative Realism: Essays and Lectures* (UK: Zero Books, John Hunt Publishing, 2010). This collection of essays and lectures recounts the evolution of Harman's object-oriented metaphysics.

Harman, Graham, *The Quadruple Object* (UK: Zero Books, John Hunt Publishing, 2010). Harman critiques philosophies that reject objects as a primary reality and replaces them with concepts of sensual and real objects.

Latour, Bruno, *Pandora's Hope: Essays on the Reality of Science Studies* (Cambridge, MA: Harvard University Press, 1999). *Pandora's Hope* argues for understanding the reality of science in practical terms. Through the case studies of scientists, Latour shows how events in the material world are transformed into items of scientific knowledge.

LeCain, Timothy J., *The Matter of History: How Things Create the Past* (Cambridge, UK: Cambridge University Press, 2017). This book combines scientific and humanistic ideas to develop a post-anthropocentric understanding of the past that reveals how organisms and things create social and cultural affairs.

Malafouris, Lambros, *How Things Shape the Mind: A Theory of Material Engagement* (Cambridge, MA: MIT Press, 2013). Malafouris draws on recent developments in cognitive science and philosophy of mind to construct Material Engagement Theory (MET). This cross-disciplinary study investigates the ways 'things' become cognitive extensions of the human body.

Miller, Adam S., *Speculative Grace: Bruno Latour and Object-Oriented Theology*, Perspectives in Continental Philosophy (New York: Fordham University Press, 2013). This book relates object-oriented philosophy to religion in which grace is embodied in objects and religious practice involves the flesh.

Morton, Timothy, *Hyperobjects: Philosophy and Ecology after the End of the World* (Minneapolis, MN: University of Minnesota Press, 2013). 'Hyperobjects' are entities of such vast temporal and spatial dimensions that they defeat traditional ideas about what a thing is in the first place. Morton explains their impact on how we experience our politics, ethics, and art.

Morton, Timothy, *The Ecological Thought* (Cambridge, MA: Harvard University Press, 2010). Morton argues that all forms of life are connected in a vast, entangling mesh. This interconnectedness penetrates all dimensions of life. 'Nature' does not exist as an entity.

Woodward, Ben, *On an Ungrounded Earth: Towards a New Geophilosophy* (Brooklyn, NY: Punctum Books, 2013). In continental philosophy, the Earth is only an object of interest when it attracts human attention and stimulates thought. This book constructs an alternative – geophilosophy – by referring to digging machines, cyclones, nuclear waste, and worms, as well as horror films and video games.

Eco-Art

Barrett, Estelle and Barbara Bolt, *Carnal Knowledge: Towards a 'New Materialism' Through the Arts* (London: I. B. Tauris, 2013). *Carnal Knowledge* assembles essays that explore the material nature of artistic practice and the notion of 'truth to materials'. The book spans dance, music, film, fashion, design, photography, literature, painting, and virtual reality.

Blanc, Natalie and Barbara Benish, *Form, Art and the Environment: Engaging in Sustainability* (London and New York: Routledge, 2016). This book examines the contributions of the arts in promoting sustainable development and culture at a grassroots level and its potential as a catalyst for social change. It considers the ecological crisis on a macro level, as well as personal lifestyles and beliefs.

Bergmann, Sigurd, Irmgard Blindow, and Konrad Ott (eds), *Aesth/Ethics in Environmental Change: Hiking Through the Arts, Ecology, Religion and Ethics of the Environment,*

Studies in Religion and the Environment/Studien zur Religion und Umwelt 7 (Berlin, Münster, Wien, Zürich, and London: LIT Verlag, 2013). How can the arts widen our perception of nature and deepen environmental ethics? This question is explored from the angles of arts, environmental ethics, ecology, religious studies, theology, art history, and philosophy. The book prompts discussion about the aesthetic and spiritual dimensions in the environmental humanities.

Brady, James. (ed.), *Elemental: An Arts and Ecology Reader* (Manchester: Gaia Project Press and Cornerhouse Press, 2016). *Elemental* presents essays by artists, activists, curators, and writers currently working in the arts and ecology. The book presents critical reflections and philosophies on a variety of eco-art practices and methodologies. Subject areas include New Materialism, socially engaged ecosystem restoration, the legal 'Rights of Nature', and ecology in theatre and performance art.

Carruthers, Beth, *Arts in Ecology: Mapping the Terrain of Contemporary Eco-ART Practice and Collaboration* (UNESCO, 2006). A research report on science–arts collaboration in sustainability commissioned by the Canadian Commission for UNESCO. Published online by UNESCO (http://unesco.ca/~/media/pdf/unesco/ bethcarruthersartinecologyresearchreportenglish.pdf). Accessed November 10, 2016.

Cox, Christoph, Jenny Jaskey, and Suhail Malik (eds), *Realism Materialism Art* (Berlin: Sternberg Press: Annandale-on-Hudson: Bard College, 2015). This book delves into Neo Materialtheories, object-oriented ontologies, and neo-rationalist philosophies. The texts address art theories, not artworks.

Dessein, J., K. Soini, G. Fairclough, and L. Horlings (eds), *Culture in, for and as Sustainable Development*, Conclusions from the COST Action IS1007: Investigating Cultural Sustainability (Helsinki: University of Jyväsky, 2015). Investigating Cultural Sustainability is a European research network focused in a multidisciplinary perspective on the relationship between culture and sustainable development and has provided policy makers with instruments for integrating culture into sustainable development.

Harrison, Helen and Newton Harrison, *The Time of the Force Majeure: After 45 Years Counterforce Is on the Horizon* (New York, NY: Prestel, 2016). Pioneers of the eco-art movement, Helen and Newton Harrison, established a worldwide network among biologists, ecologists, architects, urban planners, politicians, and other artists to initiate collaborative dialogues about ideas and solutions that support biodiversity and community development.

Ingold, Tim, *Making: Anthropology, Archaeology, Art and Architecture* (Abingdon and New York: Routledge, 2013). *Making* reflects on what it means to create things, on materials and form, the meaning of design, landscape perception, animate life, personal knowledge, and the work of the hand. It combines social and cultural anthropology, archaeology, architecture, art and design, visual studies, and material culture.

Kagan, Sacha and Volker Kirchberg (eds), *Sustainability: A New Frontier for the Arts and Cultures* (Frankfurt: VAS, 2011). This book diverges from approaching sustainability challenges as scientific problems. It presents art as offering a different intellectual, creative, and social space that can offer innovative solutions to sustainability problems.

Kagan, Sacha, *Art and Sustainability: Connecting Patterns for a Culture of Complexity* (New Brunswick: Transaction Publishers, 2011). This book explores 'culture(s) of sustainability', 'aesthetics of sustainability', and 'art and sustainability' as expressed by four decades of ecological art.

Kirksey, Eben, *The Multispecies Salon* (New York: Duke University Press, 2014). Multispecies ethnographers collaborate with artists and biological scientists to illuminate how diverse organisms are entangled in political, economic, and cultural systems.

Lippard, Lucy, *Undermining: A Wild Ride Through Land Use, Politics and Art in the Changing West* (New York: New Press, 2013). Lippard weaves together fracking, mining, land art, adobe buildings, ruins, Indian land rights, the Old West, tourism, photography, and water to illuminate the relationship between culture and the land.

Moore, Sadhbh and Alison Tickell, *D'Art Report 34b: The Arts and Environmental Sustainability: An International Overview* (Sydney: Julie's Bicycle and the International Federation of Arts Councils and Culture Agencies, 2014) (http://www.ifacca.org/topic/ecological-sustainability, accessed December 6, 2014). This report compiles surveys and interviews regarding sustainable art practices.

Sidford, Holly and Alexis, Frasz, *Beyond Green: The Arts as a Catalyst for Sustainability*, Session Report 561, February 19–24, 2016 (Salzburg: Salzburg Global Seminar, 2016) (http://www.salzburgglobal.org/fileadmin/user_upload/Documents/2010-2019/2016/Session_561/SalzburgGlobal_Report_561__online_.pdf, accessed September 24, 2016). *Beyond Green* explores the arts' role in advancing sustainability at international, national, and local levels.

Sigurjonsdottir, A. and O. Jonsson (eds), *Art, Ethics and Environment* (Newcastle-upon-Tyne: Cambridge Scholars Press, 2006). This collection gathers different trends in thinking about nature and value through art, and juxtaposes them with some other ways of thinking about these issues, such as economics and religion.

Solnit, Rebecca, *As Eve Said to the Serpent: On Landscape, Gender and Art* (Athens: University of Georgia Press, 2003). To Solnit, the word 'landscape' implies not only literal places, but also the ground on which we invent our lives, confront our innermost desires, and construct the social, political, and philosophical landscapes we inhabit.

Spade, Sue. (ed.), *Ecovention: Current Art to Transform Ecologies* (Cincinnati: Contemporary Arts Centre, Eco-artspace and Green Museum, 2002). This text explores various forms of activism that address ecological issues and the monitoring of ecological problems.

Strelow, Heike. (ed.), *Ecological Aesthetics: Art in Environmental Design: Theory and Practice*, trans. H. Prigann and V. David (Basel, Berlin, and Boston: Birkhauser, 2004). Ecological and artistic approaches to designing urban spaces, gardens, or landscapes raise important questions addressed in this book, such as whether to conserve or to intervene, to rely on natural order or apply man-made means, and whether to lend nature a romantic or a technocratic appearance.

Van Boeckel, Jan, *At the Heart of Art and Earth: An Exploration of Practices in Arts-Based Education* (Aalto: Aalto University Press, 2014). In this explorative study, the author provides catalysts to evoke transformative learning through art-making by leading participants away from consumer and digital habits.

Weintraub, Linda, *CycleLogical Art: Recycling Matters for Eco-Art (Rhinebeck, Avant-Guardians: Textlets in Art and Ecology)* (Rhinebeck: Artnow Publications, 2007). *CycleLogical Art* assembles eleven pioneering artists who have designed ingenious schemes to divert waste and cast-offs from the purgatory of landfills and grant them everlasting value.

Weintraub, Linda, *EcoCentric Topics: Pioneering Themes for Eco-Art (Avant-Guardians: Textlets in Art and Ecology)* (Rhinebeck: Artnow Publications, 2006). 'Ecocentric' describes thoughts and behaviors that are habitat-centered, as opposed to species-centered (anthropocentric) and self-centered (egocentric). It applies ecocentrism to such topics as 'nature', 'desire', 'globalism', 'power', and 'death'.

Weintraub, Linda, *EnvironMentalities: Twenty-Two Approaches to Eco-Art (Avant-Guardians: Textlets in Art and Ecology)* (Rhinebeck: Artnow Publications, 2007). *EnvironMentalities* presents chapter-long discussions of contemporary artists representing such environmental doctrines as preservation, conservation, deep ecology, ecofeminism, sustainism, urban ecology, etc.

Weintraub, Linda, *To Life! EcoArt in Pursuit of a Sustainable Planet* (Berkeley and Los Angeles: University of California Press, 2012). *To Life!* is the first college textbook that integrates environmental studies, studio art practice, and contemporary art history. The forty artists it presents are considered in all three contexts. These diverse examples of contemporary eco-art provide an overview of the current eco art movement.

Eco-Art Studio Practice

Edwards, Lynn, *The Natural Paint Book* (Allentown, PA: Rodale Books, 2003)Explains the difference between conventional and eco-friendly paints, with instruction on how to make natural paints.

Harkness, Rachel (ed.) *An Unfinished Compendium of Materials* (Aberdeen: University of Aberdeen, 2017). This book offers a comprehensive survey of contemporary art from the vantage of materials such as asphalt, concrete, and bees wax.

Hiebert, Helen, *Papermaking with Garden Plants & Common Weeds* (North Adams, MA: Storey Publishers, 2006). Step-by-step instructions on the use of plant stalks, bark, petals, pine needles, and more.

Ingold, Tim, *Correspondences* (Aberdeen: University of Aberdeen, 2017) This text situates matter and materialisms within the context of 'world', 'lines', 'words', and 'conversations'.

Ingold, Tim, *Making: Anthropology, Archaeology, Art and Architecture* (London and New York: Routledge Press, 2013). This book explores 'thinking through making' as a guiding principle for anthropology, archaeology, art, and architecture.

Manco, Tristan, *Raw+Material=Art: Found, Scavenged and Upcycled* (London: Thames & Hudson, 2012). Developments in the use of salvaged and repurposed materials by contemporary artists.

McCann, Mike, and Angela Babin, *Health Hazards Manual for Artists* (Guilford, CT: Lyons Press, 2003). This offers a practical guide to health hazards from painting, photography, ceramics, sculpture, printmaking, woodworking, and textiles.

Michel, Karen, *Green Guide for Artists: Nontoxic Recipes, Green Art Ideas & Resources for the Eco Conscious Artist* (Beverely, MA: Quarry Books, 2009). This book contains DIY recipes for making nontoxic paint, mediums, and adhesives using recycled materials, and a resource guide.

Miodownik, Mark, *Stuff Matters: Exploring the Marvelous Materials that Shape Our Man-Made World* (Boston and New York: Houghton Mifflin Harcourt, 2014). The author is a materials scientist. He describes the science of such materials as paper, glass, and concrete.

Neddo, Nick, *The Organic Artist: Make Your Own Paint, Paper, Pigments, Prints and More from Nature* (Beverly, MA: Quarry Books, 2015). Neddo demonstrates how the natural landscape can serves as a source of art supplies.

Skinner, Tina, *Found Object Art II* (Atglen, PA: Schiffer Publishing, 2009). Eighty artists from the United States and Europe are represented. Many of the artworks convey environmental messages.

Spencer, Dorothy, *Found Object Art I* (Atglen, PA: Schiffer Publishing, 2007). This book contains many examples of artworks created from trash and found objects.

Webster, Sandy, *Earthen Pigments: Hand-Gathering & Using Natural Colors in Art* (Atglen, PA: Schiffer Publishing, 2013). This book explains how to collect, process, and use pigments from the earth.